T0233676

Introductory Circuit Theory

D. Sundararajan

Introductory Circuit Theory

 Springer

D. Sundararajan
Formerly at Concordia University
Montreal, QC, Canada

Additional material to this book can be downloaded from https://Springer.com.

ISBN 978-3-030-31987-8   ISBN 978-3-030-31985-4   (eBook)
https://doi.org/10.1007/978-3-030-31985-4

This Springer imprint is published by the registered company Springer Nature Switzerland AG.
The registered company address is: Gewerbestrasse 11, 6330 Cham, Switzerland

Preface

Circuit theory is fundamental to the study of important engineering applications such as power systems and communication. As such, it forms part of the first course to engineering and science students in several disciplines. This book is written for engineering, computer science and physics students, and engineers and scientists. Both the theory and practice by programming and in laboratory are two important aspects of the learning process of circuit theory. The objective in writing this book is to present the fundamentals of circuit theory systematically in a clear and concise textbook and provide the required programming part online. Hopefully, this approach is expected to improve the readability and understandability of the theory without clutter. Plenty of examples, figures, tables, programs, analogy, and physical explanations make it easy for the reader to get a good grounding in the basics of circuit theory and some of its applications.

The learning of the circuit theory requires calculus, linear algebra, transform analysis, programming, and laboratory practice. These tools are also required for other courses and in professional career later. However, these topics are difficult for new students. But, through the methods suggested, students can learn well this all-important subject with sufficient practice. Further, the learning process for this subject will certainly help the students in their ensuing study of the other subjects.

Learning of the circuit theory consists of four aspects: (1) systematic presentation of the mathematical methods in the text; (2) verification of the analytically obtained results by coding; (3) verification of the results by simulation; and (4) verification of the results using actual components in laboratory experiments. The last component gives physical appreciation of the circuit elements and their values in practice, the voltages, currents, frequencies, measuring instruments, and all other practical aspects involved. Students must take a coordinated laboratory course. Both coding and simulation of the analysis of circuits are presented in the online programming part. Each student can practice using these four methods as much as it is required for a good understanding of the subject.

This book is intended to be a textbook for the first course in circuit theory to the new undergraduate students in several disciplines of engineering and science, which includes the primary disciplines of electrical and electronics engineering. For engineering professionals, this book will be useful for self-study. In addition, this book will be a reference for anyone, student or professional, specializing in practical applications of circuit theory. The prerequisite for reading this book is a good knowledge of physics and calculus at the high school level.

As mentioned already, programming is an important component in learning and practicing circuit theory, as well as other subjects. While several software packages are available, it is better to use a popular general-purpose software package that the students are likely to use in their other courses and ensuing professional carrier in several areas of engineering and science. Therefore, learning of only one package is required. The programming part is presented using the popular, user-friendly and widely used, both in universities and industries, MATLAB® software package. While the use of a software package is inevitable in most applications, it is better to use the software in addition to self-developed programs. The effective use of a software package or to develop own programs requires a good grounding in the basic principles of the circuit theory. Answers to selected exercises marked ∗ are given at the end of the book. A Solutions Manual and slides are available for instructors at the website of the book.

I assume the responsibility for all the errors in this book and would very much appreciate receiving readers' suggestions and pointing out any errors (email:d_sundararajan@yahoo.com). I am grateful to my Editor and the rest of the team at Springer for their help and encouragement in completing this project. I thank my family for their support during this endeavor.

D. Sundararajan

Contents

1 Basic Concepts 1
   1.1 Ohm's Law 3
   1.2 Resistors Connected in Series 4
   1.3 Resistors Connected in Parallel 7
   1.4 Resistors Connected in Series and Parallel 9
   1.5 Kirchhoff's Voltage and Current Laws 11
   1.6 Applications 13
   1.7 Summary 13

2 DC Circuits 19
   2.1 Mesh Analysis 20
   2.2 Nodal Analysis 27
   2.3 Examples 29
       2.3.1 Linearity Property of Circuits 37
       2.3.2 Analysis of a Circuit with a Controlled Voltage Source .. 40
       2.3.3 Analysis of a Circuit with a Controlled Current Source .. 44
       2.3.4 $Y - \Delta$ and $\Delta - Y$ Transformations 55
   2.4 Circuit Theorems 59
       2.4.1 Thévenin's Theorem and Norton's Theorem 59
       2.4.2 Maximum Power Transfer Theorem 65
   2.5 Application 67
       2.5.1 Strain Gauge Measurement 67
   2.6 Summary 68

3 AC Circuits 77
   3.1 Sinusoids 77
       3.1.1 The Rectangular Form of Sinusoids 79
       3.1.2 The Complex Sinusoids 79
   3.2 AC Circuit Analysis 80
       3.2.1 Time- and Frequency-Domain Representations of Circuit Elements 80
       3.2.2 Time-Domain Analysis of a Series RC Circuit 82
       3.2.3 Frequency-Domain Analysis of a RC Circuit 84
       3.2.4 Impedances Connected in Series 85
       3.2.5 Impedances Connected in Parallel 91
       3.2.6 Impedances Connected in Series and Parallel 93
       3.2.7 Analysis of Typical Circuits 94
       3.2.8 Linearity Property of Circuits 105

3.3   Circuit Theorems 122
     3.3.1   Thévenin's Theorem 122
     3.3.2   Norton's Theorem 130
     3.3.3   Maximum Average Power Transfer Theorem 131
     3.3.4   Source Transformation 133
3.4   Application 135
     3.4.1   Filters 135
3.5   Summary 137

4   Steady-State Power 149
4.1   Energy in Reactive Elements with Sinusoidal Sources 150
4.2   Power Relations in a Circuit 151
4.3   Power-Factor Correction 159
4.4   Application 173
4.5   Summary 173

5   Magnetically Coupled Circuits 177
5.1   Mutual Inductance 177
5.2   Stored Energy 178
5.3   Examples 181
5.4   Application 198
     5.4.1   Transformers 198
5.5   Summary 199

6   Three-Phase Circuits 203
6.1   Three-Phase Voltages 203
     6.1.1   The Instantaneous Power 205
6.2   The Three-Phase Balanced $Y - Y$ Circuit 205
6.3   The Three-Phase Balanced $Y - \Delta$ Circuit 207
6.4   The Three-Phase Balanced $\Delta - Y$ Circuit 210
6.5   The Three-Phase Balanced $\Delta - \Delta$ Circuit 211
6.6   Application 212
6.7   Summary 212

7   Two-Port Networks 215
7.1   Impedance Parameters 215
7.2   Admittance Parameters 221
7.3   Hybrid Parameters 224
7.4   Transmission Parameters 225
7.5   Examples 226
     7.5.1   Analysis of a π Circuit 226
     7.5.2   Analysis of a Common-Emitter Transistor Amplifier 228
     7.5.3   Analysis of a Bridge Circuit 230
     7.5.4   Ladder Circuit 231
7.6   Application 233
     7.6.1   Digital-to-Analog Converter: The $R - 2R$ Ladder Circuit 233
7.7   Summary 236

8   Transform Analysis and Transient Response 239

  8.1   Fourier Series 240

         8.1.1   Fourier Series of a Rectified Sine Wave 242

         8.1.2   Gibbs Phenomenon 246

  8.2   Fourier Transform 246

         8.2.1   The Transfer Function and the Frequency Response 248

  8.3   Transient Response 249

         8.3.1   The Unit-Impulse and Unit-Step Signals 250

  8.4   Laplace Transform 252

         8.4.1   Properties of the Laplace Transform 253

         8.4.2   Time-Differentiation 254

         8.4.3   Integration 254

         8.4.4   Initial Value 255

         8.4.5   Final Value 255

         8.4.6   Circuit Analysis in the Frequency-Domain 255

  8.5   Application 269

  8.6   Summary 271

A   Matrices 275

  A.1   Determinants 277

B   Complex Numbers 281

Answers to Selected Exercises 285

  Bibliography 291

Index 293

Abbreviations

AC	Alternating current, sinusoidally varying current or voltage
DC	Direct current, sinusoid with frequency zero, constant current or voltage
FS	Fourier series
FT	Fourier transform
Im	Imaginary part of
KCL	Kirchhoff's Current Law
KVL	Kirchhoff's Voltage Law
LTI	Linear time-invariant
pf	Power factor
RC	Circuit comprising a resistor and a capacitor in series
Re	Real part of
RL	Circuit comprising a resistor and an inductor in series
RLC	Circuit comprising a resistor, a capacitor, and an inductor in series
RMS	Root-mean-square value

Basic Concepts

1

Electrical and electronic engineering is indispensable in all applications of science and engineering, either in home, office, or industry. In applications, signals occur in different forms, such as temperature, pressure, audio, video, medical, and optical. All the signals are converted to electrical signals using appropriate transducers, which convert signals from other forms to electrical, for easier and efficient analysis, control, generation, and transmission. For example, a microphone converts sound waves into an electrical signal. A turbogenerator, a steam turbine coupled to an electric generator, produces electrical power. Windmills serve a similar purpose. Once the signals are available in electrical form, there are two major types of activities in electrical and electronic engineering those are indispensable for all applications, power systems and communication. For these and other types of applications, circuit theory is of fundamental importance. That is why, this subject is learnt by students from several disciplines, in addition to the primary disciplines of electrical and electronic engineering. By analogy, the basic principles of circuit theory are applicable to other systems, such as mechanical, optical, and acoustical.

The effects of electrical power are clearly visible in the bulb emitting light, a water heater making the water hot and a fan creating a current of air. However, the flow of electric current is invisible and this aspect makes it difficult to visualize it. Fortunately, it is similar to the flow of water in a pipe. This analogy enables us to visualize the current flow in an electrical system. The analogy is similar to the operation of a computer providing a good analogy to the working of the brain. Water head determines the flow of water through a hose and pinching it increases the resistance to the flow. Parallel pipes decrease the resistance to the flow and pipes in series increase the resistance. Water towers provide the pressure to supply water to our homes. When the tank is full, the flow is more compared with when it is near empty. Water pressure is the analog to electrical voltage and water flow is the analog to current. Figure 1.1a shows water flowing from a tap, with the tap partly open. Figure 1.1b shows water flowing from the tap, with the tap fully open. In this case, the resistance to the flow of water is relatively reduced. For the same opening of the tap, the flow can be increased by increasing the water pressure from the overhead tank.

Electricity is the movement of charged particles, such as electrons. A circuit is a closed loop, which allows the flow of charges from one place to another. Components in the circuit control the flow and use it to do work. Charge is similar to the amount of water. Voltage is similar to water pressure. Current is flow of charge. More water in the tank results in increasing the flow at the exit. Flash light gets dimmer as the batteries run down. Less pressure results in reduced water flow. The measure of water flow is cubic meter/second. The measure of current flow is coulomb/second.

© Springer Nature Switzerland AG 2020
D. Sundararajan, *Introductory Circuit Theory*,
https://doi.org/10.1007/978-3-030-31985-4_1

Fig. 1.1 (**a**) Water flow
with the tap partly open;
(**b**) more flow with the tap
fully open

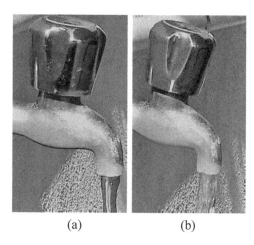

(a)   (b)

Fig. 1.2 Batteries in series

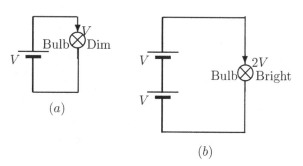

An electrical system is similar to a water supply system. Water flows through the pipes of a water
system. Electricity flows through the wires of an electrical system. Electrical pressure is measured in
volts. Electrical current flow is measured in amperes. A wire serves the purpose of a pipe. A battery
serves the purpose of a pump or a reservoir. A resistor serves the purpose of a narrow constriction in
a pipe. Current flow is similar to water flow. Current is the number of electrons moving past a point in
the circuit per second. Voltage is the pressure difference between two points, due to which the current
flows. A larger conductor offers less resistance and enables a larger current flow in the circuit with
the same voltage applied. With the resistance remaining the same, applying a higher voltage results
in a larger current flow, as shown in Fig. 1.2. In Fig. 1.2a, V volts of electrical pressure is applied
to the bulb and the resulting light is dim. In (b), two batteries connected in series produce 2 V volts
of pressure resulting in the bulb becoming bright. With a higher voltage applied to the circuit, more
current flows making the bulb emit more light. A water supply system has water tank, pumps, pipes
connected in various configurations, and valves. Similarly, an electric circuit has voltage and current
sources, wires and circuit elements, connected in various configurations.

No analogy can be exact. Here is another one. Cars cross tollgate in the highway. If there is only
one gate, certain number of cars can cross the gate per unit time. If the cars have to cross two gates,
one after another, then the rate of cars passing through the gates becomes one-half. If there are two
parallel gates, double the number of cars can pass through. If there are only few cars, then one gate
may be sufficient. But, a large number of cars forces opening of more gates resulting in a higher rate
of flow. In electrical systems, this is called Ohm's law. A higher voltage produces a larger current in
the circuit. Processing time at the gate has to be reduced by opening more gates. Similarly, a higher
voltage has to be applied for more current and more light from the bulb. Similarly, a larger conductor
offers less resistance and enables a larger current flow in the circuit with the same voltage applied.

Fig. 1.3 Resistor R in ohm or conductance G in mho

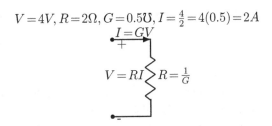

$$V = 4V, R = 2\Omega, G = 0.5\mho, I = \tfrac{4}{2} = 4(0.5) = 2A$$

$$I = GV$$

$$V = RI \qquad R = \tfrac{1}{G}$$

A voltage is developed by creating a separation of positive and negative charges. One coulomb of charge is the total charge of 6.242×10^{18} electrons. To develop 1 volt (V) between two points, 1 coulomb of charge has to be moved by applying 1 joule of energy. This is similar to pumping water to a higher level. Since potential energy is due to position, the word potential is also used to refer voltage levels. When a connection is made between two points with a potential difference, current flows. If 1 coulomb of charge passes through a point per second, then the flow of charge or current is said to be 1 ampere (A). Typical voltage sources are batteries, solar cells, and generators.

In a practical electrical circuit, the elements are physical devices. For the purpose of analysis, any system can be represented by mathematical models with acceptable tolerances. Therefore, the elements of an electric circuit are represented by mathematical definitions. The opposition to the flow of current is called as the resistance. The resistance is 1 Ω if 1 ampere current flows through it, when 1 V is applied across it. The resistor, a device used to control the flow of current, is denoted by the symbol R. The reciprocal of resistance is conductance, denoted by G. That is,

$$R = \frac{1}{G} \quad \text{and} \quad G = \frac{1}{R}.$$

The unit of measurement of resistance is Ω, called ohm. The unit of measurement of conductance is $\mho$, called mho. Figure 1.3 shows the input–output relationship of a resistor of value $R\Omega$. With 4 V applied across the resistor of value 2 Ω, the current I through it is $4/2 = 2$ A. The conductance is $1/2 = 0.5\mho$. Then, $I = 4 \times 0.5 = 2$ A, as found earlier. The resistance is similar to friction. If a surface is rough, more force is required to move an object on it. The more the roughness of the surface, the more is the heat generated. The higher the current flowing through the resistor, the higher is the voltage across it. Both the frictional devices and resistors dissipate energy, when a force is applied.

1.1 Ohm's Law

The voltage across a resistor, V, is the current, I, flowing through it times its value R, known as Ohm's law. That is,

$$I = \frac{V}{R}, \quad R = \frac{V}{I} \quad \text{and} \quad V = IR.$$

It is assumed that R is a constant to a required accuracy, since R varies with temperature, pressure, etc. This law, as is the case with most of the other laws, is applicable to both the DC and AC circuits. By convention, the current enters the positive terminal and leaves the negative terminal. That is, current flows from a higher voltage to a lower voltage. Therefore, the current direction determines the polarity of the voltage at the terminals of the resistor. If R is zero, it is called a short-circuit and the voltage across it is zero. If R is ∞, it is called an open-circuit and the current through it is zero. Resistors are essential to control the flow of current and are commonly used in such applications as controlling the speed of a fan, audio volume control in amplifiers, and emission of light by a bulb.

The power dissipated by a resistor, the rate of energy dissipation, is

$$P = VI = \frac{V^2}{R} = I^2 R.$$

The unit of measurement is joules per second or watts (W). Note that the expressions for the power are quadratic. That is, they are nonlinear. The power varies inversely as the resistance with a voltage source. The power varies proportionally as the resistance with a current source. The power consumed in a circuit is the sum of the powers consumed by all the constituent resistors, found using either of the defining expressions. Alternatively, when there is only one source, the equivalent resistance of the whole circuit can be computed. Then, the power dissipated in this resistor is the power consumed by the whole circuit. When more than one source is present, the power consumed by the circuit is the sum of the powers consumed by each resistor.

Example 1.1 Determine the resistance of a bulb, when the current through it is 0.2 A and the voltage across it is 220 V. Find the power consumed by the bulb.

Solution

$$R = \frac{220}{0.2} = 1100\Omega$$

$$P = 220 \times 0.2 = \frac{220^2}{1100} = 0.2^2 1100 = 44\text{W}.$$

■

Example 1.2 Determine the resistance of a smoothing iron, when the current through it is 10 A and the voltage across it is 110 V. Find the power consumed by the iron.

Solution

$$R = \frac{110}{10} = 11\Omega$$

$$P = 110 \times 10 = \frac{110^2}{11} = 10^2 11 = 1100\text{W}.$$

■

1.2   Resistors Connected in Series

In practice, the desired resistor is often not available and we have to use a combination of more than one resistor to make an equivalent one. Resistors can be connected in series and/or parallel configurations. Resistors have two terminals and, therefore, they come under the class of two-terminal devices or elements. In a series connection, one, and only one, terminal of a resistor is connected to adjoining resistors. Figure 1.4a shows two resistors connected in series, called a series circuit. A circuit is an interconnection of elements. The determination of current and voltages at all parts of the circuit is the essence of circuit analysis. When resistors are connected in series, the voltage across them increases, with the same current flowing through them. It is similar to connecting hoses to make

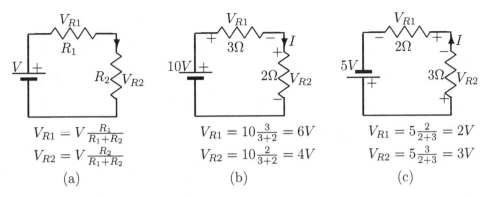

$$V_{R1} = V\frac{R_1}{R_1+R_2}$$
$$V_{R2} = V\frac{R_2}{R_1+R_2}$$
(a)

$$V_{R1} = 10\frac{3}{3+2} = 6V$$
$$V_{R2} = 10\frac{2}{3+2} = 4V$$
(b)

$$V_{R1} = 5\frac{2}{2+3} = 2V$$
$$V_{R2} = 5\frac{3}{2+3} = 3V$$
(c)

Fig. 1.4 Series circuit with a voltage source

a longer hose. The combined resistance is the sum of all the resistances. That is, with N number of resistors connected in series, the equivalent resistance R_{eq} of the series circuit is

$$R_{eq} = R_1 + R_2 + \cdots + R_N.$$

The value of R_{eq} will be larger than the largest resistor in the series connection. The same current I pass through all the resistors. Therefore, the voltage V across the series connection is

$$V = IR_1 + IR_2 + \cdots + IR_N = IR_{eq}.$$

The total resistance remains unchanged, irrespective of the order in which they are connected. Obviously, if all of them have the same value, then $R_{eq} = NR$. The source voltage applied across them gets divided in proportion to their individual values. The current through the series circuit is

$$I = \frac{V}{R_{eq}}$$

and the voltage across any resistor R_n is

$$V_{Rn} = \frac{V}{R_{eq}}R_n.$$

With just two resistors, R_1 and R_2, and V, the voltages across the series connection are

$$V_{R1} = V\frac{R_1}{R_1 + R_2} \quad \text{and} \quad V_{R2} = V\frac{R_2}{R_1 + R_2}.$$

Consider the circuit shown in Fig. 1.4b. The circuit is energized by a voltage source of 10 V. An ideal voltage source maintains a constant voltage at its terminals, irrespective of the current drawn from it. A voltage source is a constraint, clamping the voltage at a certain point in the circuit. The voltage drops across the resistors are

$$V_{R1} = \frac{V}{R_1 + R_2}R_1 = \frac{10}{3+2}3 = 6\,\text{V} \quad \text{and} \quad V_{R2} = \frac{V}{R_1 + R_2}R_2 = \frac{10}{3+2}2 = 4\,\text{V}.$$

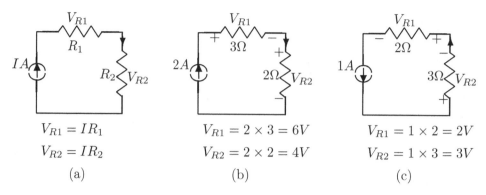

$$V_{R1} = IR_1 \qquad\qquad V_{R1} = 2 \times 3 = 6V \qquad\qquad V_{R1} = 1 \times 2 = 2V$$

$$V_{R2} = IR_2 \qquad\qquad V_{R2} = 2 \times 2 = 4V \qquad\qquad V_{R2} = 1 \times 3 = 3V$$

$$\text{(a)} \qquad\qquad\qquad\qquad \text{(b)} \qquad\qquad\qquad\qquad \text{(c)}$$

Fig. 1.5 Series circuit with a current source

The voltage drops add up to the source voltage $10 = 6 + 4$. The current through the circuit is $\frac{10}{3+2} = 2$ A. The power consumed by the circuit is

$$P = 10 \times 2 = 20\,\text{W}.$$

Consider the circuit shown in Fig. 1.4c. The circuit is energized by a voltage source of 5 V. As the polarity of the source voltage is reversed, the polarity of the voltage drops across the resistors is also reversed. The voltage drops across the resistors are

$$V_{R1} = \frac{V}{R_1 + R_2} R_1 = \frac{5}{3+2} 2 = 2\,\text{V} \quad \text{and} \quad V_{R2} = \frac{V}{R_1 + R_2} R_2 = \frac{5}{3+2} 3 = 3\,\text{V}$$

The voltage drops add up to the source voltage $5 = 2 + 3$. The current through the circuit is $-\frac{5}{3+2} = -1$ A. The power consumed by the circuit is

$$P = -5 \times -1 = 5\,\text{W}$$

Consider the circuit shown in Fig. 1.5a. The circuit is energized by a current source of $I\,A$. An ideal current source maintains a constant current at its terminals, irrespective of the voltage across its terminals. A current source is a constraint, clamping the current at a certain point in the circuit. As the same current flows through the two resistors, the respective voltage drops are

$$V_{R1} = IR_1 \quad \text{and} \quad V_{R2} = IR_2.$$

Consider the circuit shown in Fig. 1.5b. The circuit is energized by a current source of $I = 2$ A. The respective voltage drops are

$$V_{R1} = IR_1 = 2 \times 3 = 6\,\text{V} \quad \text{and} \quad V_{R2} = IR_2 = 2 \times 2 = 4\,\text{V}.$$

The power consumed by the circuit is

$$P = 10 \times 2 = 20\,\text{W}.$$

Consider the circuit shown in Fig. 1.5c. The circuit is energized by a current source of $I = 1$ A. As the direction of current flow is reversed, the polarities of the voltage drops are also reversed. The respective voltage drops are

$$V_{R1} = IR_1 = 1 \times 2 = 2\,\text{V} \quad \text{and} \quad V_{R2} = IR_2 = 1 \times 3 = 3\,\text{V}.$$

The power consumed by the circuit is

$$P = -5 \times -1 = 5\,\text{W}.$$

1.3 Resistors Connected in Parallel

In a parallel connection of elements, all of them are connected such that they have two points in common. Figure 1.6 shows two resistors connected in parallel, called a parallel circuit. When resistors are connected in parallel, the voltage across all of them remains the same, with different currents flowing through them. It is similar to connecting hoses to make a wider hose. The length remains the same, but the flowing capacity increases. The combined resistance is the reciprocal of the sum of all the conductances. That is, with N number of resistors connected in parallel, the equivalent resistance R_{eq} of the parallel circuit is

$$G_{eq} = G_1 + G_2 + \cdots + G_N \quad \text{and} \quad R_{eq} = \frac{1}{G_{eq}},$$

where $G_n = 1/R_n$. The value of R_{eq} will be smaller than the smallest resistor in the parallel connection, since the total current is more. The same voltage V is applied across all the resistors. Therefore, the total current I flowing through the parallel connection is

$$I = VG_1 + VG_2 + \cdots + VG_N = VG_{eq}.$$

The total resistance remains unchanged, irrespective of the order in which they are connected. Obviously, if all of them have the same value then $R_{eq} = R/N$. The total current gets divided in proportion to their individual conductance values. The current through the whole circuit is

$$I = VG_{eq}$$

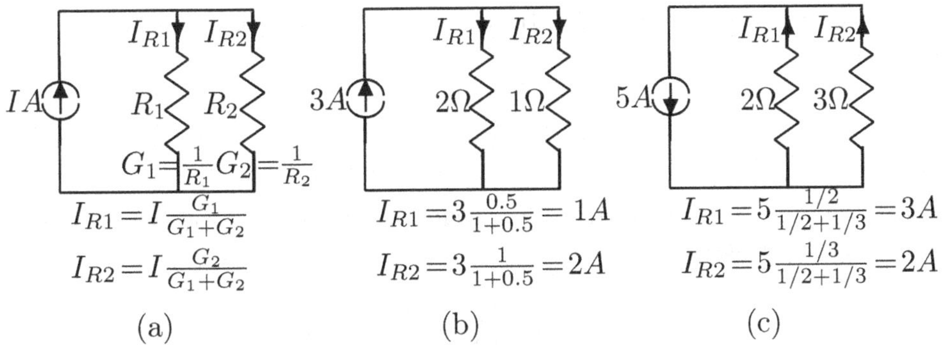

Fig. 1.6 Parallel circuit with a current source

and the current through any resistor R_n is

$$I_{Rn} = \frac{I}{G_{eq}} G_n.$$

In terms of resistance values, with just two resistors,

$$R_{eq} = R_1 \parallel R_2 = \frac{R_1 R_2}{R_1 + R_2}.$$

The parallel combination of two elements is denoted by the symbol $\parallel$.

Consider the circuit shown in Fig. 1.6a. The circuit is energized by a current source of $V = I$ A. The current through the two resistors are

$$I_{R1} = \frac{I}{G_1 + G_2} G_1 \quad \text{and} \quad I_{R2} = \frac{I}{G_1 + G_2} G_2.$$

Consider the circuit shown in Fig. 1.6b. The circuit is energized by a current source of $I = 3$ A. The current through the two resistors are

$$I_{R1} = \frac{3}{1 + 0.5} 0.5 = 1 \text{ A} \quad \text{and} \quad I_{R2} = \frac{3}{1 + 0.5} 1 = 2 \text{ A}.$$

The voltage across both the resistors are the same, $1 \times 2 = 2 \times 1 = 2$ V. The power consumed by the circuit is

$$P = 2 \times 3 = 6 \text{ W}.$$

Consider the circuit shown in Fig. 1.6c. The circuit is energized by a current source of $I = 5$ A. The current through the two resistors are

$$I_{R1} = \frac{5}{2(1/2 + 1/3)} = 3 \text{ A} \quad \text{and} \quad I_{R2} = \frac{5}{3(1/2 + 1/3)} = 2 \text{ A}.$$

As the source direction is reversed, the direction of current flow is also reversed. The voltage across both the resistors is the same, $3 \times 2 = 2 \times 3 = 6$ V. The power consumed by the circuit is

$$P = -5 \times -6 = 30 \text{ W}.$$

Consider the circuit shown in Fig. 1.7a. The circuit is energized by a voltage source of V. The currents through the resistors are

$$I_{R_1} = \frac{V}{R_1} \quad \text{and} \quad I_{R_2} = \frac{V}{R_2}.$$

Consider the circuit shown in Fig. 1.7b. The circuit is energized by a voltage source of 2 V. The currents through the resistors are

$$I_{R_1} = \frac{2}{2} = 1 \text{ A} \quad \text{and} \quad I_{R_2} = \frac{2}{1} = 2 \text{ A}.$$

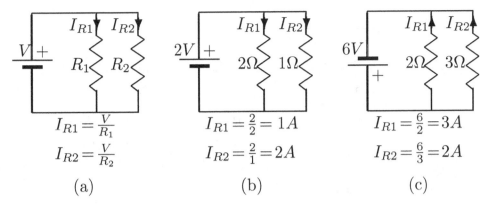

Fig. 1.7 Parallel circuit with a voltage source

The power consumed by the circuit is

$$P = 2 \times 3 = 6\,\text{W}.$$

Consider the circuit shown in Fig. 1.7c. The circuit is energized by a voltage source of 6 V. The currents through the resistors are

$$I_{R_1} = \frac{6}{2} = 3\,\text{A} \quad \text{and} \quad I_{R_2} = \frac{6}{3} = 2\,\text{A}.$$

As the polarity of the source is reversed, the direction of the current is also reversed. The power consumed by the circuit is

$$P = -5 \times -6 = 30\,\text{W}.$$

1.4 Resistors Connected in Series and Parallel

The analysis of series and parallel circuits is relatively straightforward. In general, most circuits are a combination of series and parallel circuits or connected in a random configuration in which none of the elements is in series or parallel or a combination of both. Obviously, combinations of the concepts of series and parallel circuits are used to analyze series-parallel circuits. Analysis of circuits with random configurations is presented in later chapters.

First, we have to identify the parts of the circuit with series and parallel configurations and simplify them separately. Now, the circuit gets reduced to a simpler form. These steps must be repeated until we can determine the source current. Then, using the voltage-division and current-division laws governing series and parallel circuits repeatedly, find the voltages and currents at all parts of the circuit.

Consider the circuit shown in Fig. 1.8a. The circuit is energized by a current source of $I = I_{R1} = 1\,\text{A}$. The source current flows through R_1 and, therefore, $I_{R1} = 1\,\text{A}$. The voltage across it is $1 \times 2 = 2\,\text{V}$. The source current gets divided between R_2 and R_3. Using current-division formula, we get

$$I_{R2} = \frac{I}{G_2 + G_3} G_2 = \frac{1}{1 + (1/3)} 1 = \frac{3}{4}\,A$$

$$I_{R3} = \frac{I}{G_2 + G_3} G_3 = \frac{1}{1 + (1/3)} \frac{1}{3} = \frac{1}{4}\,A$$

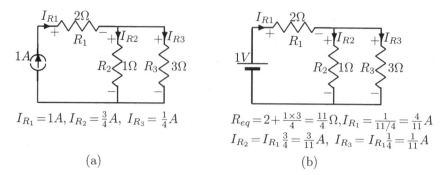

$$I_{R_1} = 1A, I_{R_2} = \tfrac{3}{4}A, \ I_{R_3} = \tfrac{1}{4}A$$

$$R_{eq} = 2 + \tfrac{1 \times 3}{4} = \tfrac{11}{4}\Omega, I_{R_1} = \tfrac{1}{11/4} = \tfrac{4}{11}A$$
$$I_{R_2} = I_{R_1}\tfrac{3}{4} = \tfrac{3}{11}A, \ I_{R_3} = I_{R_1}\tfrac{1}{4} = \tfrac{1}{11}A$$

(a)   (b)

Fig. 1.8 Resistors in series and parallel

as shown in the figure.

$$I = I_{R1} = I_{R2} + I_{R3} = \frac{3}{4} + \frac{1}{4} = 1\,\text{A}.$$

The voltage across the parallelly connected R_2 and R_3 is

$$\frac{3}{4}1 = \frac{1}{4}3 = \frac{3}{4}\,\text{V}.$$

The power dissipated in the circuit is

$$P = 1^2 \times 2 + \frac{3}{4} \times 1 = 2.75\,\text{W}.$$

Consider the circuit shown in Fig. 1.8b. The circuit is energized by a voltage source of 1 V. First, we have to find the combined resistance of the circuit, which is

$$R_{eq} = R_1 + (R_2 \parallel R_3) = \frac{11}{4}\Omega.$$

Now, the current drawn from the source is

$$I = I_{R1} = \frac{1}{11/4} = \frac{4}{11}\,A.$$

This current gets divided between R_2 and R_3. Using current-division formula, we get

$$I_{R2} = \frac{I}{G_2 + G_3}G_2 = \frac{\frac{4}{11}}{1 + (1/3)}1 = \frac{3}{11}\,A$$

$$I_{R3} = \frac{I}{G_2 + G_3}G_3 = \frac{\frac{4}{11}}{1 + (1/3)}\frac{1}{3} = \frac{1}{11}\,A$$

as shown in the figure.

$$I = I_{R1} = I_{R2} + I_{R3} = \frac{3}{11} + \frac{1}{11} = \frac{4}{11} \text{ A}$$

The voltage across the parallelly connected R_2 and R_3 is

$$\frac{3}{11} 1 = \frac{1}{11} 3 = \frac{3}{11} \text{ V}.$$

The combined voltage across the series-parallel combination is

$$2\frac{4}{11} + \frac{3}{11} = \frac{11}{11} = 1 \text{ V},$$

which is equal to the source voltage. The power dissipated in the circuit is

$$P = \frac{4}{11} \times 1 = \frac{4}{11} \text{ W}.$$

1.5 Kirchhoff's Voltage and Current Laws

Kirchhoff's voltage and current laws are two of the few fundamental laws in electrical engineering. Voltage means electric potential difference. These laws are required to find the equilibrium of the circuit in terms of its currents and voltages. A circuit can be characterized by a set of independent variables, current or voltage. If currents are chosen as the variables, then Kirchhoff's voltage law (KVL) is used to express the equilibrium of the circuit. The voltage at a point in the circuit is similar to the height of a point in a hilly terrain. Let us start at some point in a hill and climb the hill, visit the peaks, and climb down to the starting point. Then, the net height traversed by us is zero. Similarly, let us start at some point in a circuit, go through the circuit visiting junction points of elements and return back to the starting point. Then, the algebraic sum of the voltages around the loop is zero, which is KVL.

Kirchhoff's Voltage Law
The algebraic sum of voltage drops across the circuit elements around a closed path of a circuit must be zero. That is,

$$\Sigma \pm V = 0.$$

Consider the circuit shown in Fig. 1.9a. Let us start at point 1 and traverse the circuit in the clockwise direction. Voltage at point 1 minus the voltage at point 2 is the voltage drop from point 1 to 2, which is

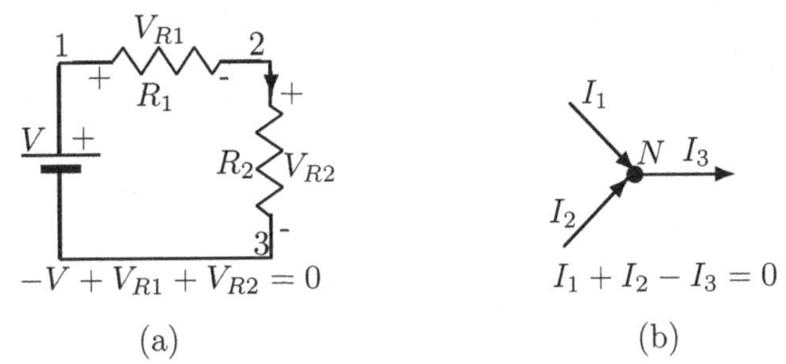

Fig. 1.9 Kirchhoff's voltage and current laws

$$-V + V_{R1} + V_{R2} = 0$$

(a)

$$I_1 + I_2 - I_3 = 0$$

(b)

positive. Similarly, the voltage drop from point 2 to 3 is also positive. However, the voltage drop from point 3 to 1 is negative. Therefore, some of the voltage drops must be negative and the rest positive so that the sum around the closed path (loop) is zero. In the circuit shown in Fig. 1.4b, we have

$$6 + 4 - 10 = 0 \text{ or } 6 + 4 = 10 \text{ or } -6 - 4 + 10 = 0.$$

In the circuit shown in Fig. 1.4c, we have

$$-2 - 3 + 5 = 0 \text{ or } 2 + 3 = 5 \text{ or } 2 + 3 - 5 = 0.$$

We can traverse the loop in the anticlockwise direction also.

Kirchhoff's Current Law (KCL)

In a market, at the end of the day, the money spent by the customers must be equal to the money received by the merchants. In a traffic junction, the number of vehicles entering must be equal to the number of vehicles leaving. At the junction of water pipes, the inflow of water must be equal to the outflow. Similarly, at the junction of several branches of a circuit, the algebraic sum of the currents must be zero. That is the sum of the incoming currents is equal to the sum of the outgoing currents.

The algebraic sum of currents flowing from branches towards a node in a circuit must be zero. That is,

$$\Sigma \pm I = 0.$$

When voltages are chosen to characterize the circuit, the equilibrium of the circuit is expressed by the KCL. In Fig. 1.9b, applying KCL, we get

$$I_1 + I_2 - I_3 = 0 \text{ or } I_1 + I_2 = I_3 \text{ or } -I_1 - I_2 + I_3 = 0.$$

The incoming currents must be assigned the opposite sign assigned to the outgoing currents.
In Fig. 1.8a,
$$I_{R1} = I_{R2} + I_{R3} = \frac{3}{4} + \frac{1}{4} = 1 \text{ A}.$$

In Fig. 1.8b,
$$I_{R1} = I_{R2} + I_{R3} = \frac{3}{11} + \frac{1}{11} = \frac{4}{11} \text{ A}.$$

Characterization of circuits on a current or voltage basis has a dual nature in that they are essentially similar with the roles of the current and voltage variables interchanged. The dual nature of the series and parallel circuits is shown in Table 1.1.

Table 1.1 The dual nature of the series and parallel circuits

Parallel circuit	Series circuit
$G_{eq} = G_1 + G_2$	$R_{eq} = R_1 + R_2$
$I = I_1 + I_2 = VG_1 + VG_2$	$V = V_1 + V_2 = IR_1 + IR_2$
$I_1 = I\frac{G_1}{G_1+G_2}, I_2 = I\frac{G_2}{G_1+G_2}$	$V_1 = V\frac{R_1}{R_1+R_2}, V_2 = V\frac{R_2}{R_1+R_2}$
$I_1 = I, G_1 \rightarrow \infty, I_2 = I, G_2 \rightarrow \infty$	$V_1 = V, R_1 \rightarrow \infty, V_2 = V, R_2 \rightarrow \infty$

1.6   Applications

In almost all electric circuits used in applications, series and parallel connection of elements will occur in some parts of the circuit. Circuit elements, such as resistors, are readily available only at certain values. It is a common practice to connect them in series and/or parallel to find the element with the required value. Another common occurrence day-to-day is the use of electric cells in series in appliances, such as torchlight. Usually, all devices, such as a motor, fan, and bulb, are connected in series with a switch to put the device on or off. Our house wiring is in parallel, connecting group of devices to each line. This prevents the shutdown of the whole power supply, when a problem occurs in a certain device.

In water heaters, the heater is connected with a thermostat, to stop and start the power supply as required to control the temperature of the water at the set level, and a switch. In electrical filters, which pass some frequency components of a signal while attenuating others, resistors, inductances, and capacitances are connected in series and/or parallel. In sensor circuits, a voltage is applied to the series connection of a standard resistor and a sensing resistor. The resistance of this resistor changes with some parameters such as pressure, temperature, etc. By measuring the change of the voltage drop across the standard resistor, the change in resistance of the sensor and, hence, the required parameter is measured. Airport runway lights are connected in series with a constant current through them. Therefore, the current, for a given voltage source, through them is kept low, requiring small conductors for the circuit. Certain mechanism is included so that the failure of one bulb will not prevent illuminating by the rest.

1.7   Summary

- Electrical and electronic engineering is indispensable in all applications of science and engineering, either in home, office, or industry.
- Electricity is the movement of charged particles, such as electrons.
- The flow of current is similar to the flow of water in a pipe.
- A circuit is a closed loop, which allows the flow of charges from one place to another. Components in the circuit control the flow and use it to do work.
- Charge is similar to the amount of water. Voltage is similar to water pressure.
- Current is the number of electrons moving past a point in the circuit per second. Voltage is the pressure difference between two points, due to which the current flows.
- The opposition to the flow of current is called the resistance. The resistance is 1 Ω if 1 ampere current flows through it, when 1 V is applied across it.
- The voltage across a resistor V is the current I flowing through it times its value R, known as Ohm's law.
- Resistors can be combined in a series and/or parallel or arbitrary configurations.
- An electric circuit is an interconnection of components, such as battery, resistors, etc.
- In a series circuit, same current flows through the circuit.
- In a parallel circuit, same voltage is applied across all the elements in the circuit.
- Kirchhoff's voltage law states that the algebraic sum of voltage drops across the circuit elements around a closed path of a circuit must be zero.
- Kirchhoff's current law states that the algebraic sum of currents flowing from branches towards a node in a circuit must be zero.

- Characterization of circuits on a current or voltage basis has a dual nature in that they are essentially similar with the roles of the current and voltage variables interchanged.
- Electric circuits are used in almost all applications of science and engineering.

Exercises

1.1 Find the current I and the voltages across the resistors,

$$V_{R1}, V_{R2}, V_{R3}, V_{R4}, V_{R5}$$

in the series circuit, shown in Fig. 1.10. Verify KVL.

1.1.1

$$V = 2\,\text{V}, \quad R_1 = 3\,\Omega, \quad R_2 = 2\,\Omega, \quad R_3 = 4\,\Omega, \quad R_4 = 1\,\Omega, \quad R_5 = 5\,\Omega.$$

* 1.1.2

$$V = -3\text{V}, \quad R_1 = 1\,\Omega, \quad R_2 = 2\,\Omega, \quad R_3 = 3\,\Omega, \quad R_4 = 4\,\Omega, \quad R_5 = 5\,\Omega.$$

1.1.3

$$V = -4\,\text{V}, \quad R_1 = 2\,\Omega, \quad R_2 = 1\,\Omega, \quad R_3 = 4\,\Omega, \quad R_4 = 3\,\Omega, \quad R_5 = 6\,\Omega.$$

1.2 Find the voltages across the resistors,

$$V_{R1}, V_{R2}, V_{R3}, V_{R4}, V_{R5}$$

in the series circuit, shown in Fig. 1.11.

1.2.1

$$I = 2\,\text{A}, \quad R_1 = 3\,\Omega, \quad R_2 = 2\,\Omega, \quad R_3 = 4\,\Omega, \quad R_4 = 1\,\Omega, \quad R_5 = 5\,\Omega.$$

1.2.2

$$I = -3\,\text{A}, \quad R_1 = 1\,\Omega, \quad R_2 = 2\,\Omega, \quad R_3 = 3\,\Omega, \quad R_4 = 4\,\Omega, \quad R_5 = 5\,\Omega.$$

* 1.2.3

$$I = -4A, \quad R_1 = 2\,\Omega, \quad R_2 = 1\,\Omega, \quad R_3 = 4\,\Omega, \quad R_4 = 3\,\Omega, \quad R_5 = 6\,\Omega.$$

1.3 Find the voltage across the resistors,

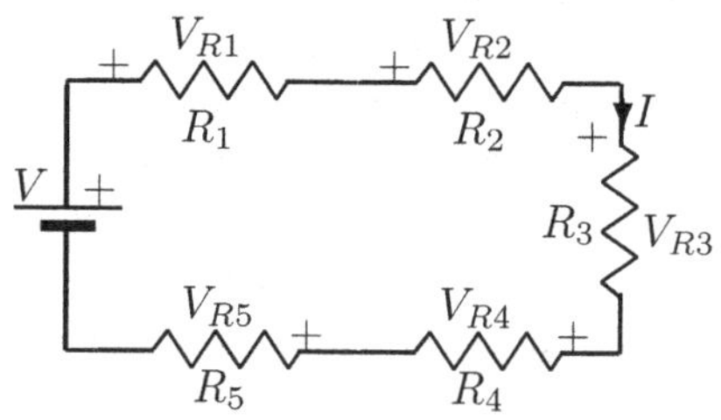

Fig. 1.10 Series circuit with a voltage source

Fig. 1.11 Series circuit
with a current source

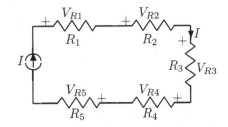

Fig. 1.12 Parallel circuit
with a current source

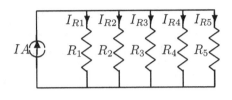

Fig. 1.13 Parallel circuit
with a voltage source

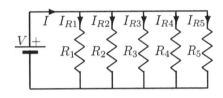

$$V_{R1}, V_{R2}, V_{R3}, V_{R4}, V_{R5}$$

in the parallel circuit, shown in Fig. 1.12. Find the currents through the resistors,

$$I_{R1}, I_{R2}, I_{R3}, I_{R4}, I_{R5}$$

Verify that the sum of the currents through the resistors is equal to the source current.

1.3.1
$$I = 2\,\text{A}, \quad R_1 = 3\Omega, \quad R_2 = 2\Omega, \quad R_3 = 4\Omega, \quad R_4 = 1\Omega, \quad R_5 = 5\Omega.$$

1.3.2

$$I = -3\,\text{A}, \quad R_1 = 1\Omega, \quad R_2 = 2\Omega, \quad R_3 = 3\Omega, \quad R_4 = 4\Omega, \quad R_5 = 5\Omega.$$

* 1.3.3

$$I = -4\,\text{A}, \quad R_1 = 2\Omega, \quad R_2 = 1\Omega, \quad R_3 = 4\Omega, \quad R_4 = 3\Omega, \quad R_5 = 6\Omega.$$

1.4 Find the current I and the voltages across the resistors,

$$V_{R1}, V_{R2}, V_{R3}, V_{R4}, V_{R5}$$

in the parallel circuit, shown in Fig. 1.13. Find the currents through the resistors,

$$I_{R1}, I_{R2}, I_{R3}, I_{R4}, I_{R5}$$

Verify that the sum of the currents through the resistors is equal to the total current, I.

1.4.1
$$V = 2\,\text{V}, \quad R_1 = 3\Omega, \quad R_2 = 2\Omega, \quad R_3 = 4\Omega, \quad R_4 = 1\Omega, \quad R_5 = 5\Omega$$

* 1.4.2

$$V = -3\,\text{V}, \quad R_1 = 1\Omega, \quad R_2 = 2\Omega, \quad R_3 = 3\Omega, \quad R_4 = 4\Omega, \quad R_5 = 5\Omega$$

1.4.3

$$V = -4\,\text{V}, \quad R_1 = 2\Omega, \quad R_2 = 1\Omega, \quad R_3 = 4\Omega, \quad R_4 = 3\Omega, \quad R_5 = 6\Omega$$

1.5 Find the current I and the voltages across the resistors,

$$V_{R1}, V_{R2}, V_{R3}, V_{R4}, V_{R5}$$

and current through

$$I_{R1}, I_{R2}, I_{R3}, I_{R4}, I_{R5}$$

in the series-parallel circuit, shown in Fig. 1.14. Verify KVL around the loops and KCL at node x.

1.5.1
$$V = 2\,\text{V}, \quad R_1 = 3\Omega, \quad R_2 = 2\Omega, \quad R_3 = 4\Omega, \quad R_4 = 1\Omega, \quad R_5 = 5\Omega.$$

* 1.5.2

$$V = -3\,\text{V}, \quad R_1 = 1\Omega, \quad R_2 = 2\Omega, \quad R_3 = 3\Omega, \quad R_4 = 4\Omega, \quad R_5 = 5\Omega.$$

1.5.3

$$V = -4\,\text{V}, \quad R_1 = 2\Omega, \quad R_2 = 1\Omega, \quad R_3 = 4\Omega, \quad R_4 = 3\Omega, \quad R_5 = 6\Omega.$$

1.6 Find the voltages across the resistors,

$$V_{R1}, V_{R2}, V_{R3}, V_{R4}, V_{R5}$$

in the series-parallel circuit, shown in Fig. 1.15. Verify KCL at node x.

1.6.1
$$I = 2\,\text{A}, \quad R_1 = 3\Omega, \quad R_2 = 2\Omega, \quad R_3 = 4\Omega, \quad R_4 = 1\Omega, \quad R_5 = 5\Omega.$$

Fig. 1.14 Series-parallel circuit with a voltage source

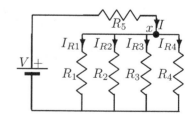

Fig. 1.15 Series-parallel
circuit with a current
source

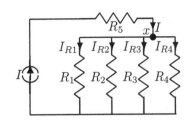

* **1.6.2**

$$I = -3\,\text{A}, \quad R_1 = 1\Omega, \quad R_2 = 2\Omega, \quad R_3 = 3\Omega, \quad R_4 = 4\Omega, \quad R_5 = 5\Omega.$$

1.6.3

$$I = -4\,\text{A}, \quad R_1 = 2\Omega, \quad R_2 = 1\Omega, \quad R_3 = 4\Omega, \quad R_4 = 3\Omega, \quad R_5 = 6\Omega.$$

DC Circuits

2

An electric circuit, for theoretical analysis, is an interconnection of idealized representation of physical components, such as voltage and current sources, switches, resistors, inductors, and capacitors. Any physical device can be approximated for practical purposes by idealized devices with sufficient accuracy. The relationship between the voltage across an element and the current through it is called its volt–ampere relationship and is assumed to be linear in the specified operating ranges. In this chapter, we consider circuits with DC sources and resistors only. DC is abbreviated form for direct current, a common designation for constant voltage or current. AC is abbreviated form for alternating current, a common designation for sinusoidally varying voltage or current.

Apart from its practical importance, the study of the basic principles of DC circuit analysis, without considering the transient response, is relatively simpler. Further, it has the priority since the mathematical process of computing the circuit response for a constant excitation is relatively easier to learn. The same basic principles can be easily adapted for AC circuit analysis. Therefore, a good understanding of the DC circuit analysis is essential and it makes the learning of the AC circuit analysis that much simpler.

With the sources and circuit elements specified, the purpose of circuit analysis is to determine the voltages and currents at all parts of the circuit. There is a voltage across any circuit element, called a branch of the circuit, and a current through it. Since the current and voltage in an element are related through its volt–ampere relationship, circuits can be analyzed in terms of the branch currents alone or branch voltages alone. A minimum set of independent branch voltages or currents has to be determined. There is no unique choice. The method of circuit analysis based on currents is called mesh or loop analysis. The other method based on voltages is called nodal analysis. Either method results in a set of equilibrium equations. The solution to these equations is obtained using linear algebra, yielding the values of the independent variables. From these values, all the values of other branch voltages and currents of the circuit can be determined, using KVL and KCL, completing the circuit analysis. The analysis presented, hereafter, is an extension of the analysis presented in the first chapter applicable to circuits with random configurations. We use the same input–output relationship of elements, Ohm's law, KVL, and KCL, but in a systematic manner as the circuit complexity is high. The equilibrium conditions for a circuit can be established in either of the two ways:

© Springer Nature Switzerland AG 2020
D. Sundararajan, *Introductory Circuit Theory*,
https://doi.org/10.1007/978-3-030-31985-4_2

Fig. 2.1 (**a**) Ideal and (**b**)
practical voltage sources
and their volt–ampere
characteristics

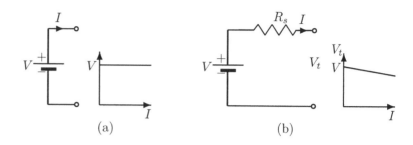

(a)   (b)

1. Through a set of N equations, using KVL, in which the mesh currents are the independent variables.
2. Through a set of M equations, using KCL, in which the branch voltages are the independent variables.

Voltage Sources

A source is required to energize a circuit. An ideal voltage source is characterized by its volt–ampere relationship of keeping its terminal voltage same irrespective of the current drawn by the load circuit connected to it, as shown in Fig. 2.1a. A practical voltage source can be approximated by an ideal voltage source in series with a resistance R_s, called the source resistance. Therefore, the voltage at the terminals of a practical voltage source, V_t, will decrease as the current drawn from it is increased, as shown in Fig. 2.1b. A voltage source is an applied constraint in the circuit. Due to this constraint, the voltages at various nodes of the circuit are fixed, depending on the characteristics of the voltage source and the circuit. The $+$ sign or an arrow indicates that a voltage raise in that direction is considered as a positive quantity.

Two voltage sources with capacities V_1 and V_2 volts, connected in series, are equivalent to a single source with capacity, which is equal to the algebraic sum of V_1 and V_2. Two voltage sources with unequal voltage capacities are not allowed to be connected in parallel, since it is a violation of KVL.

2.1   Mesh Analysis

Mesh analysis is also called loop analysis. The response of a circuit is completely characterized by the values of currents and voltages in all its branches. A circuit, geometrically, is characterized by its branches and nodes. A branch represents a single element, such as a resistor. That is, a branch is any two-terminal element. However, in certain problems, it is possible to regard series, parallel, or series-parallel combination of elements as a branch. The point where the terminals of two or more branches are connected together is called a node. Nodes, with selected independent voltages, are indicated by a dot in the circuit diagram. If two nodes are connected by a conductor (a short circuit), then the two nodes constitute a single node. A loop is a closed path in a circuit. It starts at a node, passes through a set of nodes (passing through each node only once), and returns to the starting node. **A loop is independent if it contains at least one branch that is not part of any other independent loop.** Two or more elements are connected in series if they carry the same current. Two or more elements are connected in parallel if their terminal voltages are the same.

A tree is any set of branches of the circuit that is sufficient to connect all its nodes. As the structure of the selected set of branches resembles that of a tree, it is called a tree. The number of branches forming a tree is $TB = N - 1$, where TB is the number of branches of the tree, called tree branches, and N is the number of nodes of the circuit. A tree contains no closed paths. There are several possible

Fig. 2.2 (**a**) A bridge circuit; (**b**) the circuit with branch currents

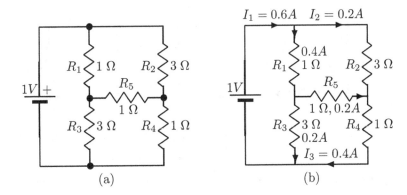

different trees for a given circuit. There is no unique tree corresponding to a circuit. But, the number of tree branches is always TB. Any one of the possible trees is sufficient for analyzing the corresponding circuit. Branches removed from a circuit, in forming a tree, are called the links. The number of links is M. Then, the total number of branches, B, in the circuit is $B = TB + M$. The equilibrium state of a circuit is determined by L independent link currents flowing through the L selected loops. That is, L branch currents only, of the B branch currents, are independent and it is the smallest number of currents in terms of which the rest can be expressed uniquely as a linear combination, using KCL at the nodes.

Given a circuit, the problem is to determine all the voltages across the branches and currents through the branches. The branch voltages are related to branch currents through volt–ampere characteristics of the particular circuit elements. Therefore, the behavior of any circuit is adequately characterized by branch currents alone or branch voltages alone. Based on this, there are two basic methods of circuit analysis: (1) mesh or loop analysis and (2) nodal analysis. In mesh analysis, the values of the independent current variables are determined. In nodal analysis, the values of the independent voltage variables are determined.

In order to reduce mistakes and confusion, the problem has to be approached systematically through several steps. Steps in establishing equilibrium equations for the mesh analysis of a circuit and finding the solution are as follows:

1. Select an appropriate number of independent current variables and the directions of current flow
2. Express the dependent current variables, by applying KCL at nodes, in terms of independent current variables
3. Apply KVL around the selected loops to set up a set of simultaneous equations
4. Solve for the independent currents and find the currents in all the branches
5. Verify the solution using KVL and KCL

The concepts are better presented through examples. Consider the analysis of the bridge circuit, shown in Fig. 2.2a. The circuit consists of 5 resistors with their values, designated R_1 to R_5, and a 1 V voltage source. Each circuit is different in the kinds of elements involved and the way they are interconnected. A resistor is characterized by its volt–ampere relationship of the voltage across it being proportional to the current through it, the constant of proportionality being its resistance value in ohms.

One possible set of independent current variables is shown in Fig. 2.2b. There are three links in the circuit and, therefore, the number of independent current variables is three, designated as

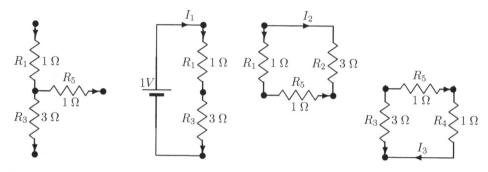

Fig. 2.3 A tree, with four nodes shown by discs, corresponding to the bridge circuit and the three independent loops

$$\{I_1, I_2, I_3\}.$$

The tree corresponding to the circuit, with four nodes shown by discs, is shown on the left side in Fig. 2.3, followed by the three loops formed by inserting the three links, in turn, in the tree. The three links are the voltage source and resistors R_2 and R_4. The three branches pertaining to the three independent currents must be the links associated with the selected tree. The insertion of one of the links in the tree must result in only one closed loop, which is different for each link. Therefore, a distinct set of loops is obtained.

Step 1 Independent and Dependent Variables

The three independent current variables are I_1, I_2, and I_3. By applying KCL at three nodes in Fig. 2.2b, the three dependent currents, flowing through resistors R_1, R_3, and R_5, are expressed in terms of the independent currents as

$$I_{R1} = I_1 - I_2$$
$$I_{R3} = I_1 - I_3$$
$$I_{R5} = I_{R1} - I_{R3} = I_3 - I_2.$$

The directions of all the branch currents, which can be arbitrary, must be assigned. If the assumed direction of the current is not correct, then the value of the current found in the analysis will be negative.

Step 2 Setting Up the Equilibrium Equations

The equilibrium conditions for a circuit can be expressed, in mesh analysis, by a set of N equations using KVL in which the N loop currents are the variables. Assuming no current sources and N independent loops, the equilibrium equations, in general, are of the form

$$Z_{11}I_1 + Z_{12}I_2 + Z_{13}I_3 + \cdots + Z_{1N}I_N = \sum V_1$$
$$Z_{21}I_1 + Z_{22}I_2 + Z_{23}I_3 + \cdots + Z_{2N}I_N = \sum V_2$$
$$\cdots \qquad\qquad = \cdots$$
$$Z_{N1}I_1 + Z_{N2}I_2 + Z_{N3}I_3 + \cdots + Z_{NN}I_N = \sum V_N,$$

where $\{I_1, I_2, \ldots, I_N\}$ are the currents in the N chosen loops of the circuit, $\{Z_{i1}, Z_{i2}, \ldots, Z_{iN}\}$ are the total impedances in the respective loops and $\{\sum V_1, \sum V_2, \ldots, \sum V_N\}$ are the algebraic sum of the source voltages in the respective loops. For example, Z_{12} is the impedance in the branch common to loops 1 and 2. It is positive or negative depending on the directions of the flow of the currents in the common branch. Impedance has a resistive part and a reactive part. In this chapter, we use only resistors. Therefore, $Z = R$. The equilibrium equations hold whether the circuit consists of resistors only or otherwise. In order to emphasize this fact and to make the understanding of the analysis of AC circuits easier, we use Z mostly instead of R.

Using vector and matrix quantities, the equilibrium equations and the solution are concisely given by

$$ZI = V \quad \text{and} \quad I = Z^{-1}V,$$

where Z is the impedance matrix, I is the current vector, and V is the voltage vector.

$$Z = \begin{bmatrix} Z_{11} & Z_{12} & Z_{13} & \cdots & Z_{1N} \\ Z_{21} & Z_{22} & Z_{23} & \cdots & Z_{2N} \\ & & \cdots \cdots & & \\ Z_{N1} & Z_{N2} & Z_{N3} & \cdots & Z_{NN} \end{bmatrix}, \quad I = \begin{bmatrix} I_1 \\ I_2 \\ I_3 \\ \vdots \\ I_N \end{bmatrix}, \quad V = \begin{bmatrix} \sum V_1 \\ \sum V_2 \\ \sum V_3 \\ \vdots \\ \sum V_N \end{bmatrix}.$$

For the circuit shown in Fig. 2.2b, applying KVL around the chosen loops, we get the three equilibrium equations.

$$Z_1(I_1 - I_2) + Z_3(I_1 - I_3) = 1$$

$$-Z_1(I_1 - I_2) - Z_5(I_3 - I_2) + Z_2 I_2 = 0$$

$$-Z_3(I_1 - I_3) + Z_5(I_3 - I_2) + Z_4 I_3 = 0.$$

These are the equations obtained traversing each loop either in the clockwise or anticlockwise direction. For example, the first equation corresponds to the loop with current I_1. Considering the direction of the current flow, the voltage drops across R_1 and R_3 add up, whereas the source voltage polarity is opposite of this drop. Therefore,

$$I_{R1}Z_1 + I_{R3}Z_3 = Z_1(I_1 - I_2) + Z_3(I_1 - I_3) = 1.$$

This is a straightforward application of KVL to loop 1. There are no sources in the other loops. Simplifying the equilibrium equations, we get

$$(Z_1 + Z_3)I_1 - Z_1 I_2 - Z_3 I_3 = 1$$

$$-Z_1 I_1 + (Z_1 + Z_2 + Z_5)I_2 - Z_5 I_3 = 0$$

$$-Z_3 I_1 - Z_5 I_2 + (Z_3 + Z_4 + Z_5)I_3 = 0.$$

Using matrices, we get

$$\begin{bmatrix} (Z_1 + Z_3) & -Z_1 & -Z_3 \\ -Z_1 & (Z_1 + Z_2 + Z_5) & -Z_5 \\ -Z_3 & -Z_5 & (Z_3 + Z_4 + Z_5) \end{bmatrix} \begin{bmatrix} I_1 \\ I_2 \\ I_3 \end{bmatrix} = \begin{bmatrix} 1 \\ 0 \\ 0 \end{bmatrix}.$$

Ensure that the correct values for the impedances and sources and the appropriate equations are used before proceeding with any numerical computation. With

$$Z_1 = 1, \quad Z_2 = 3, \quad Z_3 = 3, \quad Z_4 = 1 \text{ and } Z_5 = 1,$$

we get

$$\begin{bmatrix} 4 & -1 & -3 \\ -1 & 5 & -1 \\ -3 & -1 & 5 \end{bmatrix} \begin{bmatrix} I_1 \\ I_2 \\ I_3 \end{bmatrix} = \begin{bmatrix} 1 \\ 0 \\ 0 \end{bmatrix}.$$

The determinant of the impedance matrix must be nonzero. Otherwise, the equilibrium equations are dependent and must be checked. The determinant is 40 for this example.

Step 3 Solving the Equilibrium Equations

In order to solve for currents, we have to find the inverse of the impedance matrix. For this example, the inverse is found, as shown in Appendix A. In order to verify the inverse, it must be multiplied by the impedance matrix to get a 3×3 identity matrix, a matrix with all zeros except 1 on the main diagonal.

$$I = Z^{-1}V$$

$$\begin{bmatrix} I_1 \\ I_2 \\ I_3 \end{bmatrix} = \begin{bmatrix} 4 & -1 & -3 \\ -1 & 5 & -1 \\ -3 & -1 & 5 \end{bmatrix}^{-1} \begin{bmatrix} 1 \\ 0 \\ 0 \end{bmatrix} = \begin{bmatrix} 0.6000 & 0.2000 & 0.4000 \\ 0.2000 & 0.2750 & 0.1750 \\ 0.4000 & 0.1750 & 0.4750 \end{bmatrix} \begin{bmatrix} 1 \\ 0 \\ 0 \end{bmatrix} = \begin{bmatrix} 0.6 \\ 0.2 \\ 0.4 \end{bmatrix}$$

$$\{I_{R1} = I_1 - I_2 = 0.4 \text{ A}, \quad I_{R3} = I_1 - I_3 = 0.2 \text{ A}, \quad I_{R5} = I_3 - I_2 = 0.2 \text{ A}\}.$$

All the currents are shown in Fig. 2.2b.

Verifying the Solutions

Applying KVL to each loop, in turn, we get

$$I_{R1}R_1 + I_{R3}R_3 = (0.4)1 + (0.2)3 = 1$$

$$-I_2 R_2 + I_{R5}R_5 + I_{R1}R_1 = -(0.2)3 + (0.2)1 + (0.4)1 = 0$$

$$I_{R5}R_5 + I_3 R_4 - I_{R3}R_3 = (0.2)1 + (0.4)1 - (0.2)3 = 0.$$

Verifying the solution using KCL at the four nodes

$$I_1 - I_{R1} - I_2 = 0.6 - 0.4 - 0.2 = 0$$

$$I_{R1} - I_{R3} - I_{R5} = 0.4 - 0.2 - 0.2 = 0$$

$$I_{R5} + I_2 - I_3 = 0.2 + 0.2 - 0.4 = 0$$

$$I_{R3} - I_1 + I_3 = 0.2 - 0.6 + 0.4 = 0.$$

Symmetry of the Impedance Matrix

The procedure for loop analysis described is applicable for any circuit. However, for certain type of circuits under some conditions, the general procedure can be simplified. Of course, if a shortcut is available for the analysis of circuits it should be taken advantage of. The impedance matrix of the bridge circuit is symmetrical about its principal diagonal. That is, $Z_{ij} = Z_{ji}$. It is useful to verify the setting of the equilibrium equations. It is not an inherent property of linear circuits. This is due to the setting of KVL equations using the same loops defining the link currents and using consistently clockwise or anticlockwise direction for the independent currents in a mappable circuit. The graph of a mappable circuit does not have any branches crossing each other. Further, all the voltage sources in the circuit must be independent. In the first loop of the example circuit, shown in Fig. 2.3, the independent current I_1 flows in the same direction through all the elements. Therefore, the contribution of the voltage due to I_1 is the current multiplied by the sum of the values of all resistors in the first loop. Consequently, the first value in the main diagonal of the impedance matrix is 4. Similarly, the other values in the main diagonal are, respectively, the sum of the values of the resistors in the corresponding loops. The second and third values are 5 and 5. The other entries in the matrix are negative and symmetric. Their values are the values of the resistance common to the two loops. For example, the value of the resistor between the first and third loops is 3 and, therefore, $z_{13} = z_{31} = -3$. That is, the voltage drop in loop 1 by the current I_3 is -3 V, since it flows in the opposite direction of that of I_1. In the other cases, the values are -1. If an independent current does not flow in a certain loop, then the entry in the matrix is zero. It is possible to use two different sets of independent loops for the independent currents and the KVL equations. Then, the symmetry may not exist. Further, the determinant of the impedance matrix should be nonzero in any case.

Changed Assumption of the Current Directions
It is better to choose the current directions to be consistently clockwise or counterclockwise. However, the selection of the direction of the currents in the branches can be arbitrary. In that case, of course, the symmetry of the impedance matrix may be lost. If a current flows in the direction opposite to that assumed, the analysis result will be negative-valued. Let us do the analysis of the bridge circuit again with different direction for the currents, as shown in Fig. 2.4. Applying KCL at the nodes, we get

$$I_{R1} = -(I_1 + I_2)$$
$$I_{R3} = I_3 - I_1$$
$$I_{R5} = I_3 + I_2.$$

The equilibrium equations are

$$-Z_1(I_1 + I_2) - Z_3(I_1 - I_3) = 1$$
$$Z_1(I_1 + I_2) + Z_5(I_2 + I_3) + Z_2 I_2 = 0$$
$$Z_3(I_3 - I_1) + Z_5(I_2 + I_3) + Z_4 I_3 = 0.$$

Fig. 2.4 The bridge circuit with different current directions

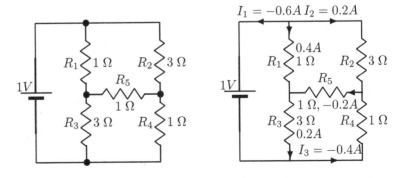

Simplifying, we get

$$-(Z_1 + Z_3)I_1 - Z_1 I_2 + Z_3 I_3 = 1$$

$$Z_1 I_1 + (Z_1 + Z_2 + Z_5)I_2 + Z_5 I_3 = 0$$

$$-Z_3 I_1 + Z_5 I_2 + (Z_3 + Z_4 + Z_5)I_3 = 0.$$

In matrix form,

$$\begin{bmatrix} -(Z_1 + Z_3) & -Z_1 & Z_3 \\ Z_1 & (Z_1 + Z_2 + Z_5) & Z_5 \\ -Z_3 & Z_5 & (Z_3 + Z_4 + Z_5) \end{bmatrix} \begin{bmatrix} I_1 \\ I_2 \\ I_3 \end{bmatrix} = \begin{bmatrix} 1 \\ 0 \\ 0 \end{bmatrix}.$$

With

$$Z_1 = 1, \quad Z_2 = 3, \quad Z_3 = 3, \quad Z_4 = 1, \quad Z_5 = 1$$

$$\begin{bmatrix} -4 & -1 & 3 \\ 1 & 5 & 1 \\ -3 & 1 & 5 \end{bmatrix} \begin{bmatrix} I_1 \\ I_2 \\ I_3 \end{bmatrix} = \begin{bmatrix} 1 \\ 0 \\ 0 \end{bmatrix}.$$

Solving the equilibrium equations, we get

$$\begin{bmatrix} I_1 \\ I_2 \\ I_3 \end{bmatrix} = \begin{bmatrix} -4 & -1 & 3 \\ 1 & 5 & 1 \\ -3 & 1 & 5 \end{bmatrix}^{-1} \begin{bmatrix} 1 \\ 0 \\ 0 \end{bmatrix} = \begin{bmatrix} -0.6000 & -0.2000 & 0.4000 \\ 0.2000 & 0.2750 & -0.1750 \\ -0.4000 & -0.1750 & 0.4750 \end{bmatrix} \begin{bmatrix} 1 \\ 0 \\ 0 \end{bmatrix} = \begin{bmatrix} -0.6 \\ 0.2 \\ -0.4 \end{bmatrix}$$

$$\{I_{R1} = 0.4, \quad I_{R3} = 0.2, \quad I_{R5} = -0.2\}.$$

A negative value for the current indicates current flow in the opposite direction from that assumed. After correcting the directions, we get the direction shown in Fig. 2.2b.

The power consumed by the circuit can be computed by multiplying both sides of the equilibrium equations, by the currents

$$\{I_1, I_1, I_3, \ldots, I_N, \}$$

respectively, and summing

$$I_1(Z_{11}I_1 + Z_{12}I_2 + Z_{13}I_3 + \cdots + Z_{1N}I_N) = \sum V_1 I_1$$

$$I_2(Z_{21}I_1 + Z_{22}I_2 + Z_{23}I_3 + \cdots + Z_{2N}I_N) = \sum V_2 I_2$$

$$\cdots \qquad \qquad = \cdots$$

$$I_N(Z_{N1}I_1 + Z_{N2}I_2 + Z_{N3}I_3 + \cdots + Z_{NN}I_N) = \sum V_N I_N.$$

For the specific example, we get

$$\begin{bmatrix} 0.6(4 & -1 & -3) \\ 0.2(-1 & 5 & -1) \\ 0.4(-3 & -1 & 5) \end{bmatrix} \begin{bmatrix} 0.6 \\ 0.2 \\ 0.4 \end{bmatrix} = \begin{bmatrix} 1(0.6) \\ 0 \\ 0 \end{bmatrix}.$$

Summing either side yields the power consumed by the circuit as 0.6 W. This procedure is useful to compute the power with multiple sources.

2.2   Nodal Analysis

The equilibrium state of a circuit can also be determined by TB independent voltages, where TB is the number of branches of the circuit forming a tree. KCL is applied at each node (except one, which becomes the ground node) of the selected tree of the circuit to set up the required equilibrium equations. The steps in nodal analysis are as follows:

1. Select an appropriate number of independent voltage variables
2. Express the dependent voltage variables, by applying KVL around the loops, in terms of independent voltage variables
3. Apply KCL at the selected nodes to set up a set of simultaneous equations
4. Solve for the independent voltages and find the voltages at all the nodes
5. Verify the solution using KVL and KCL

Prefer the node connected to the maximum number of elements and sources as the ground node. A ground node acts as a reference for voltage levels at various points in the circuit. The voltage at the ground node is assumed to be zero.

Let us analyze the same bridge circuit, shown again in Fig. 2.5a, by nodal analysis.

Step 1 Independent and Dependent Variables
The nodes are shown in Fig. 2.5b. The three independent voltages are V_1, V_2, and V_3. By applying KVL around the three loops in Fig. 2.5b, the other three dependent voltages, across the resistors R_1, R_2, and R_5, are expressed in terms of the independent voltages as

$$V_{R1} = V_1 - V_2$$

$$V_{R2} = V_1 - V_3$$

$$V_{R5} = V_2 - V_3.$$

For example, applying KVL to the top right loop, we get

$$V_{R1} + (V_2 - V_3) + (V_3 - V_1) = 0 \quad \text{and} \quad V_{R1} = V_1 - V_2.$$

Fig. 2.5 Nodal analysis of the bridge circuit

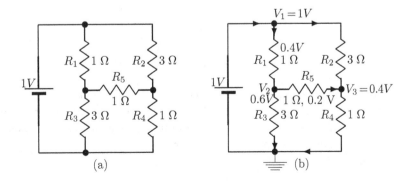

(a)

(b)

Step 2 Setting Up the Equilibrium Equations

Assuming no voltage sources and N independent nodes, applying KCL at each node in turn, the equilibrium equations, in general, are of the form

$$Y_{11}V_1 + Y_{12}V_2 + Y_{13}V_3 + \cdots + Y_{1N}V_N = \sum I_1$$

$$Y_{21}V_1 + Y_{22}V_2 + Y_{23}V_3 + \cdots + Y_{2N}V_N = \sum I_2$$

$$\cdots = \cdots$$

$$Y_{N1}V_1 + Y_{N2}V_2 + Y_{N3}V_3 + \cdots + Y_{NN}V_N = \sum I_N,$$

where $\{V_1, V_2, \ldots, V_N\}$ are the voltages at the chosen N nodes of the circuit with respect to the ground node, $\{Y_{i1}, Y_{i2}, \ldots, Y_{iN}\}$ are the admittances in the branches joining nodes i and j, and $\{\sum I_1, \sum I_2, \ldots, \sum I_N\}$ are the algebraic sum of the currents from current sources, connected to the ground node, feeding the respective nodes. Admittance has a conductive part and a susceptive part. In this chapter, we use only resistors. Therefore, $Y = G$. The equilibrium equations hold whether the circuit consists of resistors only or otherwise. In order to emphasize this fact and to make the understanding of the analysis of AC circuits easier, we use Y instead of G.

Using vector and matrix quantities, the equilibrium equations and the solution are concisely given by

$$YV = I \quad \text{and} \quad V = Y^{-1}I,$$

where V is the voltage vector, Y is the admittance matrix, and I is the current vector.

$$Y = \begin{bmatrix} Y_{11} & Y_{12} & Y_{13} & \cdots & Y_{1N} \\ Y_{21} & Y_{22} & Y_{23} & \cdots & Y_{2N} \\ & & \cdots\cdots & & \\ Y_{N1} & Y_{N2} & Y_{N3} & \cdots & Y_{NN} \end{bmatrix}, \quad V = \begin{bmatrix} V_1 \\ V_2 \\ V_3 \\ \vdots \\ V_N \end{bmatrix}, \quad I = \begin{bmatrix} \sum I_1 \\ \sum I_2 \\ \sum I_3 \\ \vdots \\ \sum I_N \end{bmatrix}.$$

Let us solve for the node voltages for the circuit shown in Fig. 2.5. Since $V_1 = 1$ is known, there are only 2 unknowns, $\{V_2, V_3\}$. When a voltage source is connected between a node and the ground node, the problem is simplified. With $Z = R$ and applying KCL at nodes with voltages V_2 and V_3, we get the equilibrium equations, respectively, at nodes 2 and 3 as

$$\frac{(V_2 - 1)}{Z_1} + \frac{(V_2 - V_3)}{Z_5} + \frac{V_2}{Z_3} = 0$$

$$\frac{(V_3 - 1)}{Z_2} + \frac{(V_3 - V_2)}{Z_5} + \frac{V_3}{Z_4} = 0.$$

In loop analysis, we can traverse the loop in the clockwise or counterclockwise direction. Similarly, in nodal analysis, we can write the KCL equations at each node so that the currents are directed out of the node or flowing towards the node. For this example, the KCL equations at each node are written so that the currents are directed out of the node.

Simplifying the equilibrium equations, we get

$$(Z_3Z_1 + Z_1Z_5 + Z_3Z_5)V_2 - Z_3Z_1V_3 = Z_3Z_5$$

$$-Z_2Z_4V_2 + (Z_2Z_4 + Z_2Z_5 + Z_4Z_5)V_3 = Z_4Z_5.$$

With

$$Z_1 = 1, \quad Z_2 = 3, \quad Z_3 = 3, \quad Z_4 = 1, \quad Z_5 = 1$$

$$7V_2 - 3V_3 = 3$$

$$-3V_2 + 7V_3 = 1.$$

In matrix form

$$\begin{bmatrix} 7 & -3 \\ -3 & 7 \end{bmatrix} \begin{bmatrix} V_2 \\ V_3 \end{bmatrix} = \begin{bmatrix} 3 \\ 1 \end{bmatrix}.$$

The determinant of the admittance matrix is 40.

Step 3 Solving the Equilibrium Equations

$$V = Y^{-1}I$$

The inverse of an arbitrary 2×2 matrix A exists if

$$|A| = (a_{11}a_{22} - a_{12}a_{21}) \neq 0.$$

Then, A^{-1} is given by

$$A = \begin{bmatrix} a_{11} & a_{12} \\ a_{21} & a_{22} \end{bmatrix} \quad \text{and} \quad A^{-1} = \frac{1}{(a_{11}a_{22} - a_{12}a_{21})} \begin{bmatrix} a_{22} & -a_{12} \\ -a_{21} & a_{11} \end{bmatrix}$$

$$\begin{bmatrix} V_2 \\ V_3 \end{bmatrix} = \begin{bmatrix} 7 & -3 \\ -3 & 7 \end{bmatrix}^{-1} \begin{bmatrix} 3 \\ 1 \end{bmatrix} = \begin{bmatrix} 0.1750 & 0.0750 \\ 0.0750 & 0.1750 \end{bmatrix} \begin{bmatrix} 3 \\ 1 \end{bmatrix} = \begin{bmatrix} 0.6 \\ 0.4 \end{bmatrix}$$

$$\{V_{R1} = 0.4, \quad V_{R2} = 0.6, \quad V_{R5} = 0.2\}.$$

These values are the same as those obtained by loop analysis.

2.3   Examples

A bridge circuit, with the voltage source in the middle, is shown in Fig. 2.6b. The tree corresponding to the circuit has 4 nodes. With the polarities marked same as in Fig. 2.2b, three independent currents

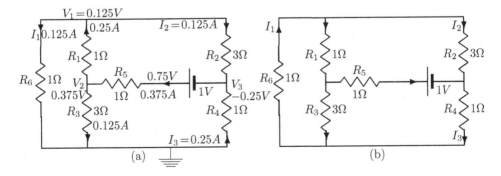

Fig. 2.6 A bridge circuit, with the voltage source in the middle

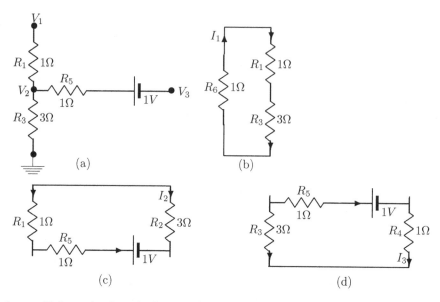

Fig. 2.7 A tree, with four nodes shown by discs, corresponding to the bridge circuit and the three independent loops

are to be found using the mesh analysis. The equilibrium equations are similar to that for the circuit in Fig. 2.2. The tree corresponding to the circuit, with four nodes shown by discs, is shown in Fig. 2.7a. Figure 2.7b–d show the three loops formed by inserting the three links, in turn, in the tree. The three links are the resistors R_6, R_2, and R_4.

The three independent current variables are I_1, I_2 and I_3. By applying KCL at three nodes in Fig. 2.6b, the other three dependent currents, flowing through resistors R_1, R_3 and R_5, are expressed in terms of the independent currents as

$$I_{R1} = I_1 - I_2$$

$$I_{R3} = I_1 - I_3$$

$$I_{R5} = I_3 - I_2.$$

The equilibrium equations are, respectively, for the loops corresponding to the independent currents, I_1, I_2, and I_3.

$$Z_1(I_1 - I_2) + Z_3(I_1 - I_3) + Z_6 I_1 = 0$$

$$-Z_1(I_1 - I_2) - Z_5(I_3 - I_2) + Z_2 I_2 = 1$$

$$-Z_3(I_1 - I_3) + Z_5(I_3 - I_2) + Z_4 I_3 = -1.$$

Simplifying, the equations in matrix form are

$$\begin{bmatrix} (Z_1 + Z_3 + Z_6) & -Z_1 & -Z_3 \\ -Z_1 & (Z_1 + Z_2 + Z_5) & -Z_5 \\ -Z_3 & -Z_5 & (Z_3 + Z_4 + Z_5) \end{bmatrix} \begin{bmatrix} I_1 \\ I_2 \\ I_3 \end{bmatrix} = \begin{bmatrix} 0 \\ 1 \\ -1 \end{bmatrix}.$$

With

$$Z_1 = 1, \quad Z_2 = 3, \quad Z_3 = 3, \quad Z_4 = 1, \quad Z_5 = 1, \quad Z_6 = 1,$$

we get

$$5I_1 - I_2 - 3I_3 = 0$$

$$-I_1 + 5I_2 - I_3 = 1$$

$$-3I_1 - I_2 + 5I_3 = -1.$$

In matrix form,

$$\begin{bmatrix} 5 & -1 & -3 \\ -1 & 5 & -1 \\ -3 & -1 & 5 \end{bmatrix} \begin{bmatrix} I_1 \\ I_2 \\ I_3 \end{bmatrix} = \begin{bmatrix} 0 \\ 1 \\ -1 \end{bmatrix}.$$

For this circuit, we can get the impedance matrix from inspection. The determinant of the impedance matrix is 64. Solving for the currents, we get

$$\begin{bmatrix} I_1 \\ I_2 \\ I_3 \end{bmatrix} = \begin{bmatrix} 5 & -1 & -3 \\ -1 & 5 & -1 \\ -3 & -1 & 5 \end{bmatrix}^{-1} \begin{bmatrix} 0 \\ 1 \\ -1 \end{bmatrix} = \begin{bmatrix} 0.375 & 0.125 & 0.250 \\ 0.125 & 0.250 & 0.125 \\ 0.250 & 0.125 & 0.375 \end{bmatrix} \begin{bmatrix} 0 \\ 1 \\ -1 \end{bmatrix} = \begin{bmatrix} -0.125 \\ 0.125 \\ -0.250 \end{bmatrix}$$

$$\{I_{R1} = -0.250, \quad I_{R3} = 0.125, \quad I_{R5} = -0.375\}.$$

Some of the currents happen to flow in the reverse direction, indicated by a negative value, to that with the initial assumption. Figure 2.6a shows the circuit with correct current directions and the node voltages. From inspection, current and voltage values satisfy KCL and KVL. For example, at the top left node, the incoming current is 0.25 and the sum of the outgoing currents is $0.125 + 0.125 = 0.25$. For the left most loop, the sum of the voltage drops across resistors R_1 and R_6 is 0.375 and the drop across R_3 is also 0.375 with the polarities reversed.

Let us do the nodal analysis for the circuit. The equilibrium equations are

$$\frac{(V_1 - V_2)}{Z_1} + \frac{V_1}{Z_6} + \frac{(V_1 - V_3)}{Z_2} = 0$$

$$\frac{(V_2 - V_1)}{Z_1} + \frac{V_2}{Z_3} + \frac{(V_2 - (V_3 + 1))}{Z_5} = 0$$

$$\frac{V_3}{Z_4} + \frac{(V_3 - V_1)}{Z_2} + \frac{((V_3 + 1) - V_2)}{Z_5} = 0.$$

Simplifying, we get

$$(Z_2 Z_6 + Z_1 Z_2 + Z_1 Z_6)V_1 - Z_2 Z_6 V_2 - Z_1 Z_6 V_3 = 0$$

$$-Z_3 Z_5 V_1 + (Z_3 Z_5 + Z_1 Z_5 + Z_1 Z_3)V_2 - Z_1 Z_3 V_3 = Z_3 Z_1$$

$$-Z_4 Z_5 V_1 - Z_2 Z_4 V_2 + (Z_2 Z_5 + Z_4 Z_5 + Z_4 Z_2)V_3 = -Z_2 Z_4.$$

With

$$Z_1 = 1, \quad Z_2 = 3, \quad Z_3 = 3, \quad Z_4 = 1, \quad Z_5 = 1, Z_6 = 1,$$

the equilibrium equations, in matrix equation form, are

$$\begin{bmatrix} 7 & -3 & -1 \\ -3 & 7 & -3 \\ -1 & -3 & 7 \end{bmatrix} \begin{bmatrix} V_1 \\ V_2 \\ V_3 \end{bmatrix} = \begin{bmatrix} 0 \\ 3 \\ -3 \end{bmatrix}.$$

The determinant of the admittance matrix is 192. Solving for the voltages, we get

$$
\begin{bmatrix} V_1 \\ V_2 \\ V_3 \end{bmatrix} = \begin{bmatrix} 7 & -3 & -1 \\ -3 & 7 & -3 \\ -1 & -3 & 7 \end{bmatrix}^{-1} \begin{bmatrix} 0 \\ 3 \\ -3 \end{bmatrix} = \begin{bmatrix} 0.2083 & 0.1250 & 0.0833 \\ 0.1250 & 0.2500 & 0.1250 \\ 0.0833 & 0.1250 & 0.2083 \end{bmatrix} \begin{bmatrix} 0 \\ 3 \\ -3 \end{bmatrix} = \begin{bmatrix} 0.125 \\ 0.375 \\ -0.250 \end{bmatrix}
$$

$$
\{V_{R1} = -0.250, \quad V_{R2} = 0.375, \quad V_{R5} = -0.375\}.
$$

Let us analyze the circuit with correct current directions.

$$
I_{R1} = I_1 + I_2
$$
$$
I_{R3} = I_3 - I_1
$$
$$
I_{R5} = I_3 + I_2.
$$

The equilibrium equations are

$$
Z_1(I_1 + I_2) + Z_3(I_1 - I_3) + Z_6 I_1 = 0
$$
$$
Z_1(I_1 + I_2) + Z_5(I_3 + I_2) + Z_2 I_2 = 1
$$
$$
Z_3(I_3 - I_1) + Z_5(I_3 + I_2) + Z_4 I_3 = 1.
$$

Simplifying, we get

$$
(Z_1 + Z_3 + Z_6)I_1 + Z_1 I_2 - Z_3 I_3 = 0
$$
$$
Z_1 I_1 + (Z_1 + Z_2 + Z_5)I_2 + Z_5 I_3 = 1
$$
$$
-Z_3 I_1 + Z_5 I_2 + (Z_3 + Z_4 + Z_5)I_3 = 1.
$$

With

$$
Z_1 = 1, \quad Z_2 = 3, \quad Z_3 = 3, \quad Z_4 = 1, \quad Z_5 = 1, \quad Z_6 = 1, \quad V = 1,
$$

we get

$$
5I_1 + I_2 - 3I_3 = 0
$$
$$
I_1 + 5I_2 + I_3 = 1
$$
$$
-3I_1 + I_2 + 5I_3 = 1.
$$

In matrix form,

$$
\begin{bmatrix} 5 & 1 & -3 \\ 1 & 5 & 1 \\ -3 & 1 & 5 \end{bmatrix} \begin{bmatrix} I_1 \\ I_2 \\ I_3 \end{bmatrix} = \begin{bmatrix} 0 \\ 1 \\ 1 \end{bmatrix}.
$$

Solving for the currents, we get

$$
\begin{bmatrix} I_1 \\ I_2 \\ I_3 \end{bmatrix} = \begin{bmatrix} 5 & 1 & -3 \\ 1 & 5 & 1 \\ -3 & 1 & 5 \end{bmatrix}^{-1} \begin{bmatrix} 0 \\ 1 \\ 1 \end{bmatrix} = \begin{bmatrix} 0.3750 & -0.1250 & 0.2500 \\ -0.1250 & 0.2500 & -0.1250 \\ 0.2500 & -0.1250 & 0.3750 \end{bmatrix} \begin{bmatrix} 0 \\ 1 \\ 1 \end{bmatrix} = \begin{bmatrix} 0.1250 \\ 0.1250 \\ 0.2500 \end{bmatrix}
$$

$$
\{I_{R1} = 0.2500, \quad I_{R3} = 0.1250, \quad I_{R5} = 0.3750\}.
$$

Let us compute the power consumed by the circuit. The resistors are

$$R_1 = 1, R_2 = 3, R_3 = 3, R_4 = 1, R_5 = 1, R_6 = 1.$$

The currents through the resistors due to the source, respectively, are

$$I_1 = 0.25, I_2 = 0.125, I_3 = 0.125, I_4 = 0.25, I_5 = 0.375, I_6 = 0.125.$$

The power consumed is 0.375 W. The current supplied by the voltage source is 0.375. Therefore, the power supplied is 0.375 W, as obtained before.

Circuit Analysis with a Supernode

If a voltage source, independent or dependent, is not connected to the ground node, then its two nodes and any elements connected in parallel with it is called a supernode. The current through that source has to be found in order to solve the problem by nodal analysis. The voltage difference between the terminals of the source is the given constraint. Sufficient number of KVL and KCL equations has to be set up and solved to find the current through the voltage source between the two nonreference nodes.

Consider the bridge circuit shown in Fig. 2.8. The voltage source is in the middle of the bridge. Same nodes and links are assumed as in earlier bridge circuit analysis. Usually, two equations using KCL is set up at the two ends of the supernode. Let the current through the source be i. At the left side node, we get

$$\frac{(V_2 - V_1)}{1} + \frac{V_2}{3} = i. \tag{2.1}$$

At the right side node, we get

$$\frac{(V_1 - (V_2 - 1))}{3} + \frac{(1 - V_2)}{1} = i. \tag{2.2}$$

Eliminating i and simplifying, we get

$$V_1 - 2V_2 = -1. \tag{2.3}$$

Applying KCL at node 1, we get

$$\frac{V_1}{1} + \frac{(V_1 - V_2)}{1} + \frac{(V_1 - (V_2 - 1))}{3} = 0.$$

Fig. 2.8 A bridge circuit with the voltage source in the middle

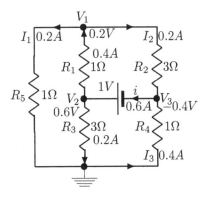

Note that the current in the source is flowing towards node 2. Since there is no other source in the circuit, this current gets split up at node 1. Simplifying, we get

$$7V_1 - 4V_2 = -1. \tag{2.4}$$

Solving Eqs. (2.3) and (2.4), we get

$$V_1 = 0.2 \quad \text{and} \quad V_2 = 0.6$$

$V_3 = V_2 - 1 = 0.6 - 1 = -0.4$ V. From Eq. (2.1), we get the current through the source as

$$\frac{(0.6 - 0.2)}{1} + \frac{0.6}{3} = i = 0.6.$$

Since we have determined the two independent voltages, there is no need to use this current for the analysis of this circuit.

$$\{V_{R1} = -0.4, \quad V_{R2} = 0.6\}.$$

An alternative method to avoid processing a supernode is that we connect a resistance in series with the source, as presented in an earlier example. As its value gets reduced compared with other resistances in the circuit, the circuit becomes more closer with a zero resistance series circuit and the result becomes closer to the exact values. A suitable small series resistance R is to be selected. For the example circuit with $R = 0.1\ \Omega$, we get, by simulation, for V_1, V_2, V_3

$$\{0.1887, \ 0.566, \ -0.3774\}.$$

With $R = 0.01\ \Omega$, we get

$$\{0.1988, \ 0.5964, \ -0.3976\}.$$

With $R = 0.001\ \Omega$, we get

$$\{0.1999, \ 0.5966, \ -0.3998\}.$$

With $R = 0.0001\ \Omega$, we get

$$\{0.2, \ 0.6, \ -0.4\}.$$

Mesh Analysis
The loop current directions are assumed to be as in Fig. 2.2b. The equilibrium equations are similar to that in Fig. 2.6.

$$Z_1(I_1 - I_2) + Z_3(I_1 - I_3) + Z_5 I_1 = 0$$

$$-Z_1(I_1 - I_2) + Z_2 I_2 = 1$$

$$-Z_3(I_1 - I_3) + Z_4 I_3 = -1.$$

In matrix form,

$$\begin{bmatrix} (Z_1 + Z_3 + Z_5) & -Z_1 & -Z_3 \\ -Z_1 & (Z_1 + Z_2) & 0 \\ -Z_3 & 0 & (Z_3 + Z_4) \end{bmatrix} \begin{bmatrix} I_1 \\ I_2 \\ I_3 \end{bmatrix} = \begin{bmatrix} 0 \\ 1 \\ -1 \end{bmatrix}.$$

With

$$Z_1 = 1, \quad Z_2 = 3, \quad Z_3 = 3, \quad Z_4 = 1, \quad Z_5 = 1, \quad V = 1,$$

we get

$$\begin{bmatrix} 5 & -1 & -3 \\ -1 & 4 & 0 \\ -3 & 0 & 4 \end{bmatrix} \begin{bmatrix} I_1 \\ I_2 \\ I_3 \end{bmatrix} = \begin{bmatrix} 0 \\ 1 \\ -1 \end{bmatrix}.$$

For this circuit, we can get the impedance matrix from inspection. The determinant of the impedance matrix is 40.

$$\begin{bmatrix} I_1 \\ I_2 \\ I_3 \end{bmatrix} = \begin{bmatrix} 5 & -1 & -3 \\ -1 & 4 & 0 \\ -3 & 0 & 4 \end{bmatrix}^{-1} \begin{bmatrix} 0 \\ 1 \\ -1 \end{bmatrix} = \begin{bmatrix} 0.4000 & 0.1000 & 0.3000 \\ 0.1000 & 0.2750 & 0.0750 \\ 0.3000 & 0.0750 & 0.4750 \end{bmatrix} \begin{bmatrix} 0 \\ 1 \\ -1 \end{bmatrix} = \begin{bmatrix} -0.2 \\ 0.2 \\ -0.4 \end{bmatrix}.$$

Figure 2.8 shows the circuit with correct current directions and the node voltages.
Let us compute the power consumed by the circuit. The resistors are

$$R_1 = 1, R_2 = 3, R_3 = 3, R_4 = 1, R_5 = 1.$$

The currents through the resistors due to the source, respectively, are

$$I_1 = 0.4, I_2 = 0.2, I_3 = 0.2, I_4 = 0.4, I_5 = 0.2.$$

The power consumed is 0.6 W. The current supplied by the voltage source is 0.6. Therefore, the power supplied is 0.6 W, as obtained before.

Current Sources

Without sources, a circuit is dead with no currents through the elements and no voltage across the elements. In practice, there is no such thing as an ideal device. However, we analyze engineering systems with ideal elements first, since it simplifies the analysis. Then, the ideal element is modified to represent actual elements. An ideal current source is characterized by its volt–ampere relationship of keeping its terminal current same irrespective of the voltage across the load circuit connected to it, as shown in Fig. 2.9a. A practical current source can be approximated by an ideal current source with a shunt resistance in parallel with its terminals, as shown in Fig. 2.9b. Therefore, the current supplied by a practical current source will decrease as the terminal voltage is increased. A current source is an applied constraint in the circuit. Due to this constraint, the currents through various branches of the circuit are fixed depending on the characteristics of the current source and the circuit. An arrow indicates that a current raise in that direction is considered as a positive quantity. A voltage source appearing in series with a branch or a current source appearing in parallel with a branch does not

Fig. 2.9 (a) Ideal and (b) practical current sources and their volt–ampere characteristics

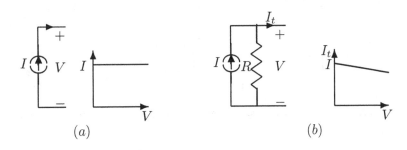

(a) (b)

affect the structure of the circuit. That is, the number of independent voltages and currents remains the same.

Two current sources with capacities I_1 A and I_2 A, connected in parallel, are equivalent to a single source with current capacity, which is equal to the algebraic sum of I_1 and I_2. Two unequal current sources are not allowed to be connected in series, since it is a violation of KCL.

Circuit Analysis with Current and Voltage Sources

Consider a circuit energized by current and voltage sources, shown in Fig. 2.10a. One possible tree is shown in Fig. 2.10b. With two voltages known and the current constrained in one branch, both the nodal and loop analysis reduce to two variable problems. Let us do the nodal analysis. $V_1 = 1$ is known. Therefore, we have to determine V_2 and V_3 only. The equilibrium equations at nodes 2 and 3, respectively, are

$$\frac{(V_2 - 1)}{1} + \frac{V_2}{3} = 1 \quad \text{or} \quad 4V_2 = 6, \; V_2 = 1.5 \, \text{V}$$

$$\frac{(V_3 + 1)}{3} + \frac{V_3}{4} = -1 \quad \text{or} \quad 7V_3 = -16, \; V_3 = -\frac{16}{7} \, \text{V}.$$

Mesh Analysis

Applying KVL around the left side loop, we get, with $I_2 = -1$,

$$(I_1)1 + 3(I_1 - I_2) = 1 \text{ or } 4I_1 = -2, \; I_1 = -0.5 \, \text{A}.$$

Applying KVL around the right side loop, we get, with $I_2 = -1$,

$$-(I_2 - I_3)3 + 4I_3 = -1 \text{ or } 7I_3 = -4, \; I_3 = -\frac{4}{7} \, \text{A}.$$

A supermesh occurs when a current source is connected between two nonreference nodes without a parallel resistance. We apply KVL to the neighboring loops to find the voltage across the current source. An alternative method to avoid processing a supermesh is that we connect a resistance in parallel with the current source. As the value gets increased compared with other resistances in the circuit, the circuit becomes more closer with an infinite resistance parallel circuit and the result

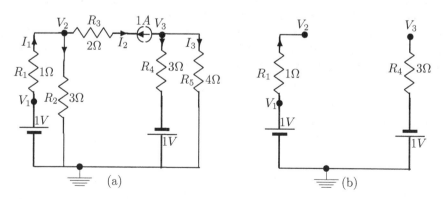

Fig. 2.10 (a) A circuit energized by current and voltage sources and (b) a possible tree

becomes closer to the exact values. A suitable high parallel resistance R is to be selected. For the example circuit, with $R = 1000\,\Omega$, we get, by simulation,

$$\{I_1 = -0.496\ \text{A}, \quad I_2 = -0.9942\ \text{A}, \quad I_3 = -0.569\ \text{A}\}.$$

With $R = 10000\,\Omega$, we get

$$\{-0.5\ \text{A}, \quad -0.9994\ \text{A}, \quad -0.5713\ \text{A}\}.$$

2.3.1 Linearity Property of Circuits

This property is called the superposition theorem. When circuits consist of linear elements, such as resistors, inductors, capacitors or, equivalently, circuits characterized by a linear differential equation with constant coefficients, the circuits are linear circuits. Most of the systems, although to some extent are nonlinear, can be approximated adequately by linear systems. For example, in theoretical analysis, we assume a pure resistor with no other parasitic elements. A resistor may be nonlinear to some extent and may have some inductive effect. It is assumed that the resistor is sufficiently pure for the accuracies required in practical applications. As the assumed linearity provides a simpler analysis, linear systems analysis is widely used in the study of signals and systems. Similarly, signals, usually, have arbitrary amplitude profile and are decomposed into a linear combination of well-defined basic signals in transform analysis, as presented later.

Mathematically, the response of a circuit to a linear combination of inputs is linear if the output is also the same linear combination of the individual outputs to the respective inputs. For example, let the response of a circuit to 1 V input is 1 A. Then, the output must be 2 A for the input 2 V. Therefore, the analysis becomes somewhat simpler with a normalized input voltage of 1 V, although the electric power is supplied at 220/110 V. And it is much higher in power transmission systems. Further, typical resistor values are in kiloohms or megaohms. These aspects are to be learnt in a concurrent laboratory course. Another implication of linearity is that the output of a circuit, with a number of sources, can be obtained by finding the output to each source separately and adding all the responses. Voltages and currents due to each source and their total in the circuit, shown in Fig. 2.10, are shown in Table 2.1. Determining the response to each source is much simpler. The use of this property may result in simpler analysis, compared with nodal and loop methods. The circuit has two voltage sources and one current source. When the response to one source is considered, the rest of the voltage sources are replaced by short-circuits and the current sources are replaced by open-circuits. Let us consider the response to the left voltage source alone, case (a). The right voltage source is short-circuited and the current source is open-circuited. The response is shown in the left side of Fig. 2.11a. Let us consider the response to the right voltage source alone, case (b). The left voltage source is short-circuited and the current source is open-circuited. The response is shown in the right side of Fig. 2.11a. Let us consider the response to the current source alone. The two voltage sources are short-circuited. The

Table 2.1 Voltages and currents due to each source and their total in the circuit shown in Fig. 2.10

Case	V_1	V_2	V_3	I_1	I_2	I_3
(a)	1	0.75	0	0.25	0	0
(b)	0	0	$-\frac{4}{7}$	0	0	$-\frac{1}{7}$
I_2 only	0	0.75	$-\frac{12}{7}$	-0.75	-1	$-\frac{3}{7}$
Total	1	1.5	$-\frac{16}{7}$	-0.5	-1	$-\frac{4}{7}$

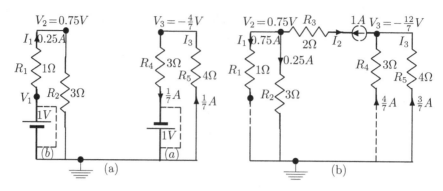

Fig. 2.11 Analysis of a circuit using the linearity property

response is shown in Fig. 2.11b. The currents are obtained by using the current-division formula for resistors connected in parallel.

Let us compute the power consumed by the circuit. The resistors are

$$R_1 = 1, R_2 = 3, R_3 = 2, R_4 = 3, R_5 = 4.$$

The net currents through the resistors due to all the sources, respectively, are

$$I_1 = 0.5, I_2 = 0.5, I_3 = 1, I_4 = \frac{3}{7}, I_5 = \frac{4}{7}.$$

The power consumed is

$$0.25 + 0.75 + 2 + \frac{27}{49} + \frac{64}{49} = 4.8571 \text{ W}.$$

The equation defining the power is nonlinear. Therefore, the total power supplied may not be equal to the sum of the powers supplied by the sources acting alone. Let the currents supplied by two sources acting alone be I_1 and I_2. Then, the sum of the powers supplied is proportional to $I_1^2 + I_2^2$. The power supplied by the sources acting together is proportional to $(I_1 + I_2)^2$ and

$$(I_1^2 + I_2^2) \neq (I_1 + I_2)^2.$$

The total power is always equal to the sum of the powers consumed by the separate branches of the circuit, irrespective of the number of sources in the circuit. When all the sources acting simultaneously, the total power supplied by the sources is the sum of the powers supplied by the individual sources. The power supplied by a source is the product of its terminal voltage and current. The voltage across a current source depends jointly upon the intensities of all the sources and vice versa for a voltage source. Therefore, the power supplied by a source can vary, when it is acting alone or jointly with other sources.

With all the sources acting, the net current through the left voltage source is -0.5. Therefore, it consumes power, which is equal to -0.5 W. Similarly, the right voltage source consumes $-3/7$ W. The voltage across the current source is $1.5 + 2 + 16/7$, when it is acting jointly. Therefore, the total power is

$$-0.5 - 3/7 + (1.5 + 2 + 16/7) = 4.8571 \text{ W},$$

as obtained before.

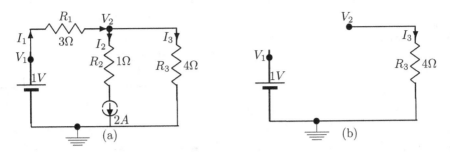

Fig. 2.12 (**a**) A circuit with a current source between two loops and (**b**) a possible tree

Circuit Analysis with the Current Source Between Two Loops

Consider the circuit with a current source between two loops, shown in Fig. 2.12a. A possible tree is shown in Fig. 2.12b. With 3 nodes and one independent voltage and one independent current given, the analysis reduces to a one-variable problem.

Mesh Analysis

With $I_1 - I_3 = 2$ A and applying KVL, we get

$$3I_1 + 4I_3 = 1$$
$$I_1 - I_3 = 2.$$

Solving, we get

$$\left\{ I_1 = \frac{9}{7} = 1.2857, \quad I_3 = -\frac{5}{7} = -0.7143 \right\}$$

and $V_2 = -20/7$ V. Current in the rightmost resistor flows opposite to the direction of initial assumption.

Nodal Analysis

$$\frac{V_2}{4} + \frac{V_2 - 1}{3} = -2 \text{ and } V_2 = -\frac{20}{7} \text{ V.}$$

Let us compute the power consumed by the circuit. The resistors are

$$R_1 = 3, R_2 = 1, R_3 = 4.$$

The net currents through the resistors due to all the sources, respectively, are

$$I_1 = 1.2857, I_2 = 2, I_3 = 0.7143.$$

The power consumed is
$$1.2857^2 (3) + 2^2 + 0.7143^2 (4) = 11 \text{ W.}$$

The net current through the voltage source is 9/7. The voltage across the current source is $(-20/7-2)$, when it acts jointly. Therefore, the total power is

$$9/7 + (-2)(-34/7) = 11 \text{ W}$$

as obtained before.

Fig. 2.13
Voltage-controlled voltage
source and its volt–ampere
characteristics

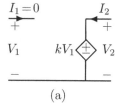

(a)   (b) I_2

Fig. 2.14
Current-controlled current
source and its volt–ampere
characteristics

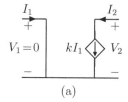

(a)   (b) V_2

Controlled Sources

Sources can both supply and absorb power. In charging, a battery absorbs power. It delivers power
when connected to a device, such as a light bulb. Batteries and generators are commonly used to
energize circuits and those are approximations to ideal sources. The voltage or current provided by
a controlled or dependent source, a 3-terminal device, is controlled by another voltage or current at
some other part of the circuit. The 3 terminals, with one common terminal, form two pairs, one pair
for input and the other for the output. Four possible types are as follows:

1. voltage-controlled voltage source
2. current-controlled voltage source
3. voltage-controlled current source
4. current-controlled current source

Dependent sources are necessary to model devices such as transistors.

   Voltage-controlled voltage source and its volt–ampere characteristics are shown in Fig. 2.13. It is
a 3-terminal device. The source voltage is k, a constant, times the control voltage V_1, irrespective of
the value of I_2. Current-controlled current source and its volt–ampere characteristics are shown in
Fig. 2.14. The source current is k, a constant, times the control current I_1, irrespective of the value
of V_2.

2.3.2   Analysis of a Circuit with a Controlled Voltage Source

A circuit with a voltage-controlled voltage source is shown in Fig. 2.15. A possible tree corresponding
to the circuit is shown in Fig. 2.16. Loops corresponding to the currents I_1, I_2, and I_3 are shown in
Figs. 2.17 and 2.18. The voltage-controlled voltage source generates a voltage of $2(V_1 - V_4)$. The
circuit has five nodes. As there is a dependent source and a branch current is known, the analysis
involves only three independent current or voltage variables. Let us analyze the circuit using the loop
method. Every element has to be included in a loop. At least one element in each loop is not part of
any other independent loop. The equilibrium equations corresponding to the middle, left, and right
loops, respectively, are

Fig. 2.15 A circuit with a voltage-controlled voltage source

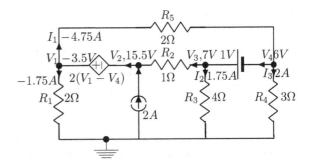

Fig. 2.16 A possible tree corresponding to the circuit in Fig. 2.15

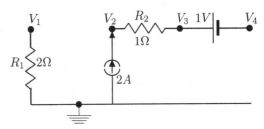

Fig. 2.17 The loop corresponding to current I_1

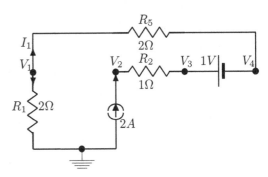

Fig. 2.18 Loops corresponding to the currents I_2 and I_3

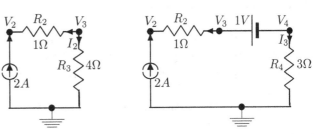

$$(Z_5 I_1 + Z_4 I_3 - 2Z_5 I_1) + (Z_2(I_1 - I_2 - I_3)) - Z_3 I_2 = 0$$

$$-Z_1(I(2) + I(3) - 2) = Z_5 I_1 + Z_4 I_3$$

$$Z_3 I_2 - Z_4 I_3 = 1.$$

The first equation corresponds to the middle loop involving resistors Z_2 and Z_3. To write the KVL equation around the loop, we need the voltage V_2.

$$V_1 = Z_5 I_1 + Z_4 I_3 \quad \text{and} \quad V_1 - V_4 = Z_5 I_1.$$

Therefore, the voltage across the controlled voltage source is $2Z_5I_1$, with opposite polarity to that of V_1. Consequently,

$$V_2 = (Z_5I_1 + Z_4I_3 - 2Z_5I_1).$$

The current I_{Z2} is $(I_1 - I_2 - I_3)$.

The second equation corresponds to the leftmost loop involving resistor Z_1. We have the equation

$$V_1 = Z_5I_1 + Z_4I_3.$$

The current entering Z_1 is $(I(2)+I(3)-2)$ from the ground side. Therefore, we get another expression for V_1 involving the current source.

$$V_1 = -Z_1(I(2) + I(3) - 2).$$

Equating these two equations yields an equilibrium equation. The last equation corresponds to the rightmost loop involving resistors Z_3 and Z_4. Simplifying, the three equilibrium equations are

$$(-Z_5 + Z_2)I_1 - (Z_2 + Z_3)I_2 + (Z_4 - Z_2)I_3 = 0$$

$$Z_5I_1 + Z_1I_2 + (Z_1 + Z_4)I_3 = 4$$

$$0I_1 + Z_3I_2 - Z_4I_3 = 1.$$

With

$$Z_1 = 2, \quad Z_2 = 1, \quad Z_3 = 4, \quad Z_4 = 3, \quad Z_5 = 2,$$

we get

$$\begin{bmatrix} -1 & -5 & 2 \\ 2 & 2 & 5 \\ 0 & 4 & -3 \end{bmatrix} \begin{bmatrix} I_1 \\ I_2 \\ I_3 \end{bmatrix} = \begin{bmatrix} 0 \\ 4 \\ 1 \end{bmatrix}.$$

The determinant of the impedance matrix is 12.

$$\begin{bmatrix} I_1 \\ I_2 \\ I_3 \end{bmatrix} = \begin{bmatrix} -1 & -5 & 2 \\ 2 & 2 & 5 \\ 0 & 4 & -3 \end{bmatrix}^{-1} \begin{bmatrix} 0 \\ 4 \\ 1 \end{bmatrix} = \begin{bmatrix} -2.1667 & -0.5833 & -2.4167 \\ 0.5000 & 0.2500 & 0.7500 \\ 0.6667 & 0.3333 & 0.6667 \end{bmatrix} \begin{bmatrix} 0 \\ 4 \\ 1 \end{bmatrix} = \begin{bmatrix} -4.75 \\ 1.75 \\ 2 \end{bmatrix}.$$

The other currents and voltages are shown in Fig. 2.15.

Nodal Analysis

The voltage source is not connected to the ground node. Therefore, we have to find the current through it. Setting up two KCL equations at the terminals of the source, we get

$$\frac{(V_4 - V_1)}{Z_5} + \frac{V_4}{Z_4} + \frac{(V_3 - V_2)}{Z_2} + \frac{V_3}{Z_3} = 0.$$

The sum of the first two terms is the current leaving from node 4 towards node 3, which is the negative of the current leaving from node 3 towards node 4. The dependent voltage source is also not connected to the ground node. Therefore, we have to find the current through it. Setting up two KCL equations at the terminals of the source, we get

$$-\frac{(V_2 - V_3)}{Z_2} + (2) - \frac{(V_1 - V_4)}{Z_5} - \frac{V_1}{Z_1} = 0.$$

Further,

$$V_1 - V_2 = 2(V_1 - V_4) \quad \text{or} \quad V_1 + V_2 - 2V_4 = 0.$$

The equilibrium equations are

$$-\frac{(V_2 - V_3)}{Z_2} + (2) - \frac{(V_1 - V_4)}{Z_5} - \frac{V_1}{Z_1} = 0 \tag{2.5}$$

$$\frac{(V_4 - V_1)}{Z_5} + \frac{V_4}{Z_4} + \frac{(V_3 - V_2)}{Z_2} + \frac{V_3}{Z_3} = 0 \tag{2.6}$$

$$V_1 + V_2 - 2V_4 = 0. \tag{2.7}$$

Replacing

$$V_3 = (V_4 + 1),$$

we get

$$\frac{((V_4 + 1) - V_2)}{Z_2} + (2) - \frac{(V_1 - V_4)}{Z_5} - \frac{V_1}{Z_1} = 0 \tag{2.8}$$

$$\frac{(V_4 - V_1)}{Z_5} + \frac{V_4}{Z_4} - \frac{(V_2 - (V_4 + 1))}{Z_2} + \frac{(V_4 + 1)}{Z_3} = 0 \tag{2.9}$$

$$V_1 + V_2 - 2V_4 = 0. \tag{2.10}$$

Simplifying, we get

$$-(Z_1 Z_2 + Z_2 Z_5)V_1 - Z_1 Z_5 V_2 + (Z_1 Z_2 + Z_1 Z_5)V_4 = -Z_1 Z_5 - 2Z_1 Z_2 Z_5$$

$$-Z_2 Z_3 Z_4 V_1 - Z_3 Z_4 Z_5 V_2 + (Z_2 Z_3 Z_4 + Z_2 Z_3 Z_5 + Z_3 Z_4 Z_5 + Z_2 Z_4 Z_5)V_4 = -Z_2 Z_4 Z_5 - Z_3 Z_4 Z_5$$

$$V_1 + V_2 - 2V_4 = 0.$$

With

$$Z_1 = 2, \quad Z_2 = 1, \quad Z_3 = 4, \quad Z_4 = 3, \quad Z_5 = 2,$$

we get

$$\begin{bmatrix} -4 & -4 & 6 \\ -12 & -24 & 50 \\ 1 & 1 & -2 \end{bmatrix} \begin{bmatrix} V_1 \\ V_2 \\ V_4 \end{bmatrix} = \begin{bmatrix} -12 \\ -30 \\ 0 \end{bmatrix}.$$

The determinant of the admittance matrix is -24.

$$\begin{bmatrix} V_1 \\ V_2 \\ V_4 \end{bmatrix} = \begin{bmatrix} -4 & -4 & 6 \\ -12 & -24 & 50 \\ 1 & 1 & -2 \end{bmatrix}^{-1} \begin{bmatrix} -12 \\ -30 \\ 0 \end{bmatrix}$$

$$= \begin{bmatrix} 0.0833 & 0.0833 & 2.3333 \\ -1.0833 & -0.0833 & -5.3333 \\ -0.5000 & 0 & -2.0000 \end{bmatrix} \begin{bmatrix} -12 \\ -30 \\ 0 \end{bmatrix} = \begin{bmatrix} -3.5 \\ 15.5 \\ 6 \end{bmatrix}.$$

The voltages are the same as those found using loop analysis.

The resistors and the respective currents are

$$R1 = 2, R2 = 1, R3 = 4, R4 = 3, R5 = 2$$

and

$$I1 = 1.75, I2 = 8.5, I3 = 1.75, I4 = 2, I5 = 4.75.$$

The power consumed is 147.75 W.

$$P = I1^2 R1 + I2^2 R2 + I3^2 R3 + I4^2 R4 + I5^2 R5 = 147.75 \text{ W}$$

The power supplied is

$$(-6.5)(-19) - (1)(6.75) + 2(15.5) = 147.75,$$

as obtained above.

2.3.3 Analysis of a Circuit with a Controlled Current Source

Nodal Analysis
A circuit with a current-controlled current source is shown in Fig. 2.19. A tree corresponding to the circuit is shown in Fig. 2.20. Since $V_3 = 1$ V is known, there are only 2 unknowns, $\{V_1, V_2\}$. The two equilibrium equations are

$$\frac{(V_2 - V_1)}{Z_2} + \frac{(V_2 - 1)}{Z_4} + \frac{V_2}{Z_3} = 0$$

Fig. 2.19 A circuit with a current-controlled current source

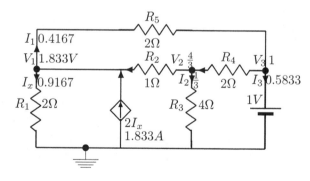

Fig. 2.20 A possible tree corresponding to the circuit in Fig. 2.19

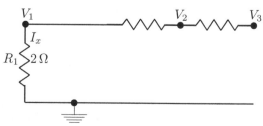

$$\frac{V_1}{Z_1} + \frac{(V_1 - 1)}{Z_5} - 2\frac{V_1}{Z_1} + \frac{(V_1 - V_2)}{Z_2} = 0.$$

The first equation is due to the application of KCL at node with voltage V_2. The second equation is due to the application of KCL at node with voltage V_1. Simplifying, we get

$$-Z_3 Z_4 V_1 + (Z_3 Z_4 + Z_2 Z_4 + Z_2 Z_3) V_2 = Z_2 Z_3$$

$$(-Z_2 Z_5 + Z_1 Z_5 + Z_1 Z_2) V_1 - Z_1 Z_5 V_2 = Z_1 Z_2.$$

With
$$Z_1 = 2, \quad Z_2 = 1, \quad Z_3 = 4, \quad Z_4 = 2, \quad Z_5 = 2,$$

we get

$$-4V_1 + 7V_2 = 2$$

$$2V_1 - 2V_2 = 1.$$

Solving for the voltages, we get

$$\begin{bmatrix} V_1 \\ V_2 \end{bmatrix} = \begin{bmatrix} \frac{11}{6} \\ \frac{4}{3} \end{bmatrix}.$$

All the currents and voltages are shown in Fig. 2.19.

Mesh Analysis
The three loops are shown in Fig. 2.21. In a controlled source, the output current does not have any influence on the input. Therefore,

$$I_{Z_1} = I_2 + I_3.$$

The equilibrium equations are

$$Z_1(I_2 + I_3) + Z_2(I_1 - I_2 - I_3) - Z_3 I_2 = 0$$

$$Z_3 I_2 + Z_4(I_1 - I_3) = 1$$

$$Z_1(I_2 + I_3) - Z_5 I_1 = 1.$$

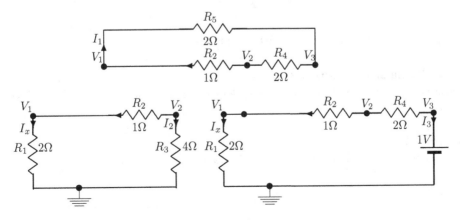

Fig. 2.21 A circuit with a controlled current source, the three loops

The first equation is obtained by equating the voltages on both sides of the current source. The second equation is obtained by applying KVL to the rightmost loop involving the voltage source. The third equation is obtained by applying KVL to the loop involving the voltage source and impedances Z_1 and Z_5. Voltage source is 1 V. Simplifying, we get

$$Z_2 I_1 + (Z_1 - Z_2 - Z_3) I_2 + (Z_1 - Z_2) I_3 = 0$$

$$Z_4 I_1 + Z_3 I_2 - Z_4 I_3 = 1$$

$$-Z_5 I_1 + Z_1 I_2 + Z_1 I_3 = 1.$$

With

$$Z_1 = 2, \quad Z_2 = 1, \quad Z_3 = 4, \quad Z_4 = 2, \quad Z_5 = 2,$$

we get

$$\begin{bmatrix} 1 & -3 & 1 \\ 2 & 4 & -2 \\ -2 & 2 & 2 \end{bmatrix} \begin{bmatrix} I_1 \\ I_2 \\ I_3 \end{bmatrix} = \begin{bmatrix} 0 \\ 1 \\ 1 \end{bmatrix}.$$

The determinant of the impedance matrix is nonzero, 24.

$$\begin{bmatrix} I_1 \\ I_2 \\ I_3 \end{bmatrix} = \begin{bmatrix} 1 & -3 & 1 \\ 2 & 4 & -2 \\ -2 & 2 & 2 \end{bmatrix}^{-1} \begin{bmatrix} 0 \\ 1 \\ 1 \end{bmatrix}$$

$$= \begin{bmatrix} 0.5000 & 0.3333 & 0.0833 \\ 0 & 0.1667 & 0.1667 \\ 0.5000 & 0.1667 & 0.4167 \end{bmatrix} \begin{bmatrix} 0 \\ 1 \\ 1 \end{bmatrix} = \begin{bmatrix} 0.4167 \\ 0.3333 \\ 0.5833 \end{bmatrix}.$$

The resistors and the respective currents are

$$R1 = 2, R2 = 1, R3 = 4, R4 = 2, R5 = 2$$

and

$$I1 = 0.9167, I2 = -0.4997, I3 = 0.3333, I4 = -0.1666, I5 = 0.4167.$$

The power consumed is 2.7776 W.

$$P = I1^2 R1 + I2^2 R2 + I3^2 R3 + I4^2 R4 + I5^2 R5 = 2.7777 \text{ W}$$

The power supplied by the sources is $-0.5833 + (11/6)^2 = 2.7777$.

Circuits with One or Two Variables

Now, we are going to present the analysis of simpler circuits with one or two variables, which can be solved manually.

Circuit in Fig. 2.22

The circuit shown in Fig. 2.22 has two voltage sources.

Nodal Analysis

$$Z_1 = 2, \quad Z_2 = 4, \quad Z_3 = 3, \quad V_1 = 1, \quad V_3 = 2$$

Fig. 2.22 A circuit with
two voltage sources

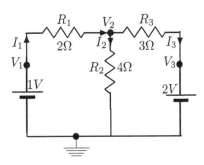

Voltage V_2 is the only unknown. Therefore, applying KCL at node 2, we get

$$\frac{(V_2 - 1)}{Z_1} + \frac{(V_2 - 2)}{Z_3} + \frac{V_2}{Z_2} = 0.$$

Solving for V_2, we get

$$V_2 = \frac{Z_2 Z_3 + 2 Z_1 Z_2}{Z_1 Z_2 + Z_2 Z_3 + Z_1 Z_3} = \frac{28}{26} = \frac{14}{13} \text{ V}.$$

The currents are

$$I_1 = -\frac{(V_2 - 1)}{Z_1} = -\frac{1}{26} \text{ A}, \quad I_2 = \frac{(V_2)}{Z_2} = \frac{14}{52} = \frac{7}{26} \text{ A}, \quad I_3 = \frac{(V_2 - 2)}{Z_3} = -\frac{4}{13} \text{ A}.$$

With $-I_1 + I_2 + I_3 = 0$, KCL is satisfied.

The total power dissipated by the resistors is

$$\frac{1}{26^2}(2 + 49 \times 4 + 64 \times 3) = \frac{390}{26^2} = \frac{15}{26} = 0.5769 \text{ W}.$$

The total power supplied by the sources is

$$-\frac{1}{26} + \frac{16}{26} = \frac{15}{26} = 0.5769 \text{ W}.$$

Current enters the positive terminal of the 1 V source, while current leaves in the 2 V source. Therefore, 1 V source absorbs power, while 2 V source delivers power.

If the current flow in a voltage source is into the positive terminal, then it absorbs power and the power is negative. If the current flow in a voltage source is out of the positive terminal, then it delivers power and the power is positive.

While the current from an ideal current source is fixed, the voltage across it may be positive or negative, depending on the circuit constraints. If the current flow in a current source is into the positive voltage terminal, then it absorbs power and the power is negative. If the current flow in a current source is out of the positive voltage terminal, then it delivers power and the power is positive.

Mesh Analysis
$I_2 = I_1 - I_3.$

$$Z_1 I_1 + Z_2 (I_1 - I_3) = 1$$

$$Z_3 I_3 - Z_2 (I_1 - I_3) = -2.$$

Simplifying, we get

$$(Z_1 + Z_2)I_1 - Z_2 I_3 = 1$$
$$-Z_2 I_1 + (Z_2 + Z_3)I_3 = -2.$$

Substituting the numerical values, we get

$$6I_1 - 4I_3 = 1$$
$$-4I_1 + 7I_3 = -2.$$

Solving the equations, we get the same values of currents, as found earlier.

Let us do the problem by superposition method. Let us find the response to 1 V source. Then, we replace 2 V source by a short-circuit. Current I_1 is

$$I_1 = \frac{1}{Z_1 + (Z_2 || Z_3)} = \frac{7}{26}.$$

By current division, we get

$$I_2 = I_1 \frac{Z_3}{Z_2 + Z_3} = \frac{3}{26}, \; I_3 = I_1 - I_2 = \frac{4}{26}.$$

Let us find the response to 2 V source. Then, we replace 1 V source by a short-circuit. Current I_3 is

$$I_3 = -\frac{2}{Z_3 + (Z_2 || Z_1)} = -\frac{6}{13}.$$

By current division, we get

$$I_1 = I_3 \frac{Z_2}{Z_2 + Z_1} = -\frac{4}{13}, \; I_2 = I_1 - I_3 = \frac{2}{13}.$$

Adding the corresponding two responses, we get the same values for the currents.

Circuit in Fig. 2.23

The circuit shown in Fig. 2.23 has voltage and current sources.

Fig. 2.23 A circuit with voltage and current sources

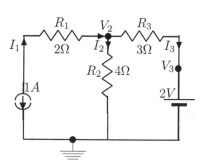

Nodal Analysis

$$Z_1 = 2, \quad Z_2 = 4, \quad Z_3 = 3, \quad I_1 = -1, \quad V_3 = 2.$$

Voltage V_2 is the only unknown. Therefore, applying KCL at node 2, we get

$$\frac{(V_2 - 2)}{Z_3} + \frac{V_2}{Z_2} = -1.$$

Solving for V_2, we get

$$V_2 = \frac{-Z_2 Z_3 + 2Z_2}{Z_2 + Z_3} = -\frac{4}{7} \, \text{V}.$$

The currents are

$$I_1 = -1 \, \text{A}, \quad I_2 = \frac{V_2}{Z_2} = -\frac{1}{7} \, \text{A}, \quad I_3 = \frac{(V_2 - 2)}{Z_3} = -\frac{6}{7} \, \text{A}.$$

With $I_1 - I_2 - I_3 = 0$, KCL is satisfied.

The total power dissipated by the resistors is

$$2 + \frac{1}{7^2}(1 \times 4 + 36 \times 3) = 2 + \frac{112}{7^2} = \frac{30}{7} = 4.2857 \, \text{W}.$$

The total power supplied by the sources is

$$\left(2 + \frac{4}{7}\right) + 2\frac{6}{7} = \frac{30}{7} = 4.2857 \, \text{W}.$$

Both sources deliver power.

Mesh Analysis
$I_2 = I_1 - I_3, \; I_1 = -1.$

$$Z_3 I_3 - Z_2(I_1 - I_3) = -2.$$

Solving for I_3, we get

$$(Z_3 + Z_2)I_3 = -6, \quad I_3 = -\frac{6}{7} \, \text{A}.$$

We get the same values of currents, as found earlier.

Let us solve the problem by superposition method. Let us find the response due to voltage source. The current source is open-circuited. Then, $I_3 = -2/7$, $I_2 = 2/7$. Let us find the response due to current source. The voltage source is short-circuited. Then, by current division, $I_3 = -4/7$, $I_2 = -3/7$. Adding the corresponding two responses, we get the same values for the currents.

Circuit in Fig. 2.24
The circuit shown in Fig. 2.24 has voltage and current sources. Source voltage is 2 V. Source current is 1 A.

$$Z_1 = 2, \quad Z_2 = 4, \quad Z_3 = 3, \quad I_3 = 1.$$

Fig. 2.24 A circuit with
voltage and current sources

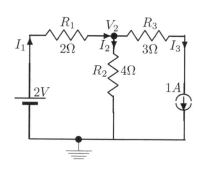

Mesh Analysis

$$I_1 - I_3 = I_2$$

$$I_1 Z_1 + I_2 Z_2 = I_1 Z_1 + (I_1 - I_3) Z_2 = 2, \quad I_1 = \frac{2 + I_3 Z_2}{Z_2 + Z_1} = 1, \quad I_2 = 0.$$

With $-I_1 + I_2 + I_3 = 0$, KCL is satisfied.

The total power dissipated by the resistors is

$$3 + 2 = 5\,\text{W}.$$

The total power supplied by the sources is

$$3 + 2 = \frac{30}{7} = 5\,\text{W}.$$

Both sources deliver power.

Nodal Analysis

$$\frac{V_2}{Z_2} + \frac{(V_2 - 2)}{Z_1} = -1.$$

Solving, $V_2 = 0$ and $V_{R3} = 3$ V. The voltage across the current source is -3 V.

Let us do the problem by superposition method. Consider the response due to the voltage source alone. The current source is open-circuited. Then,

$$I_1 = I_2 = \frac{1}{3}.$$

Consider the response due to the current source alone. The voltage source is short-circuited. Then, by current division, we get

$$I_1 = \frac{2}{3}, \quad I_2 = -\frac{1}{3}.$$

Adding the partial currents, we get the same results.

Circuit in Fig. 2.25

The circuit shown in Fig. 2.25 has current sources. Source currents 2 and 1 A.

$$Z_1 = 2, \quad Z_2 = 4, \quad Z_3 = 3.$$

Fig. 2.25 A circuit with
current sources

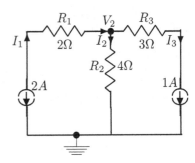

Nodal Analysis

$$\frac{V_2}{Z_2} = -3 \text{ and } V_2 = -12 \text{ V}$$

$$I_1 = -2, \ I_2 = -\frac{12}{4} = -3 \text{ A}, \ I_3 = 1 \text{ A}$$

With $-I_1 + I_2 + I_3 = 0$, KCL is satisfied.

Mesh Analysis

$$I_3 = 1, \quad I_1 = -2, \quad I_2 = (I_1 - I_3) = -3 \text{ A}$$

Let us do the problem by superposition method. Consider the response due to the 2 A current source alone. The other current source is open-circuited. Then, we get

$$I_1 = -2, \quad I_2 = -2, \quad I_3 = 0$$

Consider the response due to the 1 A current source alone. The other current source is open-circuited. Then, by current division, we get

$$I_1 = 0, \quad I_2 = -1, \quad I_3 = 1.$$

Adding the two partial results, we get the same currents.
The total power dissipated by the resistors is

$$8 + 36 + 3 = 47 \text{ W}.$$

The total power supplied by the sources is

$$16 \times 2 + 15 = 47 \text{ W}.$$

Both sources deliver power.

Circuit in Fig. 2.26

A circuit with a voltage-controlled voltage source is shown in Fig. 2.26. Source voltage is 1 V.

$$Z_1 = 2, \quad Z_2 = 4, \quad Z_3 = 3.$$

Fig. 2.26 A circuit with a
voltage-controlled voltage
source

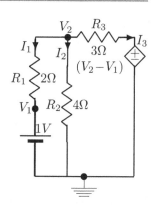

Nodal Analysis

$$\frac{(V_2 - 1)}{2} + \frac{V_2}{4} + \frac{V_2 - (V_2 - 1)}{3} = 0.$$

Solving, we get $V_2 = 2/9$. The currents are

$$I_1 = -\frac{7}{18} \text{ A}, \quad I_2 = -(I_3 + I_1) = \frac{1}{18} \text{ A}, \quad I_3 = \frac{1}{3} \text{ A}.$$

With $I_1 + I_2 + I_3 = 0$, KCL is satisfied.

The total power dissipated by the resistors is

$$\frac{1}{18^2}(98 + 4 + 108) = \frac{35}{54} = 0.6481 \text{ W}.$$

The total power supplied by the sources is

$$\frac{7}{18} + \frac{(-7)}{9}\frac{(-1)}{3} = \frac{35}{54} = 0.6481 \text{ W}.$$

Both sources deliver power.

Mesh Analysis

$$2I_1 + 4(I_3 + I_1) = -1, \quad \text{or} \quad 6I_1 + 4I_3 = -1$$
$$3I_3 + 2I_1 + 4(I_3 + I_1) = 0, \quad \text{or} \quad 6I_1 + 7I_3 = 0.$$

Solving, we get the same currents obtained earlier.

Circuit in Fig. 2.27

A circuit with a current-controlled voltage source is shown in Fig. 2.27. Source voltage is 1 V.

$$Z_1 = 2, \quad Z_2 = 4, \quad Z_3 = 3.$$

Fig. 2.27 A circuit with a current-controlled voltage source

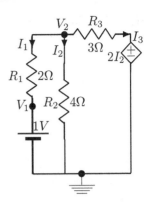

Nodal Analysis

Since $4I_2 = V_2, 2I_2 = 0.5V_2$.

$$\frac{(V_2 - 1)}{2} + \frac{V_2}{4} + \frac{V_2 - 0.5V_2}{3} = 0.$$

Solving, we get $V_2 = 6/11$. The currents are

$$I_1 = -\frac{5}{22} \text{ A}, \quad I_2 = \frac{3}{22} \text{ A}, \quad I_3 = \frac{1}{11} \text{ A}.$$

With $I_1 + I_2 + I_3 = 0$, KCL is satisfied.

The total power dissipated by the resistors is

$$\frac{1}{22^2}(50 + 36 + 12) = \frac{98}{22^2} = 0.2025 \text{ W}.$$

The total power supplied by the sources is

$$\frac{5}{22} + \frac{3}{11}\frac{(-1)}{11} = 0.2025 \text{ W}.$$

The controlled source absorbs power.

Mesh Analysis

$$2I_1 + 4(I_3 + I_1) = -1, \quad \text{or} \quad 6I_1 + 4I_3 = -1$$
$$3I_3 - 2(I_3 + I_1) + 4(I_3 + I_1) = 0, \quad \text{or} \quad 2I_1 + 5I_3 = 0.$$

Solving, we get the same currents obtained earlier.

Circuit in Fig. 2.28

A circuit with a voltage-controlled current source is shown in Fig. 2.28. Source current is 1 A.

$$Z_1 = 2, \quad Z_2 = 4, \quad Z_3 = 3.$$

Fig. 2.28 A circuit with a
voltage-controlled current
source

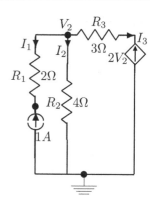

Nodal Analysis

$$-1 - 2V_2 + \frac{V_2}{4} = 0.$$

Solving, we get $V_2 = -4/7$. The currents are

$$I_1 = -1 \text{ A}, \quad I_2 = -\frac{1}{7} \text{ A}, \quad I_3 = \frac{8}{7} \text{ A}.$$

With $I_1 + I_2 + I_3 = 0$, KCL is satisfied.
    The total power dissipated by the resistors is

$$2 + \frac{4}{49} + \frac{192}{49} = 6 \text{ W}.$$

The total power supplied by the sources is

$$\frac{10}{7}(1) + (-4)\frac{(-8)}{7} = \frac{42}{7} = 6 \text{ W}.$$

Both sources deliver power.

Mesh Analysis

$$4(2V_2 + 1) = V_2, \quad \text{or} \quad V_2 = -4/7 \text{ V}$$

as before.

Circuit in Fig. 2.29
A circuit with a current-controlled current source is shown in Fig. 2.29. Source current is 1 A.

$$Z_1 = 2, \quad Z_2 = 4, \quad Z_3 = 3.$$

Nodal Analysis

$$-1 - 0.5V_2 + \frac{V_2}{4} = 0.$$

Fig. 2.29 A circuit with a current-controlled current source

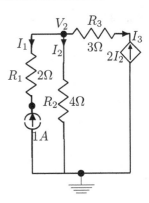

Solving, we get $V_2 = -4$. The currents are

$$I_1 = -1 \, \text{A}, \quad I_2 = -1 \, \text{A}, \quad I_3 = 2 \, \text{A}.$$

With $I_1 + I_2 + I_3 = 0$, KCL is satisfied.

The total power dissipated by the resistors is

$$2 + 4 + 12 = 18 \, \text{W}.$$

The total power supplied by the sources is

$$(-2)(1) + (-2)(-10) = 18 \, \text{W}.$$

Current source 1 A absorbs power.

Mesh Analysis
Since $4I_2 = V_2, 2I_2 = 0.5V_2$.

$$4(0.5V_2 + 1) = V_2, \quad \text{or} \quad V_2 = -4 \, \text{V}$$

as before.

2.3.4 $Y - \Delta$ and $\Delta - Y$ Transformations

In major applications, such as three-phase power systems and electrical filters, certain configurations of circuits often appear and it is required to transform one form from another for simpler circuit analysis. One is called the Y (wye) circuit, since it can be drawn resembling the letter Y, shown in Fig. 2.30a. That is, it has a common point between its three branches. The other one is called the Δ (delta) circuit, since it can be drawn resembling the Greek letter Δ, shown in Fig. 2.30b. That is, it has no common point and its three branches are circularly connected in series. They are also called star-delta and T-π configurations. The transformation is such that the impedance between each pair of terminals remains the same before and after transformation.

$\Delta - Y$ Transformation
The impedance between terminals a and b in the Y circuit is $Z_a + Z_b$. The impedance between terminals a and b in the Δ circuit is the parallel connection of Z_{ab} and $Z_{ac} + Z_{bc}$. Since the values of the impedances between the terminals have to be same in both the configurations, we get

Fig. 2.30 $Y - \Delta$ circuits

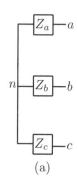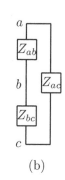

(a)   (b)

$$Z_a + Z_b = \frac{Z_{ab}(Z_{ac} + Z_{bc})}{Z_{ac} + Z_{ab} + Z_{bc}}.$$

Similarly, we get

$$Z_a + Z_c = \frac{Z_{ac}(Z_{ab} + Z_{bc})}{Z_{ac} + Z_{ab} + Z_{bc}}$$

$$Z_b + Z_c = \frac{Z_{bc}(Z_{ab} + Z_{ac})}{Z_{ac} + Z_{ab} + Z_{bc}}.$$

Solving the three equations, we get

$$Z_a = \frac{Z_{ab} Z_{ac}}{Z_{ac} + Z_{ab} + Z_{bc}}$$

$$Z_b = \frac{Z_{ba} Z_{bc}}{Z_{ac} + Z_{ab} + Z_{bc}}$$

$$Z_c = \frac{Z_{ca} Z_{cb}}{Z_{ac} + Z_{ab} + Z_{bc}}.$$

In words, the Y-impedances are given by the product of the two Δ-impedances connected at the corresponding node divided by the sum of the three Δ-impedances. We can solve the first set of equations to get the $Y - \Delta$ transformations. However, it is simpler to get this using duality property.

Duality

A number and its logarithm are equivalent representation of the same entity. However, the logarithmic representation reduces the difficult multiplication problem into the simpler addition problem. We use the more useful representation of the two to solve a problem. Similarly, transform methods reduce the problem of solving a differential equation into solving algebraic equations.

The formulation of circuit analysis problem using loop analysis is of the form

$$IZ = V,$$

where I is the current variable, V is the voltage variable, and Z is the resistance in the case of DC circuits. Let us replace I by V, V by I, and Z by $1/Z$. Then, the equation becomes

$$\frac{V}{Z} = I = VG,$$

Table 2.2 Dual concepts
and variables

Voltage	Current
Loop	Node
Resistance	Conductance
Inductance	Capacitance
Voltage source	Current source
KVL	KCL
Short-circuit	Open-circuit
Parallel paths	Series paths

where G is the conductance if Z is the resistance. This equation is another form of formulation of the circuit analysis problem using nodal analysis. The roles played by the voltage and current variables are interchanged. Since the variables play a dual role, they are called as dual variables. We know the nodal analysis is simpler for some circuits and vice versa. Obviously, we use the simpler one. Voltage is the dual of current and conductance is the dual of resistance.

Consider again the finding of the equivalent resistor R_{eq} of connecting resistors in series and parallel. If the resistors R_1 and R_2 are connected in series, we get

$$R_{eq} = R_1 + R_2.$$

If the resistors are connected in parallel, we get

$$\frac{1}{R_{eq}} = \frac{1}{R_1} + \frac{1}{R_2} \quad \text{or} \quad R_{eq} = \frac{R_1 R_2}{R_1 + R_2} = \frac{1}{G_1 + G_2}.$$

It is obvious that use of conductance is more convenient when analyzing parallel circuits. Some of the dual concepts and variables are shown in Table 2.2.

$Y - \Delta$ Transformation
Using the duality and assuming that the circuit consists of resistors, we get

$$\frac{1}{Z_{bc}} = \frac{\frac{1}{Z_b} \times \frac{1}{Z_c}}{\frac{1}{Z_a} + \frac{1}{Z_b} + \frac{1}{Z_c}} = G_{bc} = \frac{G_b G_c}{G_a + G_b + G_c} \quad \text{or} \quad Z_{bc} = \frac{Z_a Z_b + Z_b Z_c + Z_c Z_a}{Z_a}$$

$$\frac{1}{Z_{ac}} = \frac{\frac{1}{Z_a} \times \frac{1}{Z_c}}{\frac{1}{Z_a} + \frac{1}{Z_b} + \frac{1}{Z_c}} = G_{ac} = \frac{G_a G_c}{G_a + G_b + G_c} \quad \text{or} \quad Z_{ac} = \frac{Z_a Z_b + Z_b Z_c + Z_c Z_a}{Z_b}$$

$$\frac{1}{Z_{ab}} = \frac{\frac{1}{Z_a} \times \frac{1}{Z_b}}{\frac{1}{Z_a} + \frac{1}{Z_b} + \frac{1}{Z_c}} = G_{ab} = \frac{G_a G_b}{G_a + G_b + G_c} \quad \text{or} \quad Z_{ab} = \frac{Z_a Z_b + Z_b Z_c + Z_c Z_a}{Z_c}.$$

In words, the Δ-impedance connecting nodes a and b is obtained by dividing the sum of the pairwise products of the three Y-impedances by the Y-impedance at the opposite node, Z_c. Similar interpretations hold for the other two impedances. If all the impedances are equal, it is called a balanced circuit. Then,

$$Z_Y = \frac{Z_\Delta}{3} \quad \text{and} \quad Z_\Delta = 3Z_Y.$$

The difference in the conversion formulas for balanced circuits is due to the fact that Y-connection is similar to a series circuit while the other is similar to a parallel circuit.

Fig. 2.31 A bridge circuit

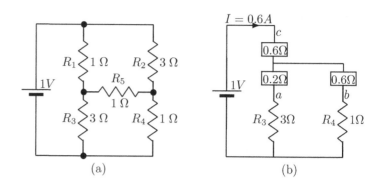

(a)   (b)

Example

Find the current I drawn by the bridge circuit, shown in Fig. 2.31a, from the source.

The circuit can be considered as the combination of two Δ circuits with a shared resistor. To make the analysis simpler, we have to convert one of them into an equivalent circuit. Replacing the top half of the Δ circuit, we get, with

$$Z_{bc} = 3, \quad Z_{ac} = 1, \quad Z_{ab} = 1$$

$$Z_a = \frac{1 \times 1}{3 + 1 + 1} = 0.2$$

$$Z_b = \frac{3 \times 1}{3 + 1 + 1} = 0.6$$

$$Z_c = \frac{3 \times 1}{3 + 1 + 1} = 0.6.$$

The transformed circuit is shown in Fig. 2.31b. Now,

$$Z_{eq} = 0.6 + \frac{(0.6 + 1)(3 + 0.2)}{(0.6 + 1 + 3 + 0.2)} = \frac{5}{3}.$$

Therefore,

$$I = \frac{V}{Z_{eq}} = (1)\frac{3}{5} = 0.6 \text{ A},$$

which is the same as found by nodal and loop analyses.

Let us get back the Δ circuit from the Y circuit.

$$Z_{bc} = \frac{Z_a Z_b + Z_b Z_c + Z_c Z_a}{Z_a} = \frac{(0.2)(0.6) + (0.2)(0.6) + (0.6)(0.6)}{0.2} = 3$$

$$Z_{ac} = \frac{Z_a Z_b + Z_b Z_c + Z_c Z_a}{Z_b} = \frac{(0.2)(0.6) + (0.2)(0.6) + (0.6)(0.6)}{0.6} = 1$$

$$Z_{ab} = \frac{Z_a Z_b + Z_b Z_c + Z_c Z_a}{Z_c} = \frac{(0.2)(0.6) + (0.2)(0.6) + (0.6)(0.6)}{0.6} = 1.$$

2.4 Circuit Theorems

2.4.1 Thévenin's Theorem and Norton's Theorem

If the interest is to find the current and voltage in a restricted part of a circuit, then simplified procedures can be used rather than complete nodal or mesh analysis. In one of the two procedures, using Thévenin's theorem, the rest of the circuit is replaced by an equivalent voltage source acting at the point of interest. Thévenin's theorem is convenient in such applications as to find the load for maximum average power transfer in a circuit. In the other procedure using Norton's theorem, the rest of the circuit is replaced by an equivalent current source acting at the point of interest. The procedures are, respectively, special cases of nodal and loop analysis.

Thévenin's Theorem

Any linear combination of voltage and current sources, independent or dependent, and resistors with two terminals can be replaced by a fixed voltage source V_{oc}, called the Thévenin equivalent voltage, in series with a resistor R_{eq}, called Thévenin equivalent resistance. The Thévenin equivalent circuit has an equivalent volt–ampere relationship only from the point of view of the load. It is a restricted kind of source equivalence. The equivalent source is derived as follows:

1. Remove the load circuit, any combination of independent and dependent voltage and current sources, and linear and nonlinear resistors. Find the open-circuit voltage, called V_{oc}, across the load circuit terminals.
2. Find the resistance across the load terminals, after short-circuiting all independent voltage sources and open-circuiting all independent current sources, called R_{eq}. This step can also be carried out by finding the short-circuit current I_{sc} through the load circuit terminals and $R_{eq} = V_{oc}/I_{sc}$.

Thévenin equivalent circuit is shown in Fig. 2.32a. The equivalent voltage source, replacing a circuit, is shown on the left side of the two nodes a and b. The constraint is that the current through the load circuit, represented by the resistor R_L in the figure, is the same as that produced by the actual circuit.

The bridge circuit, we analyzed earlier in this chapter, is shown in Fig. 2.33. The load resistor is R_5. The problem is to find the load current in the actual circuit using Thévenin's theorem.

Solution: Thévenin's Theorem Method

First, we have to find the open-circuit voltage $V_{oc} = V_{ab}$. With the load resistor removed, the current in each of the two parallel branches is $1/(3 + 1) = 0.25$ A. Therefore,

$$V_a = 0.75, \quad V_b = 0.25, \quad V_{ab} = V_{oc} = 0.5\,\text{V}.$$

Fig. 2.32 Thévenin and Norton equivalent circuits

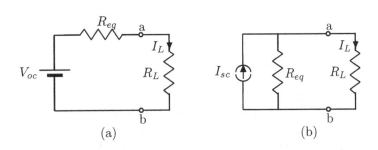

(a)   (b)

Fig. 2.33 A bridge circuit
with the load resistor R_5

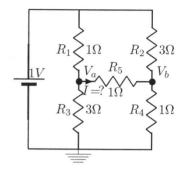

Fig. 2.34 Equivalent
resistance from ab and the
Thévenin equivalent circuit
replacing the actual circuit

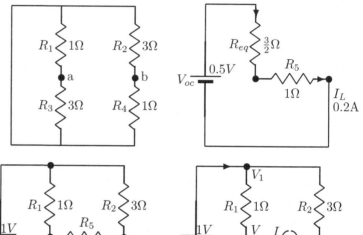

Fig. 2.35 A bridge circuit
and the load resistor
replaced by a current
source supplying current I

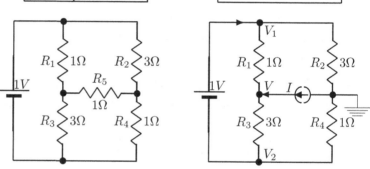

Now, we have to find the equivalent resistance, R_{eq}. For that purpose, the load resistor has to
be removed and the voltage source short-circuited, as shown on the left side of Fig. 2.34. The net
resistance at the terminal pair ab is the sum of the resistances of the pair of parallelly connected
resistors 1 and 3 Ω. That is,

$$R_{eq} = \frac{1 \times 3}{1 + 3} + \frac{1 \times 3}{1 + 3} = \frac{3}{2}\,\Omega.$$

The right side of the Fig. 2.34 shows the load resistor connected to the Thévenin equivalent circuit
replacing the actual circuit. The current in the load resistor is

$$I_L = \frac{0.5}{1.5 + 1} = 0.2\text{ A},$$

as found in our earlier analysis by the loop method.

Nodal Analysis

The load current supplied by a circuit, to be replaced by an equivalent circuit, was open-circuited to
find the Thévenin equivalent circuit in the first approach. Alternatively, the same effect can be achieved
by inserting a bucking voltage to make the current through the load circuit zero, as shown in Fig. 2.35.
Thévenin equivalent circuit is characterized by the equation

$$V = I R_{eq} + V_{oc},$$

where I is the load current. If $I = 0$, $V = V_{oc}$, the open-circuit voltage. If $V = 0$, $I_{sc} = -\frac{V_{oc}}{R_{eq}}$, the short-circuit current.

Applying KVL and KCL at the top and bottom nodes at the right side of Fig. 2.35, the equilibrium equations are

$$V_1 - V_2 = 1$$

$$\frac{V_1}{3} + \frac{V_1 - V}{1} + \frac{V2}{1} + \frac{V_2 - V}{3} = 0.$$

The second equation is obtained using the fact that the current leaving the bottom node and entering the top node must be equal. Solving for V_1 and V_2 in terms of V, we get

$$V_1 = \frac{V + 1}{2} \quad \text{and} \quad V_2 = \frac{V - 1}{2}.$$

The equilibrium equation at node with voltage V is

$$\frac{V - V_1}{1} + \frac{V - V_2}{3} = I.$$

Substituting for V_1 and V_2 in this equation, we get

$$V = \frac{3}{2}I + \frac{1}{2} = I R_{eq} + V_{oc}.$$

The first method seems to be easier for this problem.

Norton's Theorem

Any linear combination of voltage and current sources, independent or dependent, and resistors with two terminals can be replaced by a fixed current source I_{sc}, called Norton equivalent circuit, in parallel with a resistor R_{eq}, as shown in the right side of Fig. 2.32b. The equivalent source is derived as follows:

1. Remove the load circuit, any combination of independent and dependent voltage and current sources, and linear and nonlinear resistors, and find the short-circuit current, called I_{eq} or I_{sc}, through the load circuit terminals.
2. Find the resistance across the load terminals, after short-circuiting all independent voltage sources and open-circuiting all independent current sources, called R_{eq}. This step can also be carried out by finding the short-circuit current I_{sc} through the load circuit terminals and $R_{eq} = V_{oc}/I_{sc}$.

By current division, from Fig. 2.32b, we get

$$I_L = I_{sc} \frac{R_{eq}}{R_{eq} + R_L} = \frac{V_{oc}}{R_{eq} + R_L}.$$

Fig. 2.36 The short-circuit current I_{sc} and the Norton equivalent circuit replacing the actual circuit

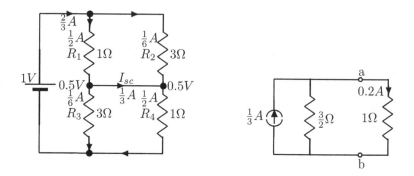

Norton equivalent circuit is characterized by the equation

$$V = I R_{eq} + V_{oc} \quad \text{or} \quad I = \frac{V}{R_{eq}} - I_{sc}.$$

For the bridge circuit, we have found $R_{eq} = \frac{3}{2}$, which is the same for both equivalent circuits. Finding the short-circuit by short-circuiting the load resistor is shown in the left side of Fig. 2.36. The total resistance, after short-circuiting the load resistor, becomes

$$1 \parallel 3 + 1 \parallel 3 = \frac{3}{2}, \quad \text{and} \quad I_{total} = 1/(3/2) = \frac{2}{3} \text{ A}.$$

This current gets divided between the resistors. The current flowing through $R_1 = (1 - 0.5)/1 = 1/2$ A. The current flowing through $R_3 = 0.5/3 = 1/6$ A. Therefore,

$$I_{sc} = \frac{1}{2} - \frac{1}{6} = \frac{1}{3} \text{ A}.$$

The load current is

$$I_L = \frac{\frac{1}{3}}{\frac{3}{2} + 1} \left(\frac{3}{2} \right) = \frac{2}{15} \left(\frac{3}{2} \right) = 0.2 \text{ A}.$$

The Norton equivalent circuit is shown on the right side of Fig. 2.36. The voltage across the load resistor in the Thévenin equivalent circuit is

$$V_L = \frac{V_{oc}}{R_{eq} + R_L} R_L.$$

The voltage across the load resistor in the Norton equivalent circuit is

$$V_L = \frac{R_{eq} I_{sc}}{R_{eq} + R_L} R_L,$$

which are the same as

$$V_{oc} = I_{sc} R_{eq}$$

A circuit with voltage and current sources is shown in Fig. 2.37. The problem is to find the current through the resistor R_4. In order to find R_{eq}, we have to short-circuit the voltage sources and open-

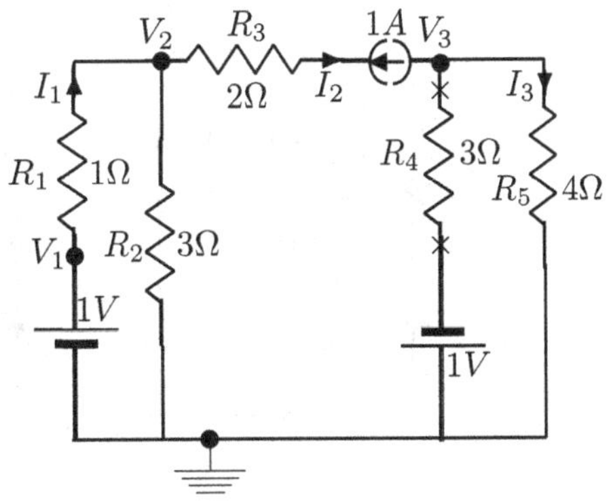

Fig. 2.37 A circuit with voltage and current sources

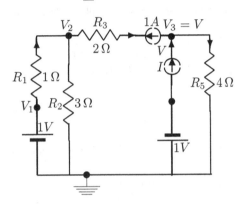

Fig. 2.38 The circuit with a bucking voltage across R_4

circuit the current source and find the resistance across the terminals marked $\times$ in the figure, with R_4 disconnected. The rightmost part of the circuit remains and the rest disconnected. Therefore, $R_{eq} = 4\,\Omega$.

Let us find the open-circuit voltage across terminals, with R_4 disconnected. At node 3, the equilibrium equation is

$$\frac{V_3}{4} = -1 \quad \text{and} \quad V_3 = -4\,\text{V}.$$

Now, $V_{oc} = -4 + 1 = -3\,\text{V}$. Therefore, the current through R_4 is

$$\frac{V_{oc}}{R_4 + R_{eq}} = \frac{-3}{4 + 3} = -\frac{3}{7}\,\text{A}.$$

The sign of V_{oc} and R_{eq} may be positive or negative, depending on the circuit constraints.

Using the alternative approach, let us insert a bucking voltage across R_4, as shown in Fig. 2.38. At the node 3, the equilibrium equation is

$$I - \frac{(V - 1)}{4} = 1.$$

Simplifying, we get

$$V = 4I - 3 \quad \text{and} \quad V = I\,R_{eq} + V_{oc}.$$

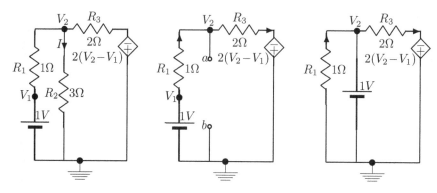

Fig. 2.39 A circuit with a controlled voltage source

A circuit with a controlled voltage source is shown in Fig. 2.39. The problem is to find the current through the resistor R_2. By nodal analysis, we get

$$\frac{(V_2 - 1)}{1} + \frac{V_2}{3} + \frac{V_2 + 2(V_2 - 1)}{2} = 0.$$

Solving, we get $V_2 = 12/17$ and the current through R_2 is $4/17$ A.

Thévenin Theorem Method
Applying KVL to the middle figure, we get

$$(1 - V_2) + 2(1 - V_2) = 1 + 2(V_2 - 1) \quad \text{and} \quad V_2 = V_{oc} = 0.8 \, \text{V}.$$

We can find the short-circuit current $I_{sc} = 1 + 1 = 2$ and $R_{eq} = 0.8/2 = 0.4\Omega$. There is another method to find R_{eq}. In order to find R_{eq} and the application of linearity, leave the dependent sources. Short-circuit independent voltage sources and open-circuit independent current sources. Then, apply a voltage source, such as 1 V, across the load terminals, as shown in the rightmost figure. Find the current through this source and the inverse of the current is R_{eq}. For the example, the current is 2.5 A. Therefore, $R_{eq} = 1/2.5 = 0.4\Omega$. Alternatively, insert a current source, such as 1 A, between the load terminals. The voltage across the source is R_{eq}. For the example, applying KCL at node 2, we get

$$\frac{V_2}{1} + \frac{3V_2}{2} = 1.$$

Solving, we get $V_2 = 0.4$ and $R_{eq} = 0.4\Omega$. The current through R_2 is

$$\frac{0.8}{3 + (0.4)} = \frac{4}{17} \, \text{A}.$$

By applying a bucking voltage across the load terminals and applying KCL, we get

$$\frac{(V - 1)}{1} - I + \frac{V + 2(V - 1)}{2} = 0 \quad \text{and} \quad V = 0.4I + 0.8.$$

Fig. 2.40 (a) A circuit with voltage sources; (b) equivalent circuit with current sources

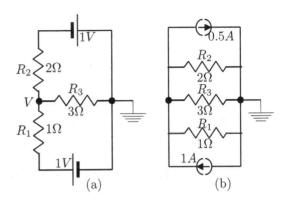

In circuits with both independent and dependent sources, the dependent source and its controlling variable must not be split when the circuit is broken to find the Thévenin or Norton equivalent.

Source Transformation

Thévenin and Norton equivalent circuits provide the same volt–ampere relationship at the terminals. It is possible to substitute one source by another, called source transformation, to simplify the circuit analysis. Source transformation is most useful if the source is localized to some portion of the circuit.

Consider the circuit shown in Fig. 2.40a with voltage sources. The problem is to find the current through R_3. Applying KCL at the middle node, we get

$$\frac{(V-1)}{1} + \frac{V}{3} + \frac{(V+1)}{2} = 0 \quad \text{and} \quad V = \frac{3}{11}\,\text{V}.$$

Consider the transformed circuit shown in Fig. 2.40b with current sources. By current division, we get

$$I_{R3} = 0.5\frac{2}{11} = \frac{1}{11}\,\text{A}$$

and the voltage across the resistor is $3/11$, as obtained above. With current sources, we got the result by using the simpler current-division formula.

2.4.2 Maximum Power Transfer Theorem

Systems absorb maximum power only for short periods. For most of the time, the load resistance is high compared with the source resistance resulting in a high efficiency. For an ideal source, $R_{eq} = 0$. It is often of interest in applications to find the condition for maximum power transfer from source to load. Using Thévenin's theorem, any linear circuit can be represented by an ideal voltage source V_{oc} in series with a resistance R_{eq}. A load resistor R_L is connected across this equivalent circuit, as shown on the left side of Fig. 2.32. The problem is to find out the value of R_L for a given R_{eq}.

The current through R_L in Fig. 2.32 is

$$I = \frac{V_{oc}}{R_L + R_{eq}}.$$

The power absorbed by R_L is

$$P_L = I^2 R_L = \frac{V_{oc}^2}{(R_L + R_{eq})^2} R_L.$$

To find the maximum power transfer, we differentiate this expression with respect to R_L and equate to zero. That is,

$$\frac{dP_L}{dR_L} = \frac{(R_L + R_{eq})^2 V_{oc}^2 - 2V_{oc}^2 R_L (R_L + R_{eq})}{(R_L + R_{eq})^4} = 0,$$

which yields $R_L = R_{eq}$. The maximum power delivered is

$$P_{\max} = \frac{V_{oc}^2}{4R_{eq}}.$$

Maximum power transfer theorem requires that maximum power is transferred between the source and load, when $R_L = R_{eq}$.

At maximum power transfer, half of the power is wasted in the source. However, circuits absorb maximum power only for short periods. For most of the time, the load resistance is high compared with the source resistance resulting in a low loss of power. For an ideal source, $R_{eq} = 0$ providing 100% efficiency of power transfer.

Example

Find the load resistance R_m in the bridge circuit, shown in Fig. 2.41a, for maximum power transfer from the source to the load.

We have to find the Thévenin's equivalent voltage and resistance across the load resistance terminals. Replacing the bottom half of the Δ circuit by Y-circuit, we get, with

$$Z_{bc} = 1, \quad Z_{ac} = 3, \quad Z_{ab} = 1$$

$$Z_a = \frac{3 \times 1}{3 + 1 + 1} = 0.6$$

$$Z_b = \frac{1 \times 1}{3 + 1 + 1} = 0.2$$

$$Z_c = \frac{3 \times 1}{3 + 1 + 1} = 0.6.$$

Fig. 2.41 (a) A bridge circuit; (b) circuit reduction using $\Delta - Y$ transformation

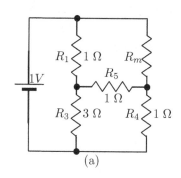

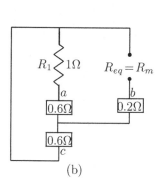

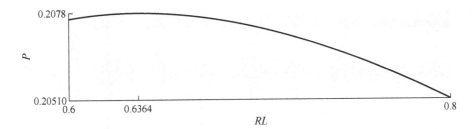

Fig. 2.42 Load resistor versus power transferred

The transformed circuit, with the voltage source short-circuited, is shown in Fig. 2.41b. Now,

$$Z_{eq} = 0.2 + (0.6 \parallel (0.6 + 1)) = 0.2 + \frac{(0.6 + 1)(0.6)}{(0.6 + 1 + 0.6)} = 0.6364.$$

The V_{oc} across the load terminals of the transformed circuit, with the voltage source inserted, is

$$V_{oc} = \frac{(1 + 0.6)1}{1 + 0.6 + 0.6} = 0.7273.$$

Therefore, the maximum power transferred is

$$P_m = \frac{V_{oc}^2}{4Z_{eq}} = \frac{(0.7273^2)}{(4 \times 0.6364)} = 0.2078 \, \text{W}.$$

Figure 2.42 shows power transferred for a range of values of the load resistance. The peak value occurs when the load resistance is equal to the Thévenin's equivalent resistance. In power system applications, efficiency of power transmission is important, while maximum power transfer is important in signal transmission.

2.5 Application

While AC power supply is advantageous for generation, transmission, and distribution, the DC power supply is equally important in utilization of electrical energy. In some industrial applications, DC supply is preferred. In electrical traction, DC supply is extensively used. In numerous electrical and electronic appliances, the circuit works with a DC supply. Invariably, the AC input is converted to provide the required DC supply. The automobile electrical system is typical one of the many systems in day-to-day usage.

2.5.1 Strain Gauge Measurement

When we apply a force to a body, the amount of deformation of it is the strain. Strain is defined as the ratio of the increment in length to the original length.

$$\varepsilon = \frac{\delta L}{L}.$$

Fig. 2.43 Wheatstone
bridge circuit for strain
measurement

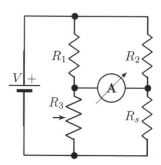

It can be tensile or compressive. The metallic strain gauge is a thin wire, a foil arranged in a grid pattern. It is bonded to a thin backing and attached to the test specimen. When the specimen experiences a strain, the electrical resistance of the gauge varies linearly. Typical value of the resistance is about $1000\,\Omega$.

The sensitivity of a strain gauge, expressed as the gauge factor (GF), is the ratio of fractional change in resistance to the fractional change in length.

$$GF = \frac{\delta R/R}{\delta L/L} = \frac{\delta R/R}{\varepsilon}.$$

Typical value for GF is 2. The strain gauge is compensated suitably to reduce its effect with temperature changes. The change in resistance is typically $0.1\,\Omega$. In order to measure this small change accurately and also compensate for temperature changes, strain gauges are usually used in bridge circuits, called Wheatstone bridge, with voltage or current source. The Wheatstone bridge circuit for strain measurement is shown in Fig. 2.43. The variable resistance R_3 consists of R and δR. Initially, $\delta R = 0$. The bridge is balanced by adjusting R so that the current through the ammeter is zero. That is

$$R_s = \frac{R_2}{R_1}R.$$

When the gauge experiences strain, its value becomes $R_s + \delta R_s$. Now, the bridge is unbalanced. By adjusting δR, the bridge is again balanced. Then,

$$R_s + \delta R_s = \frac{R_2}{R_1}(R + \delta R) \quad \text{and} \quad \delta R_s = \frac{R_2}{R_1}\delta R.$$

With δR_s, R_s and the gauge factor known, the strain is determined.

2.6   Summary

- An electric circuit, for theoretical analysis, is an interconnection of idealized representation of physical components, such as voltage and current sources, switches, resistors, inductors, and capacitors.
- The relationship between the voltage across an element and the current through it is called its volt–ampere relationship and is linear in the specified operating ranges.
- With the sources and circuit elements specified, the purpose of circuit analysis is to determine the voltages and currents at all parts of the circuit.
- Since the current and voltage in an element are related through its volt–ampere relationship, circuits can be analyzed in terms of the branch currents alone or branch voltages alone.

- The method of circuit analysis based on currents is called mesh or loop analysis. The other method based on voltages is called nodal analysis.
- The equilibrium conditions for a circuit can be established in either of the two ways: (1) through a set of N equations, using KVL, in which the mesh currents are the independent variables; (2) through a set of M equations, using KCL, in which the branch voltages are the independent variables.
- An ideal voltage source is characterized by its volt–ampere relationship of keeping its terminal voltage same irrespective of the current drawn by the load circuit connected to it.
- A circuit, geometrically, is characterized by its branches and nodes.
- A loop is a closed path in a circuit. It starts at a node, passes through a set of nodes (passing through each node only once), and returns to the starting node. A loop is independent if it contains at least one branch that is not part of any other independent loop.
- A tree is any set of branches of the circuit that is sufficient to connect all its nodes. A tree contains no closed paths. There is no unique tree corresponding to a circuit.
- Any of the left out branches of a circuit in forming a tree are called the links.
- The equilibrium state of a circuit is determined by the chosen independent link currents flowing through the selected loops.
- Procedure for circuit analysis using loops involves: (1) selection of an appropriate number of independent current variables and the directions of current flow; (2) expressing the dependent current variables, by applying KCL at nodes, in terms of independent current variables; (3) applying KVL around the loops to set up a set of simultaneous equations; (4) solving for the independent currents and finding the currents in all the branches; (5) verifying the solution using KVL and KCL.
- For certain type of circuits under some conditions, the general procedure can be simplified.
- The selection of the direction of the currents in the branches can be arbitrary. If a current flows in the direction opposite to that assumed, the analysis result will be negative-valued.
- The equilibrium state of a circuit can also be determined by a set of independent voltages, called the nodal method.
- KCL is applied at nodes of the selected tree of the circuit to find the required equilibrium equations.
- The determinant of the impedance matrix must be nonzero.
- The determinant of the admittance matrix must be nonzero.
- If a voltage source, independent or dependent, is not connected to the ground node, then its two nodes and any elements connected in parallel with it is called a supernode.
- An ideal current source is characterized by its volt–ampere relationship of keeping its terminal current same irrespective of the voltage across the load circuit connected to it.
- The response of a circuit to a linear combination of inputs is linear if the output is also the same linear combination of the individual outputs to the inputs.
- The voltage or current provided by a controlled or dependent source, a 3-terminal device, is controlled by another voltage or current at some other part of the circuit.
- Certain configurations of circuits, such as Y and Δ circuits, often appear in applications and it is required to transform one form from another for simpler circuit analysis.
- Using Thévenin's theorem, the current through a part of the circuit can be easily found by replacing the rest of the circuit by an equivalent voltage source acting at the point of interest.
- Using Norton's theorem, the current through a part of the circuit can be easily found by replacing the rest of the circuit by an equivalent current source acting at the point of interest.
- It is possible to substitute one source by another, called source transformation, to simplify the circuit analysis.
- Maximum power transfer theorem gives the condition for maximum power transfer from source to load.

Exercises

2.1 Given a circuit diagram with the independent currents, find the corresponding tree and the loops. Analyze the circuit by both nodal and loop methods to find the voltages and currents at all parts of the circuit. Verify that the results are the same by both the methods. Verify the results using KVL and KCL. Find the power consumed by the circuit.

    2.1.1 The circuit is shown in Fig. 2.44.

    * **2.1.2** The circuit is shown in Fig. 2.45.

    2.1.3 The circuit is shown in Fig. 2.46.

2.2 Given a circuit diagram with the independent currents, analyze the circuit by both nodal and loop methods to find the voltages and currents at all parts of the circuit. Verify that the results are the same by both the methods. Verify the results using KVL and KCL. Find the power consumed by the circuit.

    2.2.1 The circuit is shown in Fig. 2.47.

    * **2.2.2** The circuit is shown in Fig. 2.48.

2.3 Given a circuit diagram with the independent currents, analyze the circuit by both nodal and loop methods to find the voltages and currents at all parts of the circuit. Verify that the results are the same by both the methods. Verify the results using KVL and KCL. Find the power consumed by the circuit.

    2.3.1 The circuit is shown in Fig. 2.49.

    * **2.3.2** The circuit is shown in Fig. 2.50.

2.4 Given a circuit diagram with the independent currents, analyze the circuit by both nodal and loop methods to find the voltages and currents at all parts of the circuit. Verify that the results are the same by both the methods. Find the power consumed by the circuit. Verify the results using KVL and KCL. Analyze the circuit using the linearity property also. Verify that the power consumed is equal to the power supplied.

    2.4.1 The circuit is shown in Fig. 2.51.

    * **2.4.2** The circuit is shown in Fig. 2.52.

2.5 Given a circuit diagram with the independent currents, analyze the circuit by both nodal and loop methods to find the voltages and currents at all parts of the circuit. Verify that the results are the same by both the methods. Verify the results using KVL and KCL. Find the power consumed by the circuit. Analyze the circuit using the linearity property also.

    2.5.1 The circuit is shown in Fig. 2.53.

    * **2.5.2** The circuit is shown in Fig. 2.54.

Fig. 2.44 Circuit for Exercise 2.1.1

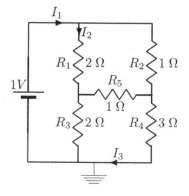

Fig. 2.45 Circuit for
Exercise 2.1.2

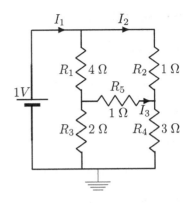

Fig. 2.46 Circuit for
Exercise 2.1.3

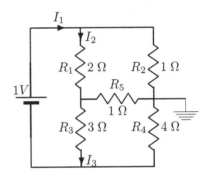

Fig. 2.47 Circuit for
Exercise 2.2.1

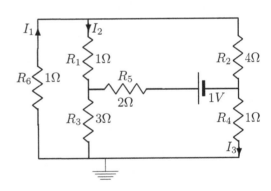

Fig. 2.48 Circuit for
Exercise 2.2.2

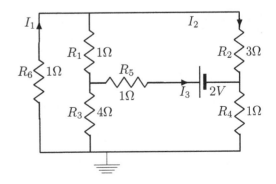

Fig. 2.49 Circuit for
Exercise 2.3.1

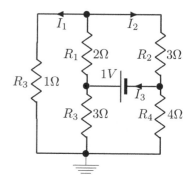

Fig. 2.50 Circuit for
Exercise 2.3.2

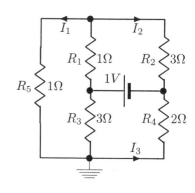

Fig. 2.51 Circuit for
Exercise 2.4.1

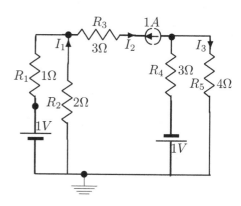

Fig. 2.52 Circuit for
Exercise 2.4.2

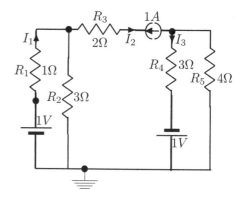

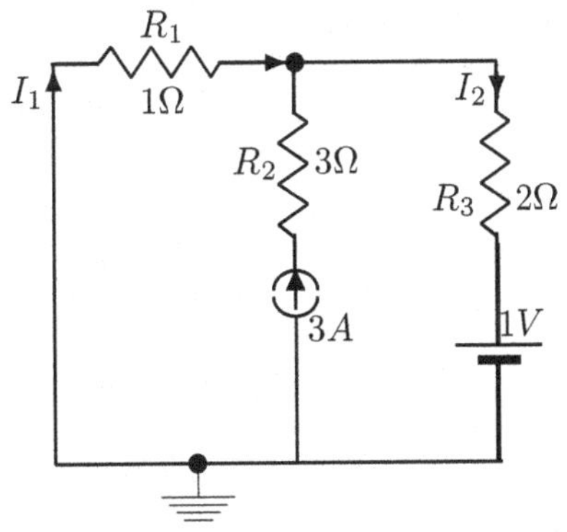

Fig. 2.53 Circuit for
Exercise 2.5.1

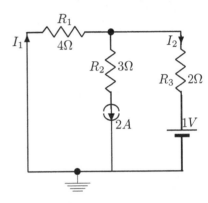

Fig. 2.54 Circuit for
Exercise 2.5.2

Fig. 2.55 Circuit for
Exercise 2.6.1

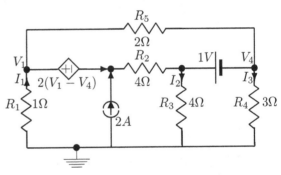

2.6 Given a circuit diagram with the independent currents, analyze the circuit by both nodal and loop methods to find the voltages and currents at all parts of the circuit. Verify that the results are the same by both the methods. Verify the results using KVL and KCL. Find the power consumed by the circuit.

    2.6.1 The circuit is shown in Fig. 2.55.

    ∗ 2.6.2 The circuit is shown in Fig. 2.56.

2.7 Given a circuit diagram with the independent currents, analyze the circuit by both nodal and loop methods to find the voltages and currents at all parts of the circuit. Verify that the results are the same by both the methods. Verify the results using KVL and KCL. Find the power consumed by the circuit.

Fig. 2.56 Circuit for
Exercise 2.6.2

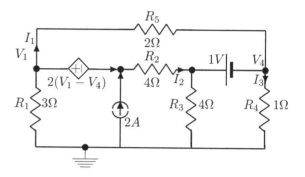

Fig. 2.57 Circuit for
Exercise 2.7.1

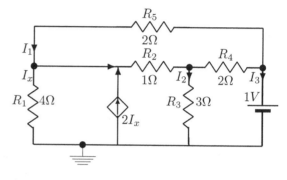

Fig. 2.58 Circuit for
Exercise 2.7.2

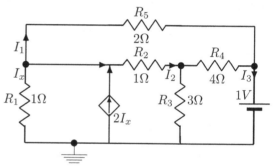

    2.7.1 The circuit is shown in Fig. 2.57.

    * 2.7.2 The circuit is shown in Fig. 2.58.

2.8 Find the Thévenin and Norton equivalent circuits and determine the current through the load resistor.

    2.8.1 The load resistor is R_1 in Fig. 2.2a.

    2.8.2 The load resistor is R_3 in Fig. 2.6b.

    2.8.3 The load resistor is R_3 in Fig. 2.8.

    2.8.4 The load resistor is R_2 in Fig. 2.10a.

    2.8.5 The load resistor is R_3 in Fig. 2.12.

    2.8.6 The load resistor is R_1 in Fig. 2.15.

    2.8.7 The load resistor is R_3 in Fig. 2.19.

2.9 Find the value of the resistor R for maximum power transfer from the source to the resistor R.

    2.9.1 The circuit is shown in Fig. 2.59.

    * 2.9.2 The circuit is shown in Fig. 2.60.

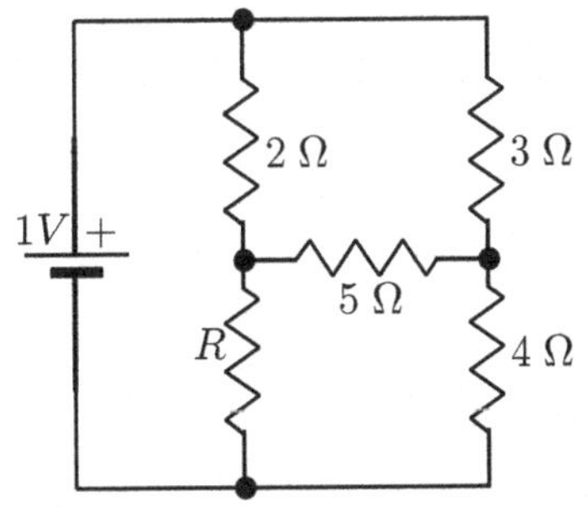

Fig. 2.59 Circuit for
Exercise 2.9.1

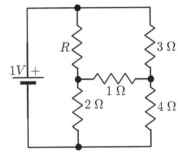

Fig. 2.60 Circuit for
Exercise 2.9.2

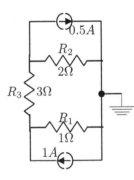

Fig. 2.61 A circuit with
current sources

Fig. 2.62 A circuit with
voltage sources

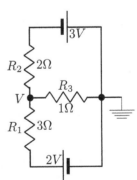

2.10 Find the current through the resistor R_3 by analyzing the circuit as shown and also after source transformation.

   * **2.10.1** The circuit is shown in Fig. 2.61.

   2.10.2 The circuit is shown in Fig. 2.62.

AC Circuits

3

Alternating current (AC) is also movement of electrical charge, but changes direction periodically. As the current is alternating, voltage must also alternate. That is, the polarity is changed in each cycle. The advantages of AC power supply are:

1. Power transmission and distribution is more efficient. Power has to be transmitted over long distances from source to consumption. Transmitting power at high voltages reduce transmission losses. But, power is utilized at low voltages. Transformers are used to provide the change in levels.
2. AC power generation is easier. It is also used in most high power applications.

However, DC is used in majority of electrical appliances and certain industrial applications, such as electric traction. Nowadays, DC power is mostly generated by batteries, solar cells, or by converting AC to DC. Therefore, the study of both DC and AC circuits is required. The basic principles are essentially the same. However, the study of DC circuit analysis, without transient analysis, is relatively simpler, as the amplitude profile of voltages and currents is a straight line. Therefore, we study DC circuit analysis first followed by AC circuit analysis. While the DC waveform is a straight line, as can be seen on the screen of an oscilloscope, the AC waveform is sinusoidal. The well-known trigonometric sine and cosine functions are the mathematical definitions of this waveform.

3.1 Sinusoids

A sinusoidal waveform is a linear combination of sine and cosine waveforms. Sinusoidal representation of signals is indispensable in the analysis of signals and systems for the following reasons. The steady-state waveform, due to an input sinusoid, in any part of a linear system is also a sinusoid of the same frequency as that of the input differing only in its magnitude and phase. In addition, the sum of any number of sinusoids of the same frequency is also a sinusoid of the same frequency. The frequency of a sinusoid remains the same in its derivative form also. Therefore, system models, such as differential equation and convolution, reduce to algebraic equations for a sinusoidal input for linear systems. Further, due to the orthogonal property, an arbitrary signal can be decomposed into a set of sinusoids easily. In addition, this decomposition can be implemented faster, in practice,

© Springer Nature Switzerland AG 2020
D. Sundararajan, *Introductory Circuit Theory*,
https://doi.org/10.1007/978-3-030-31985-4_3

using efficient numerical algorithms resulting in finding the system output faster than other methods. Physical systems also, such as a combination of an inductor and a capacitor, produce an output of sinusoidal nature. The motion of a simple pendulum is approximately sinusoidal.

The Polar Form of Sinusoids

There are two forms of representation of real sinusoids. At a given angular frequency ω, a sinusoid is characterized by its amplitude A and its phase θ (called the polar form) or by the amplitudes of its sine and cosine components (called the rectangular form). Signal amplitude can be either positive or negative. However, magnitude is always positive.

The polar form of a sinusoid is

$$x(t) = A\cos(\omega t + \theta), \qquad -\infty < t < \infty.$$

A sinusoidal waveform has a positive peak and a negative peak in each cycle. The distance of either peak of the waveform from the horizontal axis is its amplitude A. The cosine function is periodic, as

$$\cos(\omega t) = \cos(\omega t + 2\pi).$$

It repeats its values for $t = t + T = t + 2\pi/\omega$, where T is its period in seconds. Then, the cyclic frequency of the sinusoid is $f = 1/T$ Hz (cycles/second). The independent variable t, while time in most applications, can be anything else also, such as distance.

Sinusoids $x(t) = \cos(\frac{2\pi}{8}t)$ and $x(t) = 2\sin(\frac{2\pi}{8}t)$ are shown in Fig. 3.1a. Cosine and sine waveforms are special cases of a sinusoidal waveform. Cosine waveform has its peak value 1 at $t = 0$. Taking it as a reference, its phase is defined as zero radians. The radian frequency is $\omega = 2\pi/8$ rad/s. Its period is $T = 2\pi/\omega = 8$ s. That is, it makes one complete cycle in 8 s, as shown in the figure, and repeats indefinitely for $-\infty < t < \infty$. Its cyclic frequency is $f = 1/8$ Hz. The sine waveform $x(t) = 2\sin(\frac{2\pi}{8}t)$ has its peak value 2 at $t = 2$. Taking the cosine waveform as the reference, its first peak occurs at $t = 2$, after a delay of 2 s, which is one-fourth of a cycle in the period 8. Since one complete cycle corresponds to 2π radians or $360°$, its phase is defined as $-\pi/2$ radians or $-90°$. That is,

$$2\sin(\frac{2\pi}{8}t) = 2\cos(\frac{2\pi}{8}t - \frac{\pi}{2}).$$

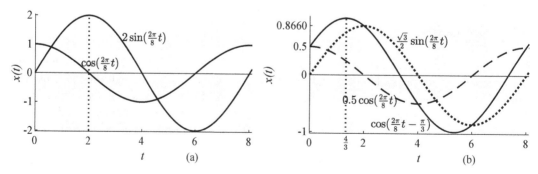

Fig. 3.1 (a) $x(t) = \cos(\frac{2\pi}{8}t)$ and $x(t) = 2\sin(\frac{2\pi}{8}t)$; (b) $x_e(t) = 0.5\cos(\frac{2\pi}{8}t)$, $x_o(t) = \frac{\sqrt{3}}{2}\sin(\frac{2\pi}{8}t)$, and $x(t) = \cos(\frac{2\pi}{8}t - \frac{\pi}{3})$

Therefore, given a sinusoidal waveform in terms of sine waveform, it can be expressed, in terms of cosine waveform as $A \sin(\omega t + \theta) = A \cos(\omega t + (\theta - \frac{\pi}{2}))$. Sinusoids remain the same by a shift of an integral number of their periods, as they are periodic. Similarly, $A \cos(\omega t + \theta) = A \sin(\omega t + (\theta + \frac{\pi}{2}))$.

3.1.1 The Rectangular Form of Sinusoids

Figure 3.1b shows the sinusoid

$$x(t) = \cos(\frac{2\pi}{8}t - \frac{\pi}{3})$$

in solid line. Its peak value of 1 occurs at $t = 4/3$ s. Therefore, its phase is $-((4/3)/8)2\pi = -\pi/3$ radians or $-60°$. Using the trigonometric subtraction formula, we get the rectangular form as

$$\cos(\frac{2\pi}{8}t - \frac{\pi}{3}) = \cos(\frac{\pi}{3})\cos(\frac{2\pi}{8}t) + \sin(\frac{\pi}{3})\sin(\frac{2\pi}{8}t)$$

$$= \frac{1}{2}\cos(\frac{2\pi}{8}t) + \frac{\sqrt{3}}{2}\sin(\frac{2\pi}{8}t).$$

The rectangular form expresses a sinusoid as the sum of its sine and cosine components, which are also, respectively, its odd and even components. The sine and cosine components are shown, respectively, by dotted and dashed lines in the figure. In general, we get

$$A \cos(\omega t + \theta) = A \cos(\theta) \cos(\omega t) - A \sin(\theta) \sin(\omega t) = C \cos(\omega t) + D \sin(\omega t),$$

where

$$C = A \cos \theta \quad \text{and} \quad D = -A \sin \theta.$$

The inverse relation is

$$A = \sqrt{C^2 + D^2} \quad \text{and} \quad \theta = \cos^{-1}(\frac{C}{A}) = \sin^{-1}(\frac{-D}{A}).$$

3.1.2 The Complex Sinusoids

While the sinusoidal waveform is used in practical systems, its mathematically equivalent form, called the complex sinusoid,

$$v(t) = Ve^{j(\omega t + \theta)} = Ve^{j\theta}e^{j\omega t}, \qquad -\infty < t < \infty$$

is found to be indispensable for analysis due to its compact form and ease of manipulation of the exponential function. $e^{j\omega t}$ is the complex sinusoid with unit magnitude and zero phase. The complex (amplitude) coefficient is $Ve^{j\theta}$. The amplitude and phase of the sinusoid is represented by the single complex number $Ve^{j\theta}$, in contrast to using two real values in the real sinusoid. Due to Euler's identity, we get

$$v(t) = \frac{V}{2}\left(e^{j(\omega t + \theta)} + e^{-j(\omega t + \theta)}\right) = V \cos(\omega t + \theta).$$

The complex exponential functions separately have no physical significance. Their sum represents a physical voltage. However, the response of a circuit to $Ve^{j\omega t}$ yields enough information with ease to deduce the response to real sinusoids. With $\omega = 0$ and $\theta = 0$,

$$v(t) = Ve^{j(\omega t + \theta)} = V.$$

Therefore, AC circuit analysis is of the same form as that of the DC. The difference is that the variables are characterized by 2 values, (V, θ) at a given frequency ω, rather than one, V, in the DC case. Note that $\omega = 0$ and $\theta = 0$ for DC.

3.2   AC Circuit Analysis

AC circuit analysis is carried out using sinusoidal input. The reason for that is the steady-state voltages and currents in any part of a linear circuit are of the same frequency as that of the source. The input–output relationship becomes algebraic and, hence, the circuit analysis. Any nonsinusoidal source can be decomposed into frequency components with different frequencies. The total output is the sum of the partial outputs due to sources with various frequencies. That is transform theory. Therefore, although most practical source waveforms are not sinusoidal, emphasis is given to analysis with sinusoidal inputs. Further, the real sinusoidal functions are difficult to manipulate. Therefore, circuit analysis boils down to analysis with sources of the form $Ae^{j(\omega t + \theta)}$, which is a mathematically equivalent representation of the real sinusoid. Then, the output to the real sinusoid is deduced from the analysis results.

3.2.1   Time- and Frequency-Domain Representations of Circuit Elements

In addition to the resistor used in DC circuit analysis, two more circuit elements, inductor and capacitor, are used in AC circuit analysis. Figure 3.2a, b show, respectively, the input–output relationship of an inductor of value L henries and a capacitor of value C farads in the time-domain. The voltage across a capacitor is the time integral of the current flowing through it times the reciprocal of its value. The voltage across an inductor is the time derivative of the current flowing through it times its value.

When the complex exponential source is used, the circuit elements, resistors, inductors, and capacitors, are to be represented appropriately. Volt–ampere relationships of circuit elements are shown in Table 3.1. The resistance R is not a function of frequency and remains the same in both the domains. For a resistor, the impedance is $Z = R$, and the admittance is the reciprocal of impedance. In the case of resistor circuits, they are also called, respectively, resistance and conductance. In the

Fig. 3.2 Input–output relationships of an inductor and a capacitor in the time-domain

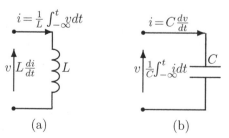

Table 3.1 Volt–ampere
relationships of circuit
elements

Element	Time-domain	Frequency-domain	Impedance	Admittance
R	$v = Ri(t)$	$V = RI$	$Z = R$	$Y = \frac{1}{R}$
L	$v = L\frac{di(t)}{dt}$	$V = j\omega LI$	$Z = j\omega L$	$Y = \frac{1}{j\omega L}$
C	$i = C\frac{dv(t)}{dt}$	$V = \frac{I}{j\omega C}$	$Z = \frac{1}{j\omega C}$	$Y = j\omega C$

case of inductors and capacitors, the volt–ampere relationships involve differential equations in the time-domain. However, for complex exponential input only, the relationships become algebraic, as in the case of the DC circuit analysis. The exponential is the only function that is its own derivative. This is the reason for the importance of decomposing an arbitrary input into frequency components, finding the output to each component, and summing all the responses to find the total output. Specifically, in the case of inductor, with the current being $e^{j\omega t}$, $V = j\omega L e^{j\omega t}$. That is, $V/I = j\omega L = Z$, the complex form of the Ohm's law. In the case of capacitor, with the voltage being $e^{j\omega t}$, $I = j\omega C e^{j\omega t}$. That is, $V/I = 1/(j\omega C) = Z$, the complex form of the Ohm's law. The equivalent impedance of series connected impedances is the sum of the individual impedances. The equivalent admittance of parallelly connected admittances is the sum of the individual admittances. In the analysis of complicated circuits, one may need to convert repeatedly from an admittance basis to an impedance basis, and vice versa. The impedance of a series connected resistor R and inductor L is $Z = R + j\omega L$. That is, the real part is a resistance. In general, the real part is called the resistive part and the imaginary part is called the reactive part. Both the parts may be combination of resistance, inductance, and capacitance elements.

The differential equation characterizing a series RL circuit, with the input $Ve^{j\omega t}$ and the response $i(t)$, is

$$L\frac{di(t)}{dt} + Ri(t) = Ve^{j\omega t}.$$

A function of the form $i(t) = Ie^{j\omega t}$ satisfies the differential equation. Therefore, the differential equation reduces to an algebraic equation, in the frequency-domain, yielding the solution

$$I = \frac{V}{R + j\omega L} = \frac{V}{Z},$$

where $Z = R + j\omega L$ is called the impedance of the circuit. The impedance is the opposition to the flow of current by the circuit for a complex exponential excitation. The admittance is the reciprocal of the impedance. The magnitude and the phase of the current, assuming the phase angle of V is zero, are, respectively,

$$\frac{V}{\sqrt{R^2 + (\omega L)^2}} \quad \text{and} \quad \theta = -\tan^{-1}(\frac{\omega L}{R}).$$

Differentiating a sinusoid any number of times results in the sinusoid of the same frequency with only changes in the amplitude and phase. Therefore, the differential and integral equations become algebraic equations, making it easy to find the solution. No other waveform has this property. Further, an arbitrary signal can be decomposed into sinusoidal components of various frequencies in transform analysis, as presented in later chapters. These are the major reasons for the importance of AC circuit analysis using sinusoids.

3.2.2   Time-Domain Analysis of a Series RC Circuit

Let us solve a simple problem of finding the steady-state current in a series RC circuit shown in Fig. 3.3, excited by a sinusoidal voltage source, $v(t) = V \cos(\omega t)$ with $V = 1$ V and $\omega = 1$. The differential equation characterizing the circuit, with input $V \cos(\omega t)$, is

$$\frac{1}{C} \int_{-\infty}^{t} i(t) + R i(t) = V \cos(\omega t).$$

Differentiating this equation and dividing both sides by R, with zero initial condition, we get

$$\frac{di(t)}{dt} + \frac{1}{RC} i(t) = -\frac{\omega V}{R} \sin(\omega t).$$

The response is expected to be of the form

$$I \cos(\omega t + \theta)$$

and its derivative with respect to t is

$$-I \omega \sin(\omega t + \theta).$$

Substituting in the differential equation, we get

$$(-\omega \sin(\omega t + \theta) + \frac{1}{RC} \cos(\omega t + \theta))I = -\frac{\omega V}{R} \sin(\omega t).$$

Expanding sine and cosine functions on the left side, we get

$$(-\omega(\sin(\omega t)\cos(\theta) + \cos(\omega t)\sin(\theta)) + \frac{1}{RC}(\cos(\omega t)\cos(\theta) - \sin(\omega t)\sin(\theta)))I$$
$$= -\frac{\omega V}{R} \sin(\omega t).$$

Combining the terms associated with $\cos(\omega t)$ and equating it to zero, we get

$$(-\omega \cos(\omega t)\sin(\theta)) + \frac{1}{RC}(\cos(\omega t)\cos(\theta)) I = 0$$

and

$$\frac{\sin(\theta)}{\cos(\theta)} = \tan(\theta) = \frac{1/RC}{\omega} = \frac{1}{\omega RC} \quad \text{and} \quad \theta = \tan^{-1}(\frac{1}{\omega RC}).$$

Fig. 3.3 A series RC circuit

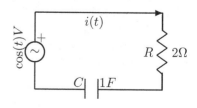

By trigonometric definition,

$$\cos(\theta) = \frac{x}{r}, \quad \sin(\theta) = \frac{y}{r} \quad \text{and} \quad r = \sqrt{x^2 + y^2}.$$

With $x = \omega$ and $y = \frac{1}{RC}$, we get

$$\cos(\theta) = \frac{\omega}{\sqrt{\omega^2 + (\frac{1}{RC})^2}} \quad \text{and} \quad \sin(\theta) = \frac{\frac{1}{RC}}{\sqrt{\omega^2 + (\frac{1}{RC})^2}}.$$

Combining the terms associated with $\sin(\omega t)$ and equating, we get

$$(\omega(\sin(\omega t)\cos(\theta)) + \frac{1}{RC}(\sin(\omega t)\sin(\theta)))I = \frac{\omega V}{R}\sin(\omega t).$$

Simplifying, we get

$$(\omega\cos(\theta) + \frac{1}{RC}\sin(\theta))I = \frac{\omega V}{R}.$$

Substituting for $\cos(\theta)$ and $\sin(\theta)$, we get

$$I = \frac{\frac{\omega V}{R}}{\sqrt{\omega^2 + (\frac{1}{RC})^2}} = \frac{V}{\sqrt{R^2 + (\frac{1}{\omega C})^2}}.$$

With

$$V = 1, R = 2, C = 1, \omega = 1, I = 0.4472, \theta = \angle(0.4636) \text{ rad}$$

For RL circuit, with

$$V = 1, R = 2, L = 1, \omega = 1, I = 0.4472, \theta = \angle(-0.4636) \text{ rad}$$

The input and the response are shown in Figs. 3.4 and 3.5. The current leads the voltage in the RC circuit. The current lags the voltage in the RL circuit.

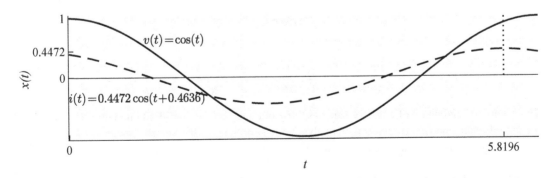

Fig. 3.4 Excitation and response of a series RC circuit

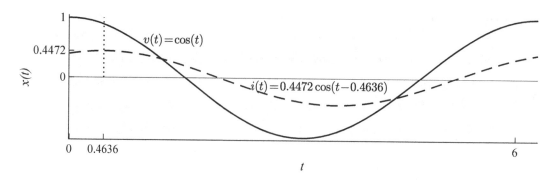

Fig. 3.5 Excitation and response of a series RL circuit

3.2.3 Frequency-Domain Analysis of a RC Circuit

When the independent variable of signals, such as time t, is not frequency, then that representation is called the time-domain representation. On the other hand, when the independent variable of signals is frequency, such as f and ω, then that representation is called frequency-domain representation. In the frequency-domain, circuit analysis becomes algebraic, rather than solving a differential equation in the time-domain. Transforms, such as Fourier transform and Laplace transform, are used to transform signals and systems from one domain into another.

The differential equation characterizing the RC circuit, with input $Ve^{j\omega t}$, is

$$\frac{1}{C}\int_{-\infty}^{t} i(t) + Ri(t) = Ve^{j\omega t}.$$

Differentiating both sides and assuming that the initial condition is zero, we get

$$\frac{di(t)}{dt} + \frac{1}{RC}i(t) = \frac{jV\omega}{R}e^{j\omega t}.$$

The response is expected to be of the form $Ie^{j\omega t}$ and its derivative is $j\omega Ie^{j\omega t}$. Substituting for $i(t)$ in the differential equation, we get

$$(\frac{1}{RC} + j\omega)Ie^{j\omega t} = \frac{jV\omega}{R}e^{j\omega t}$$

and

$$I = \frac{jV\omega}{R}\frac{RC}{1 + j\omega RC} = \frac{j\omega CV}{1 + j\omega RC} = \frac{V}{Z} = \frac{V}{R + \frac{1}{j\omega C}} = \frac{V}{R - \frac{j}{\omega C}}.$$

The magnitude and phase of I are the same as those obtained by the time-domain method. Comparing the time-domain and frequency-domain analyses, the simplicity of the latter is obvious.

The steps involved in transform analysis of AC circuits are:

1. Transform the circuit from the time-domain to the frequency-domain.
2. Find the required output using nodal or mesh analysis.
3. Transform the solution back to the time-domain.

The first task in formulating the equilibrium equations for an electric circuit is to select a set of variables, voltages or currents. They must be independent and adequate to describe the state of the network at any instant. They must be uniquely and reversibly related to all the branch variables. After selecting an appropriate set of variables, we have to use Kirchhoff's laws. The volt–ampere relations of the circuit elements are required at this stage. Of course, voltage or current sources are required to energize a circuit and their characteristics must be taken into account. Further, we used circuit theorems in analyzing circuits. All these aspects presented for DC circuits remain the same for AC circuit analysis.

One important difference in AC circuit analysis is that the independent variable is frequency. At any given frequency, a sinusoid is characterized by two parameters, amplitude and phase. These two parameters are combined into a complex quantity, called a phasor, for the convenience of analysis. A phasor is a two-element vector that represents the amplitude and phase of sinusoidal voltages and currents. Therefore, complex quantities are involved in the analysis of AC circuits. Otherwise, the analysis is similar to that of DC circuits. Same equilibrium equations are applicable for DC and AC circuits of the same structure with the circuit elements denoted by Z_1, Z_2, etc. The circuit elements for DC circuits are resistors, while those for AC circuits are resistors, inductors and capacitors. This difference comes into play only when the values of the elements are substituted in the same set of equations for the same circuit structure. We use real arithmetic in the analysis of DC circuits, while complex arithmetic is required for AC circuit analysis. Therefore, the essential part of the procedure remains the same for DC and AC circuits. Therefore, AC circuit analysis is no more complex than that of DC circuit analysis, except that of using complex arithmetic. Of course, visualization of DC voltages and currents is easier.

A suggested procedure for feeling comfortable and getting used to AC circuit analysis is as follows: Denote each circuit element by Z_1, Z_2, etc. With the same structure of the circuit, use resistors with suitable values, such as 1 and 2 Ω for Z_1, Z_2, etc. Replace the AC sources by DC sources. Write down the equilibrium equations for the type of analysis desired (nodal or mesh). Solve the equations. Verify the solution by applying KVL and KCL. Then, use the same equilibrium equations with the appropriate impedance values and AC sources. Finally, verify the solution by applying KVL and KCL. The process is similar to real and complex number arithmetic. We are more used to real arithmetic but the basic principle behind both arithmetic are the same.

Another inherent property of AC circuits is that the circuit must be analyzed at each frequency individually, if sources with more than one frequency are present in the circuit. The value of the impedance changes with the frequency of the source. The response of the circuit at all frequencies must be added to find the total response. Therefore, we analyze circuits with the same structure, studied in Chap. 2, with resistors, inductances, and capacitors, and AC sources, so that the differences in DC and AC circuit analyses are clearly highlighted.

As the sinusoidal waveform is everlasting, using a sinusoid as a source will result in the steady-state analysis. In practice, we switch on the sinusoidal source at some finite time, not from at some remote time. Due to this, there will be a transient response. For stable systems, such as the circuits, we study that the transient response will die down in a short time after switching a sinusoidal source, leaving only steady-state response. Transient response will be presented later.

For proper steady-state analysis, ensure that the effective transient response time of the circuit is short enough compared with that of the period of the excitation.

3.2.4 Impedances Connected in Series

AC circuit analysis is also based on the same concepts for the DC analysis. The difference is that the source, the sinusoid, is characterized, at a particular frequency, by two parameters, magnitude and phase, rather than the magnitude alone in DC analysis. Further, the impedances of inductors

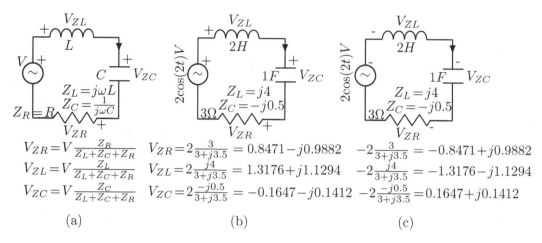

$V_{ZR} = V \frac{Z_R}{Z_L + Z_C + Z_R}$ $V_{ZR} = 2 \frac{3}{3+j3.5} = 0.8471 - j0.9882$ $-2 \frac{3}{3+j3.5} = -0.8471 + j0.9882$

$V_{ZL} = V \frac{Z_L}{Z_L + Z_C + Z_R}$ $V_{ZL} = 2 \frac{j4}{3+j3.5} = 1.3176 + j1.1294$ $-2 \frac{j4}{3+j3.5} = -1.3176 - j1.1294$

$V_{ZC} = V \frac{Z_C}{Z_L + Z_C + Z_R}$ $V_{ZC} = 2 \frac{-j0.5}{3+j3.5} = -0.1647 - j0.1412$ $-2 \frac{-j0.5}{3+j3.5} = 0.1647 + j0.1412$

(a)   (b)   (c)

Fig. 3.6 Series circuit with a voltage source

and capacitors are also characterized, at a particular frequency, by two parameters, magnitude and phase, rather than the magnitude alone for the resistor. Due to these reasons, AC circuit analysis, while based on the same concepts, requires more arithmetic. AC circuit analysis is also indispensable since AC is widely used. In addition, most of the signals occurring in practice have arbitrary amplitude profile and it is difficult to analyze circuits with them. It becomes a necessity to decompose the signals in terms of well-defined basis signals and the sinusoid is the most suitable for that purpose.

In practice, the desired circuit elements are often not available and we have to use a combination of more than one element as an equivalent one. Impedances can be combined in series and/or parallel configurations, exactly as in resistor circuits. Resistors, inductors, and capacitors have two terminals and, therefore, they come under the class of two-terminal devices or elements. In a series connection, one, and only one, terminal of an element is connected to adjoining elements. Figure 3.6a shows a resistor, an inductance and a capacitor connected in cascade, called a series circuit. A circuit is an interconnection of elements. The determination of currents and voltages at all parts of the circuit is the essence of circuit analysis. When impedances are connected in series, the voltage across them increases, with the same current flowing through them. It is similar to connecting hoses to make a longer hose. The combined impedance is the sum of all the impedances. That is, with N number of impedances connected in series, the equivalent impedance Z_{eq} of the series circuit is

$$Z_{eq} = Z_1 + Z_2 + \cdots + Z_N.$$

The same current I pass through all the impedances. Therefore, the voltage V across the series connection is

$$V = I Z_1 + I Z_2 + \cdots + I Z_N = I Z_{eq}.$$

The equivalent impedance remains unchanged, irrespective of the order in which they are connected. Obviously, if all of them have the same value, then $Z_{eq} = NZ$. The source voltage applied across them gets divided in proportion to their individual values. The current through the series circuit is

$$I = \frac{V}{Z_{eq}}$$

and the voltage across any impedance Z_n is

$$V_n = \frac{V}{Z_{eq}} Z_n.$$

With just two impedances, Z_1 and Z_2, and V, the applied voltage, in the series connection, we get

$$V_1 = \frac{V}{Z_1 + Z_2} Z_1 \quad \text{and} \quad V_2 = \frac{V}{Z_1 + Z_2} Z_2.$$

Consider the circuit in Fig. 3.6a. The circuit is energized by a voltage source V. The impedances corresponding to resistor R, inductor L, and capacitor C, respectively, are

$$Z_R = R, \quad Z_L = j\omega L, \quad Z_C = \frac{1}{j\omega C}.$$

The impedances of the inductor and capacitor are a function of the frequency. The equivalent impedance is

$$Z_{eq} = R + j\omega L + \frac{1}{j\omega C}.$$

The voltage across the impedances are, respectively,

$$V_{ZR} = V \frac{Z_R}{Z_L + Z_C + Z_R}, \quad V_{ZL} = V \frac{Z_L}{Z_L + Z_C + Z_R}, \quad V_{ZC} = V \frac{Z_C}{Z_L + Z_C + Z_R}.$$

Alternatively, we can find the current through the series circuit

$$I = \frac{V}{Z_{eq}}$$

and

$$V_{ZR} = I Z_R, \quad V_{ZL} = I Z_L, \quad V_{ZC} = I Z_C.$$

According to KVL,

$$V = V_{ZR} + V_{ZL} + V_{ZC}.$$

Consider the circuit in Fig. 3.6b. The circuit is energized by a voltage source of $V = 2\cos(2t)V$. The magnitude of the source is 2 V and its frequency $\omega = 2$ rad/s. An ideal voltage source maintains a constant voltage at its terminals, irrespective of the current drawn from it. A voltage source is a constraint, clamping the voltage at certain point in the circuit. The equivalent impedance is

$$Z_{eq} = 3 + j2 \times 2 + \frac{1}{j2 \times 1} = 3 + j4 - j0.5 = 3 + j3.5.$$

The voltage source is the real part of the complex exponential

$$2e^{j2t} = 2\cos(2t) + j2\sin(2t).$$

Therefore, with the understanding that only the real parts of resulting complex expressions in the circuit analysis are to be considered, we use the exponential $2e^{j2t}$ as the source voltage instead of $2\cos(2t)$ for mathematical convenience. As the response is of the same form throughout the circuit, the

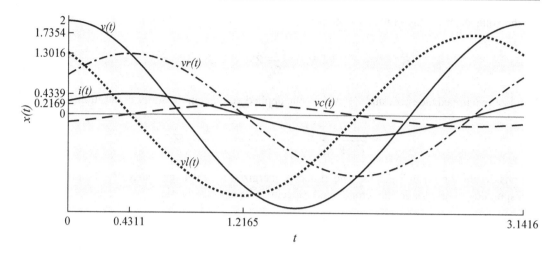

Fig. 3.7 Source and responses of a series RLC circuit

complex exponentials cancel out and we use its coefficient 2, called the phasor, in the manipulations. Phasor is a rotating vector representing an AC sinusoidal voltage or current. As it is a complex number, it may be written in exponential, polar, or rectangular form. The voltage drops across the impedances are

$$V_{ZR} = V\frac{Z_R}{Z_{eq}} = \frac{2 \times 3}{3 + j3.5} = 0.8471 - j0.9882,$$

$$V_{ZL} = V\frac{Z_L}{Z_{eq}} = \frac{2 \times j4}{3 + j3.5} = 1.3176 + j1.1294, \quad V_{ZC} = V\frac{Z_C}{Z_{eq}} = \frac{2 \times (-j0.5)}{3 + j3.5} = -0.1647 - j0.1412$$

The three voltage drops add up to 2 V, the source voltage, satisfying KVL around the only loop of the circuit. Alternatively,

$$I = \frac{V}{Z_{eq}} = \frac{2}{3 + j3.5} = 0.2824 - j0.3294$$

and

$$V_{ZR} = IZ_R = (0.2824 - j0.3294)3 = 0.8471 - j0.9882,$$

$$V_{ZL} = IZ_L = (0.2824 - j0.3294)(j4) = 1.3176 + j1.1294,$$

$$V_{ZC} = IZ_C = (0.2824 - j0.3294)(-j0.5) = -0.1647 - j0.1412$$

Figure 3.7 shows the real sinusoidal waveforms of the voltages and the current in the circuit. Although we use the complex frequency-domain representation of the circuit in its analysis, for mathematical convenience, it should be always remembered that the voltages and currents in practical circuits are real sinusoids. The real sinusoidal form of the voltages and the current are obtained by taking the real part of their corresponding form obtained in the frequency-domain analysis.

$$i(t) = |I|\cos(2t + \angle I) = 0.4339\cos(2t - 49.3987°)$$

$$v_R(t) = |V_{ZR}|\cos(2t + \angle V_{ZR}) = 1.3016\cos(2t - 49.3987°)$$

$$v_L(t) = |V_{ZL}|\cos(2t + \angle V_{ZL}) = 1.7354\cos(2t + 40.6013°)$$

$$v_C(t) = |V_{ZC}|\cos(2t + \angle V_{ZC}) = 0.2169\cos(2t - 139.3987°)$$

Fig. 3.8 The voltages and the currents represented in the complex plane

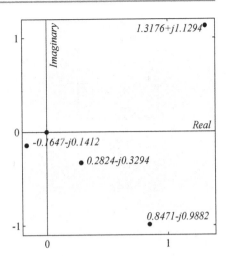

The current through and voltage across the resistor have the same phase. That is, their positive peaks occur at the same instant. The current through the inductor lags the voltage across it. Choosing the current waveform as the reference, the nearest positive peak of the voltage drop occurs earlier in a period. The current through the capacitor leads voltage across it. The positive peak of the voltage drop occurs later in a period.

The voltages and the current are represented in the complex plane in Fig. 3.8. While absolute and angle values can be computed using formulas or calling functions, it is recommended that they are visualized in the complex plane. It helps understanding and also serves as a check on numerical computation. For example, the complex value corresponding to the current is located in the middle of the bottom right quadrant. That is, the angle is about $-45°$ and the exact value is $-49.3987°$. The phase of the voltage across the resistor is also the same. The absolute value can be estimated using a scale. Similarly, the other values can be estimated. Further, the results of complex arithmetic operations can also be estimated graphically, as shown in Appendix B.

While we checked KVL using the complex values, it can be checked using the real sinusoids also. The sum of the voltages, in rectangular form of sinusoid, are

$$(0.8471 + 1.3176 - 0.1647)\cos(2t) = 2\cos(2t) \quad \text{and} \quad (0.9882 - 1.1294 + 0.1412)\sin(2t) = 0$$

That is, the sine component is zero, while the magnitude of the cosine component is 2.

Consider the circuit shown in Fig. 3.6c. The circuit is energized by the same voltage source in Fig. 3.6b, but with the polarities reversed. The voltages and the current are the same with their phases added or subtracted by π radians or $180°$.

Consider the circuit shown in Fig. 3.9a. The circuit is energized by a current source of I A. An ideal current source maintains a constant current at its terminals, irrespective of the voltage across its terminals. A current source is a constraint, clamping the current at certain point in the circuit. As the same current flows through the impedances, the respective voltage drops are

$$V_{ZR} = I Z_R, \quad \text{and} \quad V_{ZL} = I Z_L, \quad \text{and} \quad V_{ZC} = I Z_C.$$

An important step in the frequency-domain analysis of circuits is to convert the real sinusoidal sources into the mathematically equivalent complex sinusoidal sources. The conversion formulas are

$$V \cos(\omega t + \theta) \rightarrow V \angle \theta$$

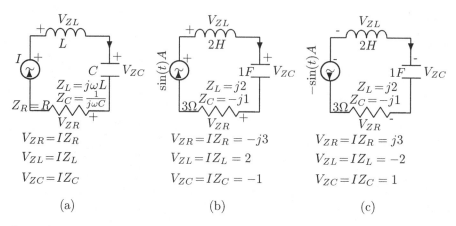

Fig. 3.9 Series circuit with a current source

and

$$V \sin(\omega t + \theta) \rightarrow V \angle (\theta - \frac{\pi}{2}).$$

These formulas hold both for current and voltage sources. For example,

$$i(t) = 2\cos(\omega t - \frac{\pi}{3}) \rightarrow 2\angle -\frac{\pi}{3}$$

and

$$v(t) = 3\sin(\omega t + \frac{\pi}{6}) \rightarrow 3\angle(\frac{\pi}{6} - \frac{\pi}{2}).$$

The last equation is due to the fact that

$$\sin(\theta) = \cos(\theta - \frac{\pi}{2}).$$

Consider the circuit in Fig. 3.9b. The circuit is energized by a current source of $i(t) = \sin(t) A$. The magnitude of the source is $1\,A$ and its frequency $\omega = 1\,\mathrm{rad/s}$. In the frequency-domain, the corresponding current source is

$$I = 1\angle -\frac{\pi}{2} = \cos(\frac{\pi}{2}) - j\sin(\frac{\pi}{2}) = -j.$$

The respective voltage drops are

$$V_{ZR} = (-j) \times 3 = -j3V \quad \text{and} \quad V_{ZL} = (-j) \times j2 = 2V$$

$$V_{ZC} = (-j) \times (-j1) = -1V.$$

In the time-domain, the values are, respectively,

$$v_R(t) = 3\sin(t), \quad v_L(t) = 2\cos(t), \quad v_C(t) = -\cos(t).$$

The equivalent impedance of the circuit is $3 + j2 - j1 = 3 + j1$. The total voltage drop across the circuit is $(-j)(3 + j1) = 1 - j3$. The sum of the individual drops $-j3 + 2 - 1 = 1 - j3$ is equal to this value.

Consider the circuit in Fig. 3.9c. The circuit is energized by a current source of $i(t) = -\sin(t)A$. As the direction of current flow is reversed, the polarities of the voltage drops are also reversed.

3.2.5 Impedances Connected in Parallel

In parallel connection, one terminal of all the elements is connected to one node and the other terminals are connected to another node. Therefore, the voltage across each element is equal. Figure 3.10 shows a resistor, an inductor, and a capacitor connected in parallel, called a parallel circuit. When elements are connected in parallel, while the voltage across all of them is the same, different currents flow through them, unless some or all of them are the same. It is similar to connecting hoses to make a wider hose. The length remains the same, but the flowing capacity increases. The combined admittance, the reciprocal of the impedance $Y = 1/Z$, is the sum of all the admittances. That is, with N number of elements connected in parallel, the equivalent admittance Y_{eq} of the parallel circuit is

$$Y_{eq} = Y_1 + Y_2 + \cdots + Y_N \quad \text{and} \quad Z_{eq} = \frac{1}{Y_{eq}},$$

where $Y_n = 1/Z_n$. The value of Z_{eq} will be smaller than the smallest of the impedances in the parallel connection, since the total current is more. The same voltage V is applied across all the elements. Therefore, the total current I flowing through the parallel connection is

$$I = VY_1 + VY_2 + \cdots + VY_N = VY_{eq}.$$

The equivalent admittance remains unchanged, irrespective of the order in which they are connected. Obviously, if all of them have the same value, then $Y_{eq} = NY$. The total current gets divided in proportion to their individual admittance values. The total current through the circuit is

$$I = VY_{eq}$$

and the current through any admittance Y_n is

$$I_n = \frac{I}{Y_{eq}}Y_n.$$

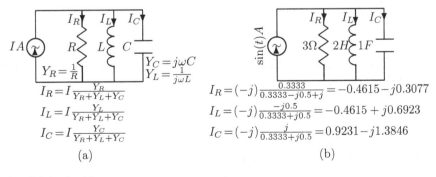

(a)   (b)

Fig. 3.10 Parallel circuit with a current source

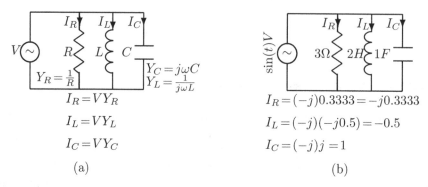

$$I_R = VY_R$$

$$I_L = VY_L$$

$$I_C = VY_C$$

(a)

$$I_R = (-j)0.3333 = -j0.3333$$

$$I_L = (-j)(-j0.5) = -0.5$$

$$I_C = (-j)j = 1$$

(b)

Fig. 3.11 Parallel circuit with a voltage source

With just two admittances, Y_1 and Y_2, and I, the total current, we get

$$I_1 = \frac{I}{Y_1 + Y_2}Y_1 \quad \text{and} \quad I_2 = \frac{I}{Y_1 + Y_2}Y_2.$$

Consider the circuit in Fig. 3.10a. The circuit is energized by a current source of $V = I\,A$. The admittances are

$$Y_R = \frac{1}{R}, \quad Y_C = j\omega C \quad Y_L = \frac{1}{j\omega L}.$$

The currents are

$$I_R = I\frac{Y_R}{Y_R + Y_L + Y_C} \quad I_L = I\frac{Y_L}{Y_R + Y_L + Y_C} \quad I_C = I\frac{Y_C}{Y_R + Y_L + Y_C}.$$

Consider the circuit in Fig. 3.10b. The circuit is energized by a current source of $I = \sin(t)A$. The magnitude of the source is 1 A and its frequency $\omega = 1$ rad/s. In the frequency-domain, the source is represented by $-j$. The currents through the three elements are shown in the figure. The total current adds up to $-j1$, which is equal to the source current. The voltages across all the elements must be the same. That is,

$$\frac{I_R}{Y_R} = \frac{I_L}{Y_L} = \frac{I_C}{Y_C} = -1.3846 - j0.9231.$$

Consider the circuit in Fig. 3.11a. The circuit is energized by a voltage source of V. The currents through the impedances are

$$I_R = VY_R, \quad I_L = VY_L, \quad I_C = VY_C.$$

Consider the circuit in Fig. 3.11b. The circuit is energized by a voltage source of $\sin(t)V$. The currents through the impedances are

$$I_R = (-j)0.3333 = -j0.3333, \quad I_L = (-j)(-j0.5) = -0.5, \quad I_C = (-j)j = 1.$$

The total admittance is $0.3333 - j0.5 + j = 0.3333 + j0.5$. The current through the circuit is $(-j1)(0.3333 + j0.5) = 0.5 - j0.3333$, which is equal to the sum of the individual currents through the elements.

3.2.6 Impedances Connected in Series and Parallel

The analysis of series and parallel circuits is relatively straightforward. In general, most circuits are a combination of series and parallel circuits or connected in a random configuration in which none of the elements is in series or parallel. Obviously, combinations of the concepts of series and parallel circuits are used to analyze series-parallel circuits. Analysis of circuits with random configurations are presented in the next section.

First, we have to identify the parts of the circuit with series and parallel configurations and simplify them separately. Now, the circuit gets reduced to a simpler form. These steps must be repeated until we can determine the source current. Then, using the voltage-division and current-division laws governing the series and parallel circuits repeatedly, we find the voltages and currents at all parts of the circuit.

Consider the circuit in Fig. 3.12a. The circuit is energized by a voltage source, $V = \cos(2t + \frac{\pi}{6})V$. The magnitude of the source is 1V and its frequency $\omega = 2$ rad/s. In the frequency-domain, it is $1\angle\frac{\pi}{6} = 0.866 + j0.5$.

First, we have to find the combined impedance of the circuit, which is

$$Z_{eq} = Z_C + (Z_R \parallel Z_L) = Z_C + \frac{1}{Y_R + Y_L} = 0.6897 - j3.2759.$$

Now, the current drawn from the source is

$$I = I_C = \frac{0.866 + j0.5}{0.6897 - j3.2759} = -0.0929 + j0.2839 A.$$

This current gets divided between Y_R and Y_L. Using current-division formula, we get

$$I_R = \frac{I}{Y_R + Y_L}Y_R = -0.1107 + j0.0071 \text{ A}$$

$$I_L = \frac{I}{Y_R + Y_L}Y_L = I - I_R = 0.0179 + j0.2768 \text{ A}.$$

Applying KCL at the common point of the three elements, we get

$$-I_C + I_L + I_R = 0.$$

The current through the circuit multiplied by the equivalent impedance is

$$(-0.0929 + j0.2839)(0.6897 - j3.2759) = 0.866 + j0.5,$$

which is equal to the source voltage.

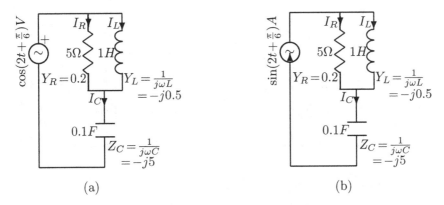

Fig. 3.12 Impedances connected in series and parallel. (a) with a voltage source; (b) with a current source

Consider the circuit in Fig. 3.12b. $I = \sin(2t + \frac{\pi}{6})A$. The magnitude of the current source is 1 and its frequency $\omega = 2$ rad/s. In the frequency-domain, it is $1\angle - \frac{\pi}{3} = 0.5 - j0.866$. The source current flows through C and, therefore, $I = I_C = (0.5 - j0.866)A$. This current gets divided between Y_R and Y_L. Using current-division formula, we get

$$I_R = \frac{I}{Y_R + Y_L} Y_R = (0.3676 + j0.0530)A$$

$$I_L = \frac{I}{Y_R + Y_L} Y_L = I - I_R = (0.1324 - j0.9190)A.$$

Applying KCL at the common point of the three elements, we get

$$-I_C + I_L + I_R = 0.$$

The voltage across R and L must be the same.

$$(0.3676 + j0.0530)/0.2 = (0.1324 - j0.9190)/(-j0.5) = 1.8380 + j0.2648$$

3.2.7   Analysis of Typical Circuits

Circuit 3.1

Consider the bridge circuit shown in Fig. 3.13a. The structure of the circuit is the same as that we analyzed in Chap. 2. The first difference is that the DC voltage source has been changed to AC source $v(t) = \cos(t + \pi/6)$. The five branches of the bridge are combinations of resistors, inductance, and capacitances rather than resistors only in the circuit for DC analysis. In AC circuit analysis, the circuit has to be transformed to the frequency-domain as shown in Fig. 3.13b so that we follow essentially the same procedure for DC analysis, although elements characterized by differential equations are part of the circuit.

The Source

The source $v(t) = \cos(t + \pi/6)$ is replaced by

$$v(t) = e^{j(t + \frac{\pi}{6})} = e^{j\frac{\pi}{6}} e^{jt}.$$

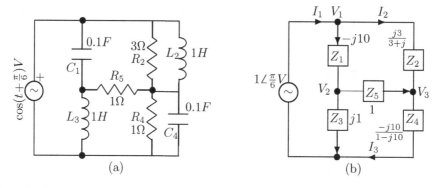

(a)   (b)

Fig. 3.13 (a) A bridge circuit; (b) its frequency-domain version

The magnitude and phase of the complex exponential is only required in the analysis, as e^{jt} remains unchanged throughout the circuit. The frequency ω is 1. The use of complex exponential source makes the analysis compact and easier. As the complex exponential has a real part and an imaginary part

$$v(t) = e^{j(t+\frac{\pi}{6})} = \cos(t + \frac{\pi}{6}) + j\sin(t + \frac{\pi}{6})$$

and our input is the cosine waveform, the real part is implied. Therefore, the complex coefficient of the exponential $V = 1\angle\frac{\pi}{6}$ represents the input source voltage in Fig. 3.13b.

The Impedances
Using the formulas in Table 3.1 for volt–ampere relationships of elements and formulas for series and parallelly connected impedances, we get, with $\omega = 1$ of the source,

$$Z_1 = \frac{1}{j\omega C_1} = -j10, \ \ Z_2 = \frac{R_2 j\omega L_2}{R_2 + j\omega L_2} = \frac{j3}{3 + j1}, \ \ Z_3 = j\omega L_3 = j1,$$

$$Z_4 = \frac{R_4 \frac{1}{j\omega C_4}}{R_4 + \frac{1}{j\omega C_4}} = \frac{-j10}{1 - j10}, \ \ Z_5 = R_5 = 1,$$

shown in Fig. 3.13b. When computing the impedance values, it is important to take into account the frequency of the voltage or current source. For each source frequency, the analysis has to be repeated.

Mesh Analysis
The impedance concept enables us to use the same methods, those are applicable to DC circuits, to analyze AC circuits in the steady state. The Ohm's and Kirchoff's laws, in terms of complex voltage, current, and impedance, are equally applicable to AC steady-state analysis. Therefore, for an AC circuit with a similar geometrical structure as that of a DC circuit, the equilibrium equations of the DC circuit, for both mesh and nodal methods of analysis, remain the same. The difference is that the elements are characterized by real quantities for DC circuits, while they are characterized by complex quantities for AC circuits. Therefore, the formal procedure of DC circuit analysis remains the same for the AC case in steady state. The resistance values are replaced by impedance values and the DC sources are replaced by AC sources.

The equilibrium equations are of the same form as those we derived for the DC bridge with the source changed and are

$$Z_1(I_1 - I_2) + Z_3(I_1 - I_3) = 1\angle\frac{\pi}{6} \tag{3.1}$$

$$-Z_1(I_1 - I_2) - Z_5(I_3 - I_2) + Z_2 I_2 = 0 \tag{3.2}$$

$$-Z_3(I_1 - I_3) + Z_5(I_3 - I_2) + Z_4 I_3 = 0. \tag{3.3}$$

Simplifying, we get

$$(Z_1 + Z_3)I_1 - Z_1 I_2 - Z_3 I_3 = \left(\frac{\sqrt{3}}{2} + j\frac{1}{2}\right) \tag{3.4}$$

$$-Z_1 I_1 + (Z_1 + Z_2 + Z_5)I_2 - Z_5 I_3 = 0 \tag{3.5}$$

$$-Z_3 I_1 - Z_5 I_2 + (Z_3 + Z_4 + Z_5)I_3 = 0. \tag{3.6}$$

The equilibrium equations, in matrix form, are

$$
\begin{bmatrix}
(Z_1 + Z_3) & -Z_1 & -Z_3 \\
-Z_1 & (Z_1 + Z_2 + Z_5) & -Z_5 \\
-Z_3 & -Z_5 & (Z_3 + Z_4 + Z_5)
\end{bmatrix}
\begin{bmatrix}
I_1 \\
I_2 \\
I_3
\end{bmatrix}
=
\begin{bmatrix}
(\frac{\sqrt{3}}{2} + j\frac{1}{2}) \\
0 \\
0
\end{bmatrix}
$$

With

$$
Z_1 = -j10, \quad Z_2 = \frac{j3}{3+j1}, \quad Z_3 = j1, \quad Z_4 = \frac{-j10}{1-j10}, \quad Z_5 = 1,
$$

the determinant of the impedance matrix is $27.8624 - j7.0762$.
Solving the equilibrium equations, we get

$$
\begin{bmatrix}
I_1 \\
I_2 \\
I_3
\end{bmatrix}
=
\begin{bmatrix}
0.0000 - j9.000 & 0.0000 + j10.000 & 0.0000 - j1.000 \\
0.0000 + j10.000 & 1.3000 - j9.100 & -1.0000 + j0.000 \\
0.0000 - j1.000 & -1.0000 + j0.000 & 1.9901 + j0.901
\end{bmatrix}^{-1}
\begin{bmatrix}
(\frac{\sqrt{3}}{2} + j\frac{1}{2}) \\
0 \\
0
\end{bmatrix}
$$

$$
=
\begin{bmatrix}
0.4750 - j0.4873 & 0.4656 - j0.5601 & 0.3813 - j0.2154 \\
0.4656 - j0.5601 & 0.4605 - j0.5259 & 0.4142 - j0.2178 \\
0.3813 - j0.2154 & 0.4142 - j0.2178 & 0.7104 - j0.2395
\end{bmatrix}
\begin{bmatrix}
(\frac{\sqrt{3}}{2} + j\frac{1}{2}) \\
0 \\
0
\end{bmatrix}
$$

$$
=
\begin{bmatrix}
0.6550 - j0.1845 \\
0.6833 - j0.2523 \\
0.4379 + j0.0041
\end{bmatrix}
$$

$$
\{I_{Z_1} = -0.0283 + j0.0677, \quad I_{Z_3} = 0.2171 - j0.1886, \quad I_{Z_5} = -0.2454 + j0.2564\}
$$

Verifying the Solutions
Applying KVL around the loops, we get

$$
I_{Z_1} Z_1 + I_{Z_3} Z_3 =
$$

$$
(-0.0283 + j0.0677)(-j10) + (0.2171 - j0.1886)(j1)
$$

$$
= (0.6774 + j0.2829) + (0.1886 + j0.2171) = (\frac{\sqrt{3}}{2} + j\frac{1}{2})
$$

$$
-I_2 Z_2 + I_{Z_5} Z_5 + I_{Z_1} Z_1 =
$$

$$
-(0.6833 - j0.2523)(\frac{j3}{3+j}) + (-0.2454 + j0.2564)(1) + (-0.0283 + j0.0677)(-j10)
$$

$$
= (-0.4320 - j0.5393) + (-0.2454 + j0.2564) + (0.6774 + j0.2829) = 0
$$

$$
I_{Z_5} Z_5 + I_3 Z_4 - I_{Z_3} Z_3 =
$$

$$
(-0.2454 + j0.2564)(1) + (0.4379 + j0.0041)(\frac{-j10}{1-j10}) - (0.2171 - j0.1886)(j1)
$$

$$
= (-0.2454 + j0.2564) + (0.4340 - j0.0393) - (0.1886 + j0.2171) = 0
$$

While these verifications in complex form are correct, as the given excitation is a real sinusoid, we have to relate the excitation to the branch voltages and currents. Consider the loop involving Z_1, Z_3 and the excitation.

$$I_{Z_1} Z_1 + I_{Z_3} Z_3 = (0.6774 + j0.2829) + (0.1886 + j0.2171) = (\frac{\sqrt{3}}{2} + j\frac{1}{2}).$$

In terms of complex exponentials, this equation corresponds to

$$0.7341 e^{j(t+0.3956)} + 0.2876 e^{j(t+0.8554)} = e^{j(t+\frac{\pi}{6})}.$$

The magnitudes and phases are the absolute values and angles of the corresponding complex coefficients. By taking the real parts, this equation corresponds to, in terms of real sinusoids,

$$0.7341 \cos(t + 0.3956) + 0.2876 \cos(t + 0.8554) = \cos(t + \frac{\pi}{6}).$$

Expressing the sinusoids in rectangular form, we get

$$(0.6774 \cos(t) - 0.2829 \sin(t)) + (0.1886 \cos(t) - 0.2171 \sin(t)) = (0.866 \cos(t) - 0.5 \sin(t))$$

In a similar manner, we can interpret the other complex quantities in terms of real sinusoids. Verifying the solution using KCL at nodes, we get

$$I_1 - I_{Z_1} - I_2 = (0.6550 - j0.1845) - (-0.0283 + j0.0677) - (0.6833 - j0.2523) = 0$$

$$I_{Z_1} - I_{Z_3} - I_{Z_5} = (-0.0283 + j0.0677) - (0.2171 - j0.1886) - (-0.2454 + j0.2564) = 0$$

$$I_{Z_5} + I_2 - I_3 = (-0.2454 + j0.2564) + (0.6833 - j0.2523) - (0.4379 + j0.0041) = 0$$

$$I_{Z_3} - I_1 + I_3 = (0.2171 - j0.1886) - (0.6550 - j0.1845) + (0.4379 + j0.0041) = 0$$

Nodal Analysis
Except for the differences pointed out for loop analysis, the equilibrium equations are of the same form as those derived for the DC bridge in Chap. 2.

$$V_1 = 1\angle\frac{\pi}{6} = (\frac{\sqrt{3}}{2} + j\frac{1}{2})$$

$$\frac{(V_2 - V_1)}{Z_1} + \frac{(V_2 - V_3)}{Z_5} + \frac{V_2}{Z_3} = 0 \tag{3.7}$$

$$\frac{V_3 - V_1}{Z_2} + \frac{(V_3 - V_2)}{Z_5} + \frac{V_3}{Z_4} = 0. \tag{3.8}$$

The first equation is the application of KCL at the left middle node. The second equation is the application of KCL at the right middle node. Simplifying, we get

$$(Z_3 Z_5 + Z_1 Z_3 + Z_1 Z_5) V_2 - Z_1 Z_3 V_3 = Z_3 Z_5 V_1 \tag{3.9}$$

$$-Z_2 Z_4 V_2 + (Z_4 Z_5 + Z_2 Z_4 + Z_2 Z_5) V_3 = Z_4 Z_5 V_1. \tag{3.10}$$

With

$$Z_1 = -j10, \quad Z_2 = \frac{j3}{3+j}, \quad Z_3 = j1, \quad Z_4 = \frac{-j10}{1-j10}, \quad Z_5 = 1$$

$$\begin{bmatrix} 10.0000 - j9.0000 & -10.0000 + j0.0000 \\ -0.3861 - j0.8614 & 1.6762 + j1.6624 \end{bmatrix} \begin{bmatrix} V_2 \\ V_3 \end{bmatrix} = \begin{bmatrix} -0.5000 + j0.8660 \\ 0.9070 + j0.4093 \end{bmatrix}.$$

The determinant of the admittance matrix is $27.8624 - j7.0762$.
Solving the equilibrium equations, we get

$$\begin{bmatrix} V_2 \\ V_3 \end{bmatrix} = \begin{bmatrix} 10.0000 - j9.0000 & -10.0000 + j0.0000 \\ -0.3861 - j0.8614 & 1.6762 + j1.6624 \end{bmatrix}^{-1} \begin{bmatrix} -0.5000 + j0.8660 \\ 0.9070 + j0.4093 \end{bmatrix}$$

$$= \begin{bmatrix} 0.0423 + j0.0704 & 0.3372 + j0.0856 \\ 0.0056 + j0.0323 & 0.4142 - j0.2178 \end{bmatrix} \begin{bmatrix} -0.5000 + j0.8660 \\ 0.9070 + j0.4093 \end{bmatrix}$$

$$= \begin{bmatrix} 0.1886 + j0.2171 \\ 0.4340 - j0.0393 \end{bmatrix}$$

$$\{V_{Z_1} = 0.6774 + j0.2829, \quad V_{Z_2} = 0.4320 + j0.5393, \quad V_{Z_5} = -0.2454 + j0.2564\}.$$

Circuit 3.2

Consider the loop analysis of the circuit shown in Fig. 3.14a. The corresponding frequency-domain circuit is shown in Fig. 3.14b. The equilibrium equations are of the same form as those derived for the corresponding to DC bridge in Chap. 2 with the source changed.

$$Z_1(I_1 - I_2) + Z_3(I_1 - I_3) + Z_6 I_1 = 0 \tag{3.11}$$

$$-Z_1(I_1 - I_2) - Z_5(I_3 - I_2) + Z_2 I_2 = 1\angle\frac{\pi}{3} \tag{3.12}$$

$$-Z_3(I_1 - I_3) + Z_5(I_3 - I_2) + Z_4 I_3 = -1\angle\frac{\pi}{3} \tag{3.13}$$

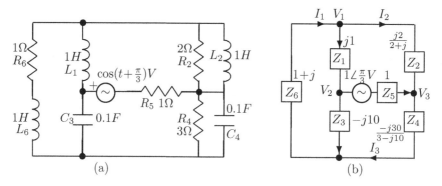

Fig. 3.14 (a) A bridge circuit, with the voltage source in the middle; (b) its frequency-domain version

Simplifying, we get, in matrix form,

$$
\begin{bmatrix}
(Z_1 + Z_3 + Z_6) & -Z_1 & -Z_3 \\
-Z_1 & (Z_1 + Z_2 + Z_5) & -Z_5 \\
-Z_3 & -Z_5 & (Z_3 + Z_4 + Z_5)
\end{bmatrix}
\begin{bmatrix} I_1 \\ I_2 \\ I_3 \end{bmatrix}
=
\begin{bmatrix} 0 \\ 1\angle\frac{\pi}{3} \\ -1\angle\frac{\pi}{3} \end{bmatrix}
$$

With

$$
Z_1 = j1, \quad Z_2 = \frac{j2}{2+j1}, \quad Z_3 = -j10, \quad Z_4 = \frac{-j30}{3-j10}, \quad Z_5 = 1, \quad Z_6 = 1 + j1
$$

$$
\begin{bmatrix}
1 - j8 & 0.0000 - j1.0000 & 0.0000 + j10.0000 \\
-j1 & 1.4000 + j1.8000 & -1.0000 + j0.0000 \\
j10 & -1.0000 + j0.0000 & 3.7523 - j10.8257
\end{bmatrix}
\begin{bmatrix} I_1 \\ I_2 \\ I_3 \end{bmatrix}
=
\begin{bmatrix} 0 \\ 0.5 + j0.866 \\ -0.5 - j0.866 \end{bmatrix}
$$

The determinant of the impedance matrix is $80.2771 - j29.1431$. Solving the equilibrium equations, we get

$$
\begin{bmatrix} I_1 \\ I_2 \\ I_3 \end{bmatrix}
=
\begin{bmatrix}
1 - j8 & 0.0000 - j1.0000 & 0.0000 + j10.0000 \\
-j1 & 1.4000 + j1.8000 & -1.0000 + j0.0000 \\
j10 & -1.0000 + j0.0000 & 3.7523 - j10.8257
\end{bmatrix}^{-1}
\begin{bmatrix} 0 \\ 0.5 + j0.866 \\ -0.5 - j0.866 \end{bmatrix}
$$

$$
=
\begin{bmatrix}
0.2949 + j0.0024 & 0.1441 - j0.0255 & 0.2501 - j0.0712 \\
0.1441 - j0.0255 & 0.3519 - j0.3810 & 0.1530 - j0.0441 \\
0.2501 - j0.0712 & 0.1530 - j0.0441 & 0.2225 - j0.0363
\end{bmatrix}
\begin{bmatrix} 0 \\ 0.5 + j0.866 \\ -0.5 - j0.866 \end{bmatrix}
$$

$$
=
\begin{bmatrix}
-0.0925 - j0.0689 \\
0.3912 + j0.0038 \\
-0.0280 - j0.0640
\end{bmatrix}
$$

$$
\{I_{Z_1} = -0.4837 - j0.0727, \; I_{Z_3} = -0.0645 - j0.0049, \; I_{Z_5} = -0.4192 - j0.0678\}
$$

Nodal Analysis

$$
\frac{(V_1 - V_2)}{Z_1} + \frac{V_1}{Z_6} + \frac{(V_1 - V_3)}{Z_2} = 0 \tag{3.14}
$$

$$
\frac{(V_2 - V_1)}{Z_1} + \frac{V_2}{Z_3} + \frac{((V_2 - 1\angle\frac{\pi}{3}) - V_3)}{Z_5} = 0 \tag{3.15}
$$

$$
\frac{V_3}{Z_4} + \frac{(V_3 - V_1)}{Z_2} + \frac{((V_3 + 1\angle\frac{\pi}{3}) - V_2)}{Z_5} = 0 \tag{3.16}
$$

$$
(Z_2 Z_6 + Z_1 Z_2 + Z_1 Z_6) V_1 - Z_2 Z_6 V_2 - Z_1 Z_6 V_3 = 0 \tag{3.17}
$$

$$
-Z_3 Z_5 V_1 + (Z_3 Z_5 + Z_1 Z_5 + Z_1 Z_3) V_2 - Z_1 Z_3 V_3 = Z_1 Z_3 1\angle\frac{\pi}{3} \tag{3.18}
$$

$$
-Z_4 Z_5 V_1 - Z_2 Z_4 V_2 + (Z_2 Z_5 + Z_4 Z_5 + Z_4 Z_2) V_3 = -Z_2 Z_4 1\angle\frac{\pi}{3}. \tag{3.19}
$$

The determinant of the admittance matrix is $-52.5644 + j55.4253$. Solving the equilibrium equations, we get

$$
\begin{bmatrix} V_1 \\ V_2 \\ V_3 \end{bmatrix} = \begin{bmatrix} -2.2000 + j2.6000 & 0.4000 - j1.2000 & 1.0000 - j1.0000 \\ 0.0000 + j10.0000 & 10.0000 - j9.0000 & -10.0000 \\ -2.7523 + j0.8257 & -1.7615 - j1.8716 & 4.9138 + j1.8459 \end{bmatrix}^{-1} \begin{bmatrix} 0 \\ 5.0000 + j8.6603 \\ 0.7401 - j2.4613 \end{bmatrix}
$$

$$
= \begin{bmatrix} -0.8561 - j0.0565 & 0.1183 + j0.0287 & 0.3395 - j0.2318 \\ -0.9594 + j0.0803 & 0.1721 + j0.0837 & 0.4501 - j0.2103 \\ -0.8306 + j0.0876 & 0.1186 + j0.0472 & 0.4925 - j0.2759 \end{bmatrix} \begin{bmatrix} 0 \\ 5.0000 + j8.6603 \\ 0.7401 - j2.4613 \end{bmatrix}
$$

$$
= \begin{bmatrix} 0.0236 + j0.1614 \\ -0.0491 + j0.6452 \\ -0.1299 - j0.1531 \end{bmatrix}
$$

Circuit 3.3

Consider the nodal analysis of the circuit shown in Fig. 3.15a. The corresponding frequency-domain circuit is shown in Fig. 3.15b. The voltage source is in the middle of the bridge. Usually, two equations using KCL is set up at the two ends of the supernode. Let the current through the source be i flowing from right to left. At the left side node, we get

$$
\frac{(V_2 - V_1)}{Z_1} + \frac{V_2}{Z_3} = i. \tag{3.20}
$$

At the right side node, we get

$$
\frac{((V_2 - 1\angle\frac{\pi}{3}) - V_1)}{Z_2} + i + \frac{(V_2 - 1\angle\frac{\pi}{3})}{Z_4} = 0. \tag{3.21}
$$

Since the currents are flowing in the opposite directions, their sum must be equal to zero.

$$
\frac{((V_2 - 1\angle\frac{\pi}{3}) - V_1)}{Z_2} + \frac{(V_2 - V_1)}{Z_1} + \frac{V_2}{Z_3} + \frac{(V_2 - 1\angle\frac{\pi}{3})}{Z_4} = 0. \tag{3.22}
$$

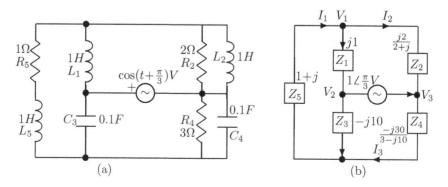

Fig. 3.15 (a) A bridge circuit with the voltage source in the middle; (b) its frequency-domain version

Eliminating i and simplifying, we get

$$-(Z_1Z_3Z_4+Z_2Z_3Z_4)V_1+(Z_1Z_3Z_4+Z_2Z_3Z_4+Z_1Z_2Z_4+Z_1Z_2Z_3)V_2 = (Z_1Z_2Z_3+Z_1Z_3Z_4)1\angle\frac{\pi}{3}$$
$$(3.23)$$

Applying at KCL at node 1, we get

$$\frac{V_1}{Z_5} + \frac{(V_1 - V_2)}{Z_1} + \frac{(V_1 - (V_2 - 1\angle\frac{\pi}{3}))}{Z_2} = 0.$$

$$(Z_1Z_2 + Z_2Z_5 + Z_1Z_5)V_1 - (Z_2Z_5 + Z_1Z_5))V_2 = (-Z_1Z_5)1\angle\frac{\pi}{3} \qquad (3.24)$$

Note that the current in the source is flowing towards node 2. Since there is no other source in the circuit, this current gets split up at node 1. With

$$Z_1 = j1, \quad Z_2 = \frac{j2}{2+j1}, \quad Z_3 = -j10, \quad Z_4 = \frac{-j30}{3-j10}, \quad Z_5 = 1+j1$$

solving for V_1 and V_2, we get

$$\begin{bmatrix} -46.2385 + j25.8716 & 48.3670 - j16.1101 \\ -2.2000 + j2.6000 & 1.4000 - j2.2000 \end{bmatrix} \cdot \begin{bmatrix} V_1 \\ V_2 \end{bmatrix} = \begin{bmatrix} 15.9839 + j27.1712 \\ 1.3660 + j0.3660 \end{bmatrix}$$

The determinant of the admittance matrix is $56.7046 - j23.2514$, which is nonzero. This is one of the checks on the problem formulation. Solving the equilibrium equations, we get

$$\begin{bmatrix} V_1 \\ V_2 \end{bmatrix} = \begin{bmatrix} -46.2385 + j25.8716 & 48.3670 - j16.1101 \\ -2.2000 + j2.6000 & 1.4000 - j2.2000 \end{bmatrix}^{-1} \begin{bmatrix} 15.9839 + j27.1712 \\ 1.3660 + j0.3660 \end{bmatrix}$$

$$= \begin{bmatrix} 0.0348 - j0.0245 & -0.8299 - j0.0562 \\ 0.0493 - j0.0256 & -0.8582 + j0.1043 \end{bmatrix} \begin{bmatrix} 15.9839 + j27.1712 \\ 1.3660 + j0.3660 \end{bmatrix}$$

$$= \begin{bmatrix} 0.1094 + j0.1714 \\ 0.2741 + j0.7585 \end{bmatrix}.$$

Substituting for V_1 and V_2, from either of the previous equations set up at the ends of the source, we get the current through the voltage source as $0.5112 - j0.1373$. Since we have determined the two independent voltages, there is no need to use this current for the analysis of this circuit. $V_3 = (0.2741 + j0.7585) - (0.5 + j0.866) = -0.2259 - j0.1075$ V.

An alternate method to avoid processing a supernode is that we insert an impedance in series with the source, as presented in an earlier example. As its value gets reduced compared with other impedances in the circuit, the circuit becomes more closer with a zero impedance series circuit and the result becomes closer to the exact values. A suitably small series impedance is to be selected. For the example circuit, an impedance of $Z = 10^{-9} + j10^{-9}$ in series with the voltage source yields almost the exact values.

Loop Analysis

The equilibrium equations are of the same form as those derived for the corresponding DC bridge in Chap. 2 with the source changed.

$$Z_1(I_1 - I_2) + Z_3(I_1 - I_3) + Z_5 I_1 = 0 \tag{3.25}$$

$$-Z_1(I_1 - I_2) + Z_2 I_2 = 1\angle\frac{\pi}{3} \tag{3.26}$$

$$-Z_3(I_1 - I_3) + Z_4 I_3 = -1\angle\frac{\pi}{3} \tag{3.27}$$

$$\begin{bmatrix} (Z_1 + Z_3 + Z_5) & -Z_1 & -Z_3 \\ -Z_1 & (Z_1 + Z_2) & 0 \\ -Z_3 & 0 & (Z_3 + Z_4) \end{bmatrix} \begin{bmatrix} I_1 \\ I_2 \\ I_3 \end{bmatrix} = \begin{bmatrix} 0 \\ 1\angle\frac{\pi}{3} \\ -1\angle\frac{\pi}{3} \end{bmatrix}.$$

With

$$Z_1 = j1, \quad Z_2 = \frac{j2}{2 + j1}, \quad Z_3 = -j10, \quad Z_4 = \frac{-j30}{3 - j10}, \quad Z_5 = 1 + j1$$

$$\begin{bmatrix} 1.0000 - j8.0000 & 0.0000 - j1.0000 & 0.0000 + j10.0000 \\ 0.0000 - j1.0000 & 0.4000 + j1.8000 & 0.0000 + j0.0000 \\ 0.0000 + j10.0000 & 0.0000 + j0.0000 & 2.7523 - j10.8257 \end{bmatrix} \begin{bmatrix} I_1 \\ I_2 \\ I_3 \end{bmatrix} = \begin{bmatrix} 0 \\ 0.5 + j0.866 \\ -0.5 - j0.866 \end{bmatrix}$$

The determinant of the impedance matrix is $68.3303 + j5.1009$. Solving the equilibrium equations, we get

$$\begin{bmatrix} I_1 \\ I_2 \\ I_3 \end{bmatrix} = \begin{bmatrix} 1.0000 - j8.0000 & 0.0000 - j1.0000 & 0.0000 + j10.0000 \\ 0.0000 - j1.0000 & 0.4000 + j1.8000 & 0.0000 + j0.0000 \\ 0.0000 + j10.0000 & 0.0000 + j0.0000 & 2.7523 - j10.8257 \end{bmatrix}^{-1} \begin{bmatrix} 0 \\ 0.5 + j0.866 \\ -0.5 - j0.866 \end{bmatrix}$$

$$= \begin{bmatrix} 0.3003 - j0.0133 & 0.1605 + j0.0283 & 0.2576 - j0.0778 \\ 0.1605 + j0.0283 & 0.1993 - j0.4955 & 0.1455 - j0.0109 \\ 0.2576 - j0.0778 & 0.1455 - j0.0109 & 0.2284 - j0.0375 \end{bmatrix} \begin{bmatrix} 0 \\ 0.5 + j0.866 \\ -0.5 - j0.866 \end{bmatrix}$$

$$= \begin{bmatrix} -0.1404 - j0.0310 \\ 0.4466 - j0.1958 \\ -0.0645 - j0.0584 \end{bmatrix}$$

Circuit 3.4

Figure 3.16a shows a circuit with a current source. The corresponding tree is shown in Fig. 3.16b. The left voltage source is $\sin(t) = \cos(t - \frac{\pi}{2})$. Therefore, $V_1 = -j$ is known and we have to solve for V_2 and V_3 only. The equilibrium equations are

$$\frac{(V_2 - (-j1))}{Z_1} + \frac{V_2}{Z_2} = -j1 \tag{3.28}$$

$$\frac{(V_3 + 2)}{Z_4} + \frac{V_3}{Z_5} = j1. \tag{3.29}$$

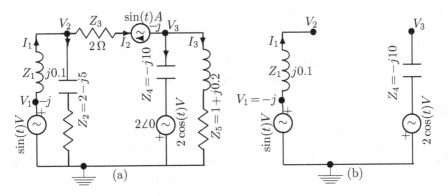

Fig. 3.16 (a) A circuit with a current source; (b) the corresponding tree

Simplifying and solving, we get

$$(Z_1 + Z_2)V_2 = (-j1)(Z_1 Z_2 + Z_2) \tag{3.30}$$

$$(Z_4 + Z_5)V_3 = j(Z_4 Z_5) - 2Z_5 \tag{3.31}$$

$$V_2 = \frac{(-j1)(Z_1 Z_2 + Z_2)}{(Z_1 + Z_2)} \tag{3.32}$$

$$V_3 = \frac{j(Z_4 Z_5) - 2Z_5}{(Z_4 + Z_5)}. \tag{3.33}$$

With

$$Z_1 = j0.1, \quad Z_2 = 2 - j5, \quad Z_3 = 2, \quad Z_4 = -j10, \quad Z_5 = 1 + j0.2$$

we get

$$\{V_2 = 0.0946 - j1.0182, \ V_3 = -0.0791 + j0.8244\}.$$

Mesh Analysis

Applying KVL around the left side loop, we get, with $I_2 = j1$,

$$I_1 Z_1 + Z_2(I_1 - I_2) = -j1$$

$$I_1(Z_1 + Z_2) = -j1 + Z_2 j1$$

$$I_1 = \frac{j1(Z_2 - 1)}{Z_1 + Z_2} = 0.1821 + j0.9461$$

$$V_2 = -I_1 Z_1 - j = 0.0946 - j1.0182$$

Applying KVL around the right side loop, we get,

$$-(I_2 - I_3)Z_4 + Z_5 I_3 = -2$$

$$I_3 = \frac{-2 + jZ_4}{Z_4 + Z_5} = 0.0824 + j0.8079$$

$$V_3 = I_3 Z_5 = -0.0791 + j0.8244.$$

An alternative method to avoid processing a supermesh is that we insert an impedance in parallel with the source. As its value gets increased compared with other impedances in the circuit, the circuit becomes more closer with an infinite impedance in parallel and the result becomes closer to the exact values. A suitably high parallel impedance is to be selected. For the example circuit, the values are almost the same as the exact ones, with parallel impedance $100 + j100$.

Circuit 3.5

In the last example, a current source exists only in one loop. Let us consider a circuit with a current source between two loops. The circuit and the tree are shown, respectively, in Fig. 3.17a, b. There are two nodes. But, one node voltage is given. Similarly, there are two loops. But, the currents are related. Therefore, the circuit analysis reduces to one variable problem for both the nodal and loop analyses. With

$$Z_1 = j0.1, \quad Z_2 = \frac{-j5}{1 - j5}, \quad Z_3 = 2, \quad V_1 = -jV.$$

With the current source supplying $2\,\text{A}$ towards the ground node, KVL around the circuit and the current source constraint yields

$$Z_1 I_1 + Z_3 I_3 = -j1.$$

Since

$$I_1 - I_3 = 2$$

$$Z_1 I_1 + Z_3 (I_1 - 2) = -j1 \quad \text{or} \quad (Z_1 + Z_3) I_1 = -j1 + 2Z_3.$$

Solving, we get $I_1 = 1.9701 - j0.5985$ and $I_3 = I_1 - 2 = -0.0299 - j0.5985$.

Nodal Analysis

$$\frac{(V_2 - (-j1))}{Z_1} + \frac{V_2}{Z_3} = -2.$$

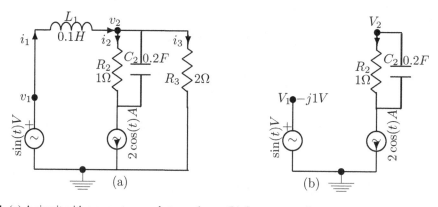

Fig. 3.17 (**a**) A circuit with a current source between loops; (**b**) the corresponding tree

Solving for V_2, we get

$$V_2 = \frac{-2Z_1Z_3 - jZ_3}{Z_1 + Z_3} = -0.0599 - j1.1970.$$

Let us do this problem considering one source at a time. When we consider the voltage source alone, the current source has to be open-circuited. Then, by voltage division, we get the voltage across the $2\,\Omega$ resistor as

$$2\frac{-j}{2 + j0.1} = -0.0499 - j0.9975.$$

When we consider the current source alone, the voltage source has to be short-circuited. Then, by current division, we get the voltage across the $2\,\Omega$ resistor as

$$-2\frac{j0.1(2)}{2 + j0.1} = -0.0100 - j0.1995.$$

By adding the partial voltages, we get

$$V_2 = -0.0599 - j1.1970$$

as obtained earlier.

3.2.8 Linearity Property of Circuits

Consider the circuit shown in Fig. 3.18. The circuit has three sources, A DC voltage source with magnitude 2 V with the positive terminal connected to the ground node. An AC voltage source $\sin(t + \frac{\pi}{3})$ V, with magnitude 1 V at frequency 1 rad/s, is placed with the negative terminal connected to the ground node. An AC current source $2\cos(5t)$ A, with magnitude 2 A at frequency 5 rad/s, is placed with the the current flowing from right to left. Since different frequencies are involved and impedances vary with frequencies, we have to use the linearity property of the circuits and the resulting superposition theorem to analyze this circuit. The total response is the sum of the individual time-domain responses. In the analysis of AC circuits with sources of different frequencies, the use of linearity is required. The transform analysis of circuits is based on linearity. For DC circuits, linearity may provide simplified analysis for some circuits.

Fig. 3.18 A circuit with voltage and current sources of different frequencies

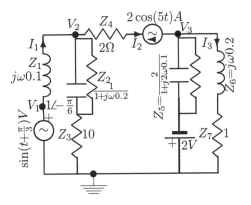

Table 3.2 Voltages due to each source and their total in circuit shown in Fig. 3.18

Source	V_1	V_2	V_3
V source, $\omega = 1$	$1\angle(-\frac{\pi}{6})$	$0.8615 - j0.5079$	0
DC, $\omega = 0$	0	0	$-\frac{2}{3}$
I source, $\omega = 5$	0	$0.0476 + j1$	-2
Total	$\cos(t - \frac{\pi}{6})$	$\cos(t - 0.5327) + 1.0011\cos(5t + 1.5232)$	$-2\cos(5t) - \frac{2}{3}$

Table 3.3 Currents due to each source and their total in circuit shown in Fig. 3.18

Source	I_1	I_2	I_3
V source	$0.0794 - j0.0449$	0	0
DC	0	0	$-\frac{2}{3}$
I source	$-(2.0000 - j0.0952)$	-2	$(-1 + j1)$
Total	$0.0912\cos(t - 0.5152) - 2.0023\cos(5t - 0.0476)$	$-2\cos(5t)$	$\sqrt{2}\cos(5t + \frac{3\pi}{4}) - \frac{2}{3}$

Consider the effect of the DC source alone. Since the frequency of DC source is zero, the reactance due to an inductance is zero and that due to a capacitance is infinity. Therefore, an inductance, in effect, is a short-circuit and a capacitance is an open-circuit. Further, an ideal current source is replaced by an open-circuit and an ideal voltage source is replaced by a short-circuit. Consequently, the circuit reduces to the DC source with amplitude -2 V and the two resistors, with values $2\,\Omega$ in Z_5 and $1\,\Omega$ in Z_7, connected in series. Therefore, $I_3 = -\frac{2}{3}$ A and $V_3 = -\frac{2}{3}$ V. All other voltages and currents are zero, as shown in Tables 3.2 and 3.3.

With

$$Z_1 = j\omega 0.1, \ Z_2 = \frac{1}{1 + j\omega 0.2}, \quad Z_3 = 10, \ Z_4 = 2, \ Z_5 = \frac{2}{1 + j2\omega 0.1}, \ Z_6 = j\omega 0.2, \ Z_7 = 1,$$

let us find the response of the circuit to the voltage source

$$\sin(t + \frac{\pi}{3}) = \cos(t - \frac{\pi}{6})$$

alone. Impedance Z_2 is due to the parallel connection of a resistor with $1\,\Omega$ and a capacitor with $C = 0.2$ F. With $\omega = 1$,

$$Z_1 = j0.1, \quad Z_2 = \frac{(1)(\frac{1}{j0.2})}{1 + \frac{1}{j0.2}} = 0.9615 - j0.1923, \quad Z_3 = 10, \ V = 1\angle -\frac{\pi}{6} = 0.8660 - j0.5$$

Impedance Z_5 is due to the parallel connection of a resistor with $2\,\Omega$ and a capacitor with $C = 0.1$ F. The right side of the circuit gets disconnected as the current source is replaced by an open-circuit. The DC source is replaced by a short-circuit. Therefore, $V_3 = 0$, $I_2 = 0$, and $I_3 = 0$. $V_1 = 1\angle(-\frac{\pi}{6})$.

$$I_1 = \frac{V_1}{Z_1 + Z_2 + Z_3} = 0.0794 - j0.0449$$

$$V_2 = I_1(Z_2 + Z_3) = V_1 - I_1 Z_1 = 0.8615 - j0.5079.$$

With $\omega = 5$,

$$Z_1 = j0.5, \quad Z_2 = 0.5 - j0.5, Z_3 = 10, Z_4 = 2, Z_5 = 1 - j1, \quad Z_6 = j1, \quad Z_7 = 1$$

Let us find the response of the circuit with the current source $2\cos(5t)$ alone. Note that the frequency ω is 5 radians, not 1 compared with the AC voltage source. Therefore, the impedance values are different. The two voltages sources are replaced by short-circuits. Applying KCL at nodes 3 and 2

$$\frac{V_3}{Z_5} + \frac{V_3}{Z_6 + Z_7} + 2 = 0$$

$$V_3 = \frac{-2(Z_5(Z_6 + Z_7))}{Z_5 + Z_6 + Z_7} = -2$$

$$\frac{V_2}{Z_1} + \frac{V_2}{Z_2 + Z_3} - 2 = 0$$

$$V_2 = \frac{2(Z_1(Z_2 + Z_3))}{Z_1 + Z_2 + Z_3} = 0.0476 + j1$$

$$-I_1 = \frac{V_2}{Z_1} = 2 - j0.0952, \quad I_2 = -2, \quad I_3 = \frac{V_3}{Z_6 + Z_7} = -1 + j1.$$

Using current-division formula also, we can find the currents I_1 and I_3. Voltages and currents due to each source and their totals are shown in Tables 3.2 and 3.3.

Circuit 3.6

Consider the circuit with a voltage-controlled voltage source, shown in Fig. 3.19. Voltage source is $2\cos(t)$ V. Current source is $\sin(t) = \cos(t - \frac{\pi}{2})$ A. In the frequency-domain, the voltage source is 2 and the current source is $-j1 = 1\angle(-\frac{\pi}{2})$. The circuit contains a controlled voltage source. Therefore, although there are five nodes, three equilibrium equations are enough. With the angular frequency of the sources being 1 radian,

$$Z_1 = 2 + j0.1, \quad Z_2 = 1, \quad Z_3 = 3 - j10, \quad Z_4 = 3, \quad Z_5 = 2.$$

Fig. 3.19 A circuit with a voltage-controlled voltage source

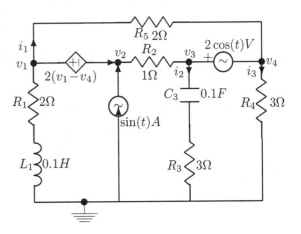

Loop Analysis

The equilibrium equations are

$$(Z_5 I_1 + Z_4 I_3 - 2Z_5 I_1) + (Z_2 (I_1 - I_2 - I_3)) - Z_3 I_2 = 0$$

$$-Z_1 (I(2) + I(3) + j1) = Z_5 I_1 + Z_4 I_3$$

$$Z_3 I_2 - Z_4 I_3 = 2.$$

The first equation corresponds to the middle loop involving resistors R_2 and R_3. To write the KVL equation around the loop, we need the voltage V_2.

$$V_1 = Z_5 I_1 + Z_4 I_3 \quad \text{and} \quad V_1 - V_4 = Z_5 I_1.$$

Therefore, the voltage across the controlled voltage source is $2Z_5 I_1$, with opposite polarity to that of V_1. Consequently,

$$V_2 = (Z_5 I_1 + Z_4 I_3 - 2Z_5 I_1).$$

The second equation corresponds to the leftmost loop involving resistor Z_1. The current entering Z_1 is $(I(2) + I(3) - (-j1)$ from the ground side. Therefore, we get another expression for V_1 involving the current source.

$$V_1 = -Z_1 (I(2) + I(3) + j1)$$

Equating these two equations yields an equilibrium equation. The last equation corresponds to the rightmost loop involving impedances Z_3 and Z_4. Rearranging the equations, we get

$$(-Z_5 + Z_2)I_1 - (Z_2 + Z_3)I_2 + (Z_4 - Z_2)I_3 = 0$$

$$Z_5 I_1 + Z_1 I_2 + (Z_1 + Z_4)I_3 = -jZ_1$$

$$0I_1 + Z_3 I_2 - Z_4 I_3 = 2.$$

Substituting the numerical values for the impedances and using matrices, we get

$$\begin{bmatrix} -1 & -4+j10 & 2 \\ 2 & 2 & 5+j0.1 \\ 0 & 3-j10 & -3 \end{bmatrix} \begin{bmatrix} I_1 \\ I_2 \\ I_3 \end{bmatrix} = \begin{bmatrix} 0 \\ -jZ_1 \\ 2 \end{bmatrix}$$

The determinant of the impedance matrix is nonzero, $10 - j29.4$. The determinant of the impedance matrix must be nonzero. Otherwise, the equations are not independent. Verify that the product of the impedance matrix and its inverse is the identity matrix of the same size. One of the equations must be based on the relation between dependent and independent source. Solving for the currents, we get

$$\begin{bmatrix} I_1 \\ I_2 \\ I_3 \end{bmatrix} = \begin{bmatrix} -1 & -4+j10 & 2 \\ 2 & 2 & 5+j0.1 \\ 0 & 3-j10 & -3 \end{bmatrix}^{-1} \begin{bmatrix} 0 \\ 0.1-j2 \\ 2 \end{bmatrix}$$

$$= \begin{bmatrix} -1.7342 - j0.1584 & -0.3671 - j0.0792 & -1.7653 - j0.2499 \\ 0.0622 + j0.1829 & 0.0311 + j0.0915 & 0.0903 + j0.2754 \\ 0.6719 - j0.0245 & 0.3360 - j0.0122 & 0.6750 - j0.0255 \end{bmatrix} \begin{bmatrix} 0 \\ 0.1 - j2 \\ 2 \end{bmatrix}$$

$$= \begin{bmatrix} -3.7257 + j0.2264 \\ 0.3666 + j0.4978 \\ 1.3591 - j0.7242 \end{bmatrix}$$

Let us find the node voltages and verify their relations.

$$V_3 = I_2 Z_3 = (0.3666 + j0.4978)(3 - j10) = 6.0774 - j2.1726$$

$$V_4 = I_3 Z_4 = (1.3591 - j0.7242)(3) = 4.0774 - j2.1726$$

$$V_1 = V_4 + I_1 Z_5 = (4.0774 - j2.1726) + (-3.7257 + j0.2264)(2) = -3.3741 - j1.7197$$

$$V_2 = 2V_4 - V_1 = 2(4.0774 - j2.1726) - (-3.3741 - j1.7197) = 11.5288 - j2.6254$$

Now,

$$V_3 - V_4 = 2$$

$$V_1 - V_2 = (-3.3741 - j1.7197) - (11.5288 - j2.6254) = -14.9028 + j0.9057$$

$$= 2I_1 Z_1 = 2(-3.7257 + j0.2264)(2 + j0.1)$$

$$V_3 - V_2 = -5.4514 + 0.4529 = (I_1 - I_2 - I_3)Z_2$$

The current through the independent voltage source flowing from left to right is

$$ir = I_1 - I_3 = -5.0848 + j0.9506.$$

The current through the dependent voltage source flowing from left to right is

$$il = I_1 - I_2 - I_3 - j = -5.4514 + j0.4529 - j = -5.4514 - j0.5471.$$

Nodal Analysis
The voltage source is not connected to the ground node. Therefore, we have to find the current through it. Setting up two KCL equations at the terminals of the source, we get

$$\frac{(V_4 - V_1)}{Z_5} + \frac{V_4}{Z_4} - \frac{(V_2 - V_3)}{Z_2} + \frac{V_3}{Z_3} = 0.$$

The dependent voltage source is also not connected to the ground node. Therefore, we have to find the current through it. Setting up two KCL equations at the terminals of the source, we get

$$\frac{(V_3 - V_2)}{Z_2} + (-j1) - \frac{(V_1 - V_4)}{Z_5} - \frac{V_1}{Z_1} = 0.$$

Now, applying KVL to the branches involving the dependent voltage source, we get the equilibrium equations as

$$\frac{(V_3 - V_2)}{Z_2} + (-j1) - \frac{(V_1 - V_4)}{Z_5} - \frac{V_1}{Z_1} = 0 \tag{3.34}$$

$$\frac{(V_4 - V_1)}{Z_5} + \frac{V_4}{Z_4} - \frac{(V_2 - V_3)}{Z_2} + \frac{V_3}{Z_3} = 0 \tag{3.35}$$

$$V_1 + V_2 - 2V_4 = 0. \tag{3.36}$$

Replacing

$$V_3 = (V_4 + 2),$$

we get

$$\frac{((V_4 + 2) - V_2)}{Z_2} + (-j1) - \frac{(V_1 - V_4)}{Z_5} - \frac{V_1}{Z_1} = 0 \tag{3.37}$$

$$\frac{(V_4 - V_1)}{Z_5} + \frac{V_4}{Z_4} - \frac{(V_2 - (V_4 + 2))}{Z_2} + \frac{(V_4 + 2)}{Z_3} = 0 \tag{3.38}$$

$$V_1 + V_2 - 2V_4 = 0 \tag{3.39}$$

Simplifying, we get

$$-(Z_1 Z_2 + Z_2 Z_5)V_1 - Z_1 Z_5 V_2 + (Z_1 Z_2 + Z_1 Z_5)V_4 = -2Z_1 Z_5 + jZ_1 Z_2 Z_5$$

$$-Z_2 Z_3 Z_4 V_1 - Z_3 Z_4 Z_5 V_2 + (Z_2 Z_3 Z_4 + Z_2 Z_3 Z_5 + Z_3 Z_4 Z_5 + Z_2 Z_4 Z_5)V_4 = -2(Z_2 Z_4 Z_5 + Z_3 Z_4 Z_5)$$

$$V_1 + V_2 - 2V_4 = 0$$

With

$$Z_1 = 2 + j0.1, \quad Z_2 = 1, \quad Z_3 = 3 - j10, \quad Z_4 = 3, \quad Z_5 = 2$$

$$\begin{bmatrix} -0.0400 - j0.0010 & -0.0400 - j0.0020 & 0.0600 + j0.0030 \\ -0.0900 + j0.3000 & -0.1800 + j0.6000 & 0.3900 - j1.1000 \\ 0.0100 + j0.0000 & 0.0100 + j0.0000 & -0.0200 + j0.0000 \end{bmatrix} \begin{bmatrix} V_1 \\ V_2 \\ V_4 \end{bmatrix} = \begin{bmatrix} -0.082 + j0.036 \\ -0.480 + j1.200 \\ 0 \end{bmatrix}$$

The determinant of the admittance matrix is $-20 + j58.8$.

$$\begin{bmatrix} V_1 \\ V_2 \\ V_4 \end{bmatrix} = \begin{bmatrix} -0.040 - j0.001 & -0.0400 - j0.002 & 0.060 + j0.003 \\ -0.090 + j0.300 & -0.180 + j0.600 & 0.390 - j1.100 \\ 0.010 + j0.000 & 0.010 + j0.000 & -0.020 + j0.000 \end{bmatrix}^{-1} \begin{bmatrix} -0.082 + j0.036 \\ -0.480 + j1.200 \\ 0 \end{bmatrix}$$

$$= 100 \begin{bmatrix} -0.1369 + j0.0976 & 0.0088 + j0.0310 & 1.4525 + j0.3903 \\ -0.8710 - j0.0609 & -0.0119 - j0.0300 & -4.4842 - j0.2435 \\ -0.5040 + j0.0184 & -0.0015 + j0.0005 & -2.0158 + 0.0734 \end{bmatrix} \begin{bmatrix} -0.082 + j0.036 \\ -0.480 + j1.200 \\ 0 \end{bmatrix}$$

$$= \begin{bmatrix} -3.3741 - j1.7197 \\ 11.5288 - j2.6254 \\ 4.0774 - j2.1726 \end{bmatrix}$$

The voltages are the same as those obtained by the loop method.

Nodal or Mesh Analysis
Basically, both types of circuit analysis are important, since one type may be conceptually and/or computationally simpler for a given circuit. In certain applications, one of the two methods is preferred or applicable. Further, the solution obtained by one method can be checked using the other method. Therefore, it is good to be familiar with both the methods. In nodal analysis, if the circuit has a voltage source connected to the ground node, then that voltage can be considered as one independent variable and the number of equilibrium equations can be reduced. Further, if a current source, without a parallel resistance, is connected between non-ground nodes, then nodal method is advantageous. Similarly, if a voltage source is connected between two non-ground nodes, then mesh analysis is easier. The solution to the last example is simpler by the mesh method. Further, if the circuit has a current source connected to the ground node, then that current can be considered as one independent variable and the number of equilibrium equations can be reduced.

Circuit 3.7
A circuit with a current-controlled current source is shown in Fig. 3.20.

Nodal Analysis
Since $V_3 = j$ V is known, there are only 2 unknowns, $\{V_1, V_2\}$. When a voltage source is connected between a reference node and a nonreference node, the problem is simplified. The equilibrium equations are

$$\frac{(V_2 - V_1)}{Z_2} + \frac{(V_2 - j)}{Z_4} + \frac{(V_2)}{Z_3} = 0 \qquad (3.40)$$

$$\frac{V_1}{Z_1} + \frac{(V_1 - j)}{Z_5} - 2\frac{V_1}{Z_1} + \frac{(V_1 - V_2)}{Z_2} = 0. \qquad (3.41)$$

The first equation is due to the application of KCL at node with voltage V_2. The second equation is due to the application of KCL at node with voltage V_1. Simplifying, we get

$$- Z_3 Z_4 V_1 + (Z_3 Z_4 + Z_2 Z_4 + Z_2 Z_3) V_2 = j Z_2 Z_3 \qquad (3.42)$$

$$(-Z_2 Z_5 + Z_1 Z_5 + Z_1 Z_2) V_1 - Z_1 Z_5 V_2 = j Z_1 Z_2. \qquad (3.43)$$

Fig. 3.20 A circuit with a current-controlled current source

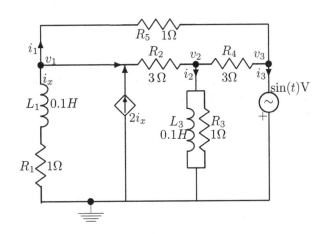

With

$$Z_1 = 1 + j0.1, \quad Z_2 = 3, \quad Z_3 = \frac{j0.1}{1 + j0.1}, \quad Z_4 = 3, \quad Z_5 = 1,$$

we get, in matrix form,

$$\begin{bmatrix} -0.0297 - j0.2970 & 9.0594 + j0.5941 \\ 1.0000 + j0.4000 & -1.0000 - j0.1000 \end{bmatrix} \begin{bmatrix} V_1 \\ V_2 \end{bmatrix} = \begin{bmatrix} -0.2970 + j0.0297 \\ -0.3 + j3 \end{bmatrix}.$$

The determinant of the admittance matrix is $-8.8218 - j3.9178$. Solving for the voltages, we get

$$\begin{bmatrix} V_1 \\ V_2 \end{bmatrix} = \begin{bmatrix} -0.0297 - j0.2970 & 9.0594 + j0.5941 \\ 1.0000 + j0.4000 & -1.0000 - j0.1000 \end{bmatrix}^{-1} \begin{bmatrix} -0.2970 + j0.0297 \\ -0.3 + j3 \end{bmatrix}$$

$$= \begin{bmatrix} 0.0989 - j0.0326 & 0.8827 - j0.3247 \\ 0.1115 - j0.0042 & 0.0153 + j0.0269 \end{bmatrix} \begin{bmatrix} -0.2970 + j0.0297 \\ -0.3 + j3 \end{bmatrix}$$

$$= \begin{bmatrix} 0.6808 + j2.7582 \\ -0.1182 + j0.0424 \end{bmatrix}$$

$$I_1 = \frac{(V_1 - j)}{Z_5} = 0.6808 + j1.7582$$

$$I_2 = \frac{V_2}{Z_3} = 0.3058 + j1.2245$$

$$I_3 = I_1 + \frac{(V_2 - j)}{Z_4} = 0.6414 + j1.4390.$$

Mesh Analysis

$$I_{Z1} = I_2 + I_3$$

$$Z_1(I_2 + I_3) + Z_2(I_1 - I_2 - I_3) - Z_3 I_2 = 0$$

$$Z_3 I_2 + Z_4(I_1 - I_3) = j$$

$$Z_1(I_2 + I_3) - Z_5 I_1 = j.$$

The first equation is obtained by equating the voltages on both sides of the current source. The second equation is obtained by applying KVL to the rightmost loop involving the voltage source. The third equation is obtained by applying KVL to the loop involving the voltage source and impedances Z_1 and Z_5. Voltage source is $\sin(t)$ V. Simplifying, we get

$$Z_2 I_1 + (Z_1 - Z_2 - Z_3)I_2 + (Z_1 - Z_2)I_3 = 0$$

$$Z_4 I_1 + Z_3 I_2 - Z_4 I_3 = j$$

$$-Z_5 I_1 + Z_1 I_2 + Z_1 I_3 = j.$$

With

$$Z_1 = 1 + j0.1, \quad Z_2 = 3, \quad Z_3 = \frac{j0.1}{1 + j0.1}, \quad Z_4 = 3, \quad Z_5 = 1$$

$$\begin{bmatrix} 3.0000 + j0.0000 & -2.0099 + j0.0010 & -2.0000 + j0.1000 \\ 3.0000 + j0.0000 & 0.0099 + j0.0990 & -3.0000 + j0.0000 \\ -1.0000 + j0.0000 & 1.0000 + 0.1000 & 1.0000 + j0.1000 \end{bmatrix} \begin{bmatrix} I_1 \\ I_2 \\ I_3 \end{bmatrix} = \begin{bmatrix} 0 \\ j \\ j \end{bmatrix}.$$

The determinant of the impedance matrix is nonzero, $2.9406 + j1.3059$.

$$\begin{bmatrix} I_1 \\ I_2 \\ I_3 \end{bmatrix} = \begin{bmatrix} 3.0000 + j0.0000 & -2.0099 + j0.0010 & -2.0000 + j0.1000 \\ 3.0000 + j0.0000 & 0.0099 + j0.0990 & -3.0000 + j0.0000 \\ -1.0000 + j0.0000 & 1.0000 + 0.1000 & 1.0000 + j0.1000 \end{bmatrix}^{-1} \begin{bmatrix} 0 \\ j \\ j \end{bmatrix}$$

$$= \begin{bmatrix} 0.9026 - j0.2648 & 0.0126 + j0.0284 & 1.7456 - j0.7093 \\ -0.0378 - j0.0852 & 0.3345 - j0.0125 & 0.8900 - j0.2932 \\ 0.9053 - 0.2664 & -0.3192 + j0.0394 & 1.7582 - j0.6808 \end{bmatrix} \begin{bmatrix} 0 \\ j \\ j \end{bmatrix}$$

$$= \begin{bmatrix} 0.6808 + j1.7582 \\ 0.3058 + j1.2245 \\ 0.6414 + j1.4390 \end{bmatrix}.$$

Circuits with One or Two Variables

Now, we are going to present the analysis of simpler circuits with one or two variables, which can be solved manually.

Circuit in Fig. 3.21

The circuit shown in Fig. 3.21 has two voltage sources.

Nodal Analysis

$$Z_1 = 2, \quad Z_2 = j2, \quad Z_3 = 3, \quad V_1 = 1, \quad V_3 = 2.$$

Voltage V_2 is the only unknown. Therefore, applying KCL at node 2, we get

Fig. 3.21 A circuit with two voltage sources

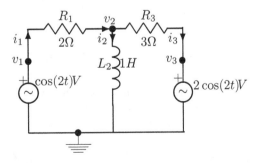

$$\frac{(V_2 - 1)}{Z_1} + \frac{(V_2 - 2)}{Z_3} + \frac{V_2}{Z_2} = 0.$$

Solving, we get

$$V_2 = \frac{Z_2 Z_3 + 2 Z_1 Z_2}{Z_1 Z_2 + Z_2 Z_3 + Z_1 Z_3} = (1.0294 + j0.6176)\,\text{V}.$$

The currents are

$$I_1 = -\frac{(V_2 - 1)}{Z_1} = (-0.0147 - j0.3088)\,\text{A}, \quad I_2 = \frac{(V_2)}{Z_2} = (0.3088 - j0.5147)\,\text{A},$$

$$I_3 = \frac{(V_2 - 2)}{Z_3} = (-0.3235 + j0.2059)\,\text{A}.$$

With $-I_1 + I_2 + I_3 = 0$, KCL is satisfied.

Mesh Analysis
$I_2 = I_1 - I_3$.

$$Z_1 I_1 + Z_2 (I_1 - I_3) = 1$$
$$Z_3 I_3 - Z_2 (I_1 - I_3) = -2.$$

Simplifying, we get

$$(Z_1 + Z_2) I_1 - Z_2 I_3 = 1$$
$$-Z_2 I_1 + (Z_2 + Z_3) I_3 = -2.$$

Substituting the numerical values, we get the impedance matrix Z as

$$Z = \begin{bmatrix} 2 + j2 & -j2 \\ -j2 & 3 + j2 \end{bmatrix}.$$

Solving the equations, we get the same values of currents, as found earlier.

Let us do the problem by superposition method. Let us find the response to the source at the left side. Then, we replace other source by a short-circuit. Current I_1 is

$$I_1 = \frac{1}{Z_1 + (Z_2 || Z_3)} = (0.2794 - j0.1324).$$

By current division, we get

$$I_2 = I_1 \frac{Z_3}{Z_2 + Z_3} = (0.1324 - j0.2206), \quad I_3 = I_1 - I_2 = (0.1471 + j0.0882).$$

Let us find the response to the response to the source at the right side. Then, we replace the other source by a short-circuit. Current I_3 is

$$I_3 = -\frac{2}{Z_3 + (Z_2 || Z_1)} = (-0.4706 + j0.1176).$$

By current division, we get

$$I_1 = I_3 \frac{Z_2}{Z_2 + Z_1} = (-0.2941 - j0.1765), \quad I_2 = I_1 - I_3 = (0.1765 - j0.2941).$$

Adding the corresponding two responses, we get the same values for the currents.

Let us find the current I_2 through Z_2 using Thevenin's theorem. The equivalent impedance, with Z_2 disconnected, is

$$Z_{eq} = Z_1 \parallel Z_2 = 2 \parallel 3 = \frac{6}{5}.$$

Due to the 1 V source alone, $V_2 = 3/5$ V. Due to the 2 V source alone, $V_2 = 4/5$ V. Therefore, $V_2 = V_{oc} = 7/5$. The current through Z_2 is

$$I_2 = \frac{V_{oc}}{Z_2 + Z_{eq}} = \frac{7/5}{6/5 + j2} = \frac{7}{6 + j10} = 0.3088 - j0.5147$$

as found earlier.

Circuit in Fig. 3.22

Figure 3.22 shows a circuit with voltage and current sources.

Nodal Analysis

$$Z_1 = 2, \quad Z_2 = j2, \quad Z_3 = 3, \quad I_1 = -1, \quad V_3 = 2.$$

Voltage V_2 is the only unknown. Therefore, applying KCL at node 2, we get

$$\frac{(V_2 - 2)}{Z_3} + \frac{(V_2)}{Z_2} = -1.$$

Solving, we get

$$V_2 = \frac{-Z_2 Z_3 + 2Z_2}{Z_2 + Z_3} = (-0.3077 - j0.4615) \, \text{V}.$$

The currents are

$$I_1 = -\frac{(V_2 - 1)}{Z_1} = (-0.0147 - j0.3088)A, \quad I_2 = \frac{(V_2)}{Z_2} = (0.3088 - j0.5147)A,$$

$$I_3 = \frac{(V_2 - 2)}{Z_3} = (-0.3235 + j0.2059)A$$

With $-I_1 + I_2 + I_3 = 0$, KCL is satisfied.

Fig. 3.22 A circuit with voltage and current sources

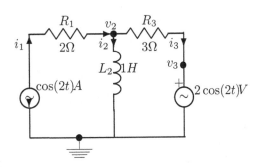

Mesh Analysis

$I_2 = I_1 - I_3, \ I_1 = -1.$

$$Z_3 I_3 - Z_2(I_1 - I_3) = -2.$$

Solving, we get

$$I_3 = \frac{-2 + I_1 Z_2}{Z_3 + Z_2}, \quad I_3 = (-0.7692 - j0.1538)\,\text{A}.$$

We get the same values of currents, as found earlier.

Let us solve the problem by superposition method. Let us find the response due to the voltage source. The current source is open-circuited. Then,

$$I_3 = -\frac{2}{Z_2 + Z_3} = (-0.4615 + j0.3077), \quad I_2 = -I_3.$$

Let us find the response due to the current source. The voltage source is short-circuited. Then, by current division,

$$I_3 = -\frac{Z_2}{Z_2 + Z_3} = (-0.3077 - j0.4615), \quad I_2 = -\frac{Z_3}{Z_2 + Z_3} = (-0.6923 + j0.4615)$$

Adding the corresponding two responses, we get the same values for the currents.

Let us find the current I_3 through Z_3 using Thévenin's theorem. The equivalent impedance, with Z_3 disconnected, is

$$Z_{eq} = Z_2 = j2.$$

Remember that the current source is open-circuited. Due to the 1 A source, $V_2 = -j2V$. Then, $V_{oc} = -j2 - 2$. Therefore, the current through Z_3 is

$$I_3 = \frac{V_{oc}}{Z_3 + Z_{eq}} = \frac{-j2 - 2}{3 + j2} = -0.7692 - j0.1538$$

as found earlier.

Circuit in Fig. 3.23

Figure 3.23 shows a circuit with voltage and current sources.

$$Z_1 = 2, \quad Z_2 = j2, \quad Z_3 = 3, \quad I_3 = 1, \quad V_1 = 2.$$

Mesh Analysis

$$I_1 - I_2 = I_3$$

$$I_1 Z_1 + I_2 Z_2 = I_1 Z_1 + (I_1 - I_3)Z_2 = 2, \quad I_1 = \frac{2 + I_3 Z_2}{Z_2 + Z_1} = 1, \quad I_2 = 0.$$

With $-I_1 + I_2 + I_3 = 0$, KCL is satisfied.

$$V_2 = I_2 Z_2 = 0.$$

Fig. 3.23 A circuit with voltage and current sources

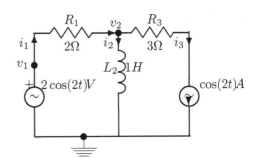

Nodal Analysis

$$\frac{V_2}{Z_2} + \frac{(V_2 - 2)}{Z_1} = -1.$$

Solving, $V_2 = 0$ and $V_{R3} = 3V$. The voltage across the current source is -3 V.

Let us do the problem by superposition method. Consider the response due to the voltage source alone. The current source is open-circuited. Then,

$$I_1 = I_2 = \frac{2}{2 + j2}.$$

Consider the response due to the current source alone. The voltage source is short-circuited. Then, by current division, we get

$$I_1 = \frac{j2}{2 + j2}, \quad I_2 = I_1 - I_3 = -\frac{2}{2 + j2}.$$

Adding the partial currents, we get the same results.

Let us find the current I_2 through Z_2 using Thevenin's theorem. The equivalent impedance, with Z_2 disconnected, is

$$Z_{eq} = Z_1 = 2.$$

Remember that the current source is open-circuited. Due to the 1 A source alone,

$$V_2 = -\frac{j2 \times 2}{2 + j2}.$$

Due to the 2 V source alone,

$$V_2 = \frac{j2 \times 2}{2 + j2}.$$

Then, $V_{oc} = 0$ and $I_2 = 0$, as found earlier.

Circuit in Fig. 3.24

Figure 3.24 shows a circuit with two current sources.

$$Z_1 = 2, \quad Z_2 = j2, \quad Z_3 = 3, \quad I_3 = 1, \quad I_1 = -2.$$

Fig. 3.24 A circuit with two current sources

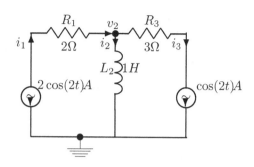

Nodal Analysis

$$\frac{V_2}{Z_2} = -3 \text{ and } V_2 = -j6V$$

$$I_1 = -2, \ I_2 = -\frac{j6}{j2} = -3A, \ I_3 = 1 \text{ A}.$$

With $-I_1 + I_2 + I_3 = 0$, KCL is satisfied.

Mesh Analysis

$$I_3 = 1, \quad I_1 = -2, \quad I_2 = (I_1 - I_3) = -3A.$$

Let us do the problem by superposition method. Consider the response due to the 2 A current source alone. The other current source is open-circuited. Then, by current division, we get

$$I_1 = -2, \quad I_2 = -2, \quad I_3 = 0.$$

Consider the response due to the 1 A current source alone. The other current source is open-circuited. Then, by current division, we get

$$I_1 = 0, \quad I_2 = -1, \quad I_3 = 1.$$

Adding the two partial results, we get the same currents.

Circuit in Fig. 3.25

A circuit with a voltage-controlled voltage source is shown in Fig. 3.25.

$$Z_1 = 2, \quad Z_2 = j1, \quad Z_3 = 3$$

By nodal analysis, we get

$$\frac{(V_2 - 1)}{Z_1} + \frac{V_2}{Z_2} + \frac{V_2 - (V_2 - 1)}{Z_3} = 0 \quad \text{or} \quad V_2 = \frac{Z_3 Z_2 - Z_1 Z_2}{Z_1 Z_3 + Z_2 Z_3} = (0.0667 + j0.1333)$$

The currents are

$$I_1 = -\frac{(V_2 - 1)}{Z_1} = (0.4667 - j0.0667)A, \quad I_2 = \frac{V_2}{Z_2} = (0.1333 - j0.0667)A, \quad I_3 = I_1 - I_2 = \frac{1}{3}A$$

With $-I_1 + I_2 + I_3 = 0$, KCL is satisfied.

Fig. 3.25 A circuit with a
voltage-controlled voltage
source

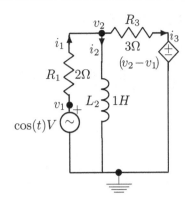

Mesh Analysis

$$Z_1(I_2 + I_3) + I_2 Z_2 = 1, \quad \text{or} \quad (Z_2 + Z_1)I_2 + Z_1 I_3 = 1$$
$$Z_3 I_3 - Z_2 I_2 = Z_1(I_3 + I_2), \quad \text{or} \quad -(Z_2 + Z_1)I_2 + (Z_3 - Z_1)I_3 = 0.$$

Solving, we get the same currents obtained earlier.
   Let us find the current I_2 through Z_2 using Thevenin's theorem.

$$\frac{(V_2 - 1)}{Z_1} + \frac{V_2 - (V_2 - 1)}{Z_3} = 0 \quad \text{or} \quad V_2 = V_{oc} = \frac{Z_3 - Z_1}{Z_3} = \frac{1}{3}.$$

Let us apply a 1 A current source instead of the load impedance and short-circuit the independent
voltage source. Then,

$$Z_{eq} = V_2 = Z_1 = 2.$$

Therefore, the current through Z_2 is

$$I_2 = \frac{V_{oc}}{Z_2 + Z_{eq}} = \frac{1/3}{2 + j1} = 0.1333 - j0.0667$$

as found earlier.

Circuit in Fig. 3.26
A circuit with a current-controlled voltage source is shown in Fig. 3.26.

$$Z_1 = 2, \quad Z_2 = j1, \quad Z_3 = 3.$$

By nodal analysis, we get

$$\frac{(V_2 - 1)}{Z_1} + \frac{V_2}{Z_2} + \frac{V_2 + 2jV_2}{Z_3} = 0.$$

Solving, we get

$$V_2 = \frac{Z_3 Z_2}{Z_3 Z_2 + Z_3 Z_1 + Z_1 Z_2 + j2 Z_1 Z_2} = (0.5172 + j0.2069) \text{ V}.$$

Fig. 3.26 A circuit with a
current-controlled voltage
source

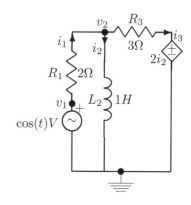

The currents are

$$I_1 = (0.2414 - j0.1034)A, \quad I_2 = (0.2069 - j0.5172)A, \quad I_3 = (0.0345 + j0.4138)A$$

With $-I_1 + I_2 + I_3 = 0$, KCL is satisfied.

Mesh Analysis

$$Z_1 I_1 + I_2 Z_2 = 1$$

$$(I_1 - I_2)Z_3 - I_2 Z_2 + 2I_2 = 0 \text{ or } Z_3 I_1 + I_2(2 - Z_3 - Z_2) = 0$$

Solving, we get the same currents obtained earlier.

Let us find the current I_3 through Z_3 using Thevenin's theorem.

$$\frac{(V_2 - 1)}{Z_1} + \frac{V_2}{Z_2} = 0 \quad \text{or} \quad V_2 = \frac{Z_2}{Z_1 + Z_2} = \frac{j1}{2 + j1}$$

$$V_{oc} = V_2 - \frac{2V_2}{Z_2} = -0.6 + j0.8.$$

Let us apply a 1 A current source instead of the load impedance and short-circuit the independent voltage source. Then, the voltage across the current source is Z_{eq}.

$$I_2 = \frac{2}{2 + j1} \text{ and } V_2 = \frac{j2}{2 + j1}$$

$$Z_{eq} = V_2 - 2I_2 = -1.2 + j1.6.$$

Therefore, the current through Z_3 is

$$I_3 = \frac{V_{oc}}{Z_3 + Z_{eq}} = \frac{(-0.6 + j0.8)}{1.8 + j1.6} = 0.0345 + j0.4138$$

as found earlier.

Circuit in Fig. 3.27

A circuit with a voltage-controlled current source is shown in Fig. 3.27.

$$Z_1 = 2, \quad Z_2 = 2 - j1, \quad Z_3 = 3.$$

By nodal analysis, we get

$$-1 - 2V_2 + \frac{V_2}{Z_2} = 0.$$

Solving, we get $V_2 = \frac{Z_2}{1 - 2Z_2} = -0.6154 - j0.0769$. The currents are

$$I_1 = -1A, \quad I_2 = \frac{V_2}{Z_2} = -0.2308 - j0.1538A, \quad I_3 = -(I_1 + I_3) = 1.2308 + j0.1538A$$

With $I_1 + I_2 + I_3 = 0$, KCL is satisfied.

Mesh Analysis

$$Z_2(2V_2 + 1) = V_2, \quad \text{or} \quad V_2 = \frac{Z_2}{1 - 2Z_2} V$$

as before.

Let us find the current I_2 through Z_2 using Thevenin's theorem.

$$-1 - 2V_2 = 0 \text{ and } V_2 = V_{oc} = \frac{-1}{2}.$$

Let us apply a 1 A current source instead of the load impedance and open-circuit the independent current source. Then,

$$1 + 2V_2 = 0 \text{ or } Z_{eq} = V_2 = -0.5.$$

Therefore, the current through Z_2 is

$$I_2 = \frac{V_{oc}}{Z_2 + Z_{eq}} = \frac{-0.5}{1.5 - j1} = -0.2308 - j0.1538$$

as found earlier.

Fig. 3.27 A circuit with a
voltage-controlled current
source

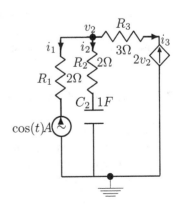

Fig. 3.28 A circuit with a
current-controlled current
source

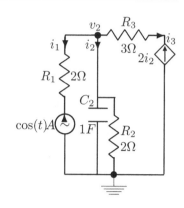

Circuit in Fig. 3.28

A circuit with a current-controlled current source is shown in Fig. 3.28.

$$Z_1 = 2, \quad Z_2 = 0.4 - j0.8, \quad Z_3 = 3$$

Nodal Analysis

$$-1 - 2\frac{V_2}{Z_2} + \frac{V_2}{Z_2} = 0.$$

Solving, we get $V_2 = -Z_2 = -0.4 + j0.8$. The currents are

$$I_1 = -1 \, \text{A}, \quad I_2 = -1 \, \text{A}, \quad I_3 = 2 \, \text{A}.$$

With $I_1 + I_2 + I_3 = 0$, KCL is satisfied.

Mesh analysis

$$Z_2(2\frac{V_2}{Z_2} + 1) = V_2, \quad \text{or} \quad V_2 = -Z_2 V$$

as before.

3.3   Circuit Theorems

Except for the same two differences between AC and DC circuit analysis, the theory remains the same as for DC circuits. The values of circuit elements become complex quantities involving complex arithmetic. The AC circuit must be analyzed separately at each frequency of interest.

3.3.1   Thévenin's Theorem

Any linear combination of voltage and current sources, independent or dependent, and impedances with two terminals can be replaced by a fixed voltage source V_{eq} or V_{oc}, called Thévenin equivalent voltage, in series with an impedance Z_{eq}, called Thévenin equivalent impedance. The equivalent source is derived as follows:

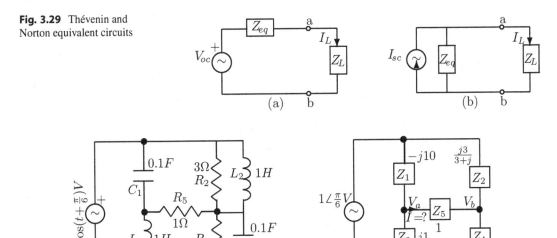

Fig. 3.29 Thévenin and Norton equivalent circuits

Fig. 3.30 A bridge circuit and its frequency-domain version

1. Remove the load circuit, any combination of independent voltage and current sources and impedances and find the open-circuit voltage, called V_{eq}, across the load circuit terminals.
2. Find the impedance across the load terminals, after short-circuiting all independent voltage sources and open-circuiting all independent current sources, called Z_{eq}. This step can also be carried out by finding the short-circuit current I_{sc} through the load circuit terminals and $Z_{eq} = V_{eq}/I_{sc}$.

Thévenin equivalent circuit is shown in Fig. 3.29a. The bridge circuit, we analyzed earlier, is shown in Fig. 3.30a. The frequency-domain version of the circuit is shown in Fig. 3.30b. The problem is to find the current I through Z_5 using Thévenin's theorem. The impedance values are

$$Z_1 = -j10, \quad Z_2 = \frac{j3}{3+j1}, \quad Z_3 = j1, \quad Z_4 = \frac{-j10}{1-j10}, \quad Z_5 = Z_L = 1.$$

With the impedance Z_5 disconnected in Fig. 3.30b, the voltages V_a, V_b and V_{ab}, the open-circuit voltage, are found as

$$V_a = V \frac{Z_3}{Z_1 + Z_3} = -0.0962 - j0.0556, \quad V_b = V \frac{Z_4}{Z_2 + Z_4} = 0.6496 - j0.0860,$$

$$V_{ab} = V_a - V_b = -0.7458 + j0.0305.$$

The equivalent impedance at terminals a and b in Fig. 3.31a, which is the series combination of parallel impedances $Z_1 \| Z_3$ and $Z_2 \| Z_4$ is determined as

$$Z_{eq} = \frac{Z_1 Z_3}{Z_1 + Z_3} + \frac{Z_2 Z_4}{Z_2 + Z_4} = 0.5152 + j1.4589.$$

Fig. 3.31 (a) Circuit to determine the equivalent impedance from ab; (b) circuit showing the determination of the load current

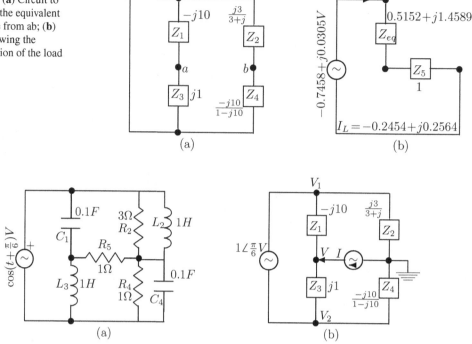

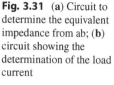

Fig. 3.32 (a) A bridge circuit; (b) with a bucking voltage V replacing the load impedance

Now, the load current I_L is determined as shown in Fig. 3.31b

$$I_L = \frac{V_{ab}}{Z_{eq} + Z_L} = \frac{-0.7458 + j0.0305}{(0.5152 + j1.4589) + 1} = -0.2454 + j0.2564,$$

which is the same as found in the complete analysis of the circuit in an earlier example.

Bucking Voltage Method

Thévenin equivalent circuit is characterized by the equation

$$V = I Z_{eq} + V_{oc}.$$

If $I = 0$, $V = V_{oc}$, the open-circuit voltage. If $V = 0$, $I_{sc} = -\frac{V_{oc}}{Z_{eq}}$, the short-circuit current. Applying KVL and KCL at the top and bottom nodes at the right side of Fig. 3.32b, the equilibrium equations are

$$V_1 - V_2 = (0.8660 + j0.5)$$

$$\frac{V_1}{Z_2} + \frac{V_1 - V}{Z_1} + \frac{V2}{Z_4} + \frac{V_2 - V}{Z_3} = 0.$$

The second equation is obtained using the fact that the current leaving the bottom node and entering the top node must be equal. Simplifying, we get

$$V_1 - V_2 = (0.8660 + j0.5)$$

$$V_1(Z_1 Z_3 Z_4 + Z_2 Z_3 Z_4) + V_2(Z_1 Z_2 Z_3 + Z_1 Z_2 Z_4) = V(Z_1 Z_2 Z_4 + Z_2 Z_3 Z_4).$$

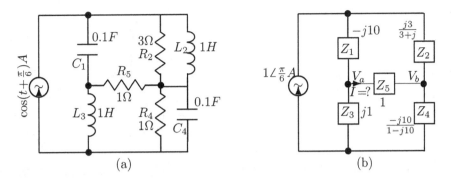

Fig. 3.33 (a) A bridge circuit with a current source; (b) its frequency-domain version

Solving for V_1 and V_2 in terms of V, we get

$V_1 = V(0.3229 - j0.2391) + (0.4499 + j0.3978)$ and $V_2 = V(0.3229 - j0.2391) - (-0.4161 - j0.1022)$

The equilibrium equation at node with voltage V is

$$\frac{V - V_1}{Z_1} + \frac{V - V_2}{Z_3} = I.$$

Substituting for V_1 and V_2 in this equation, we get

$$V = (0.5152 + j1.4589)I + (-0.7458 + j0.0305) = IZ_{eq} + V_{oc}.$$

The first method seems to be easier for this problem.

Consider the bridge circuit with a current source shown in Fig. 3.33a and its frequency-domain version shown in Fig. 3.33b.

$$I = \cos(\frac{\pi}{6}) + j\sin(\frac{\pi}{6}) = (0.8660 + j0.5)\,\text{A}$$

The problem is to find the current through Z_5. With Z_5 disconnected, the voltages V_a, $V_b V_{ab} = V_{oc}$ are computed as

$$V_a = Z_3 \frac{I(Z_2 + Z_4)}{Z_1 + Z_2 + Z_3 + Z_4} = -0.1104 - j0.1459$$

$$V_b = Z_4 \frac{I(Z_1 + Z_3)}{Z_1 + Z_2 + Z_3 + Z_4} = 1.0405 + j0.2856$$

$$V_{ab} = V_a - V_b = (-1.1509 - j0.4315)\text{V}$$

In order to find Z_{eq}, the current source is replaced by an open-circuit. Then, we get

$$Z_{eq} = \frac{(Z_1 + Z_2)(Z_3 + Z_4)}{Z_1 + Z_3 + Z_2 + Z_4} = (1.1993 + j0.8475)\,\Omega.$$

Fig. 3.34 A circuit with voltage and current sources with different frequencies

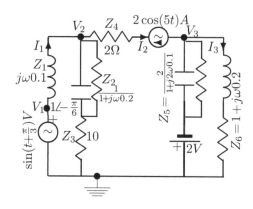

Therefore, the current through Z_5 is

$$I_L = \frac{V_{ab}}{Z_{eq} + Z_L} = \frac{-1.1509 - j0.4315}{(1.1993 + j0.8475) + 1} = -0.5215 + j0.0048.$$

Consider the circuit with voltage and current sources with different frequencies, shown in Fig. 3.34. The problem is to find the current through the impedance Z_5. Let us find the current due to the current source

$$I = 2\cos(5t) = 2\angle 0 \, \text{A}$$

alone, with the voltage sources short-circuited. With $\omega = 5$,

$$Z_5 = 1 - j1, \quad Z_6 = 1 + j1.$$

The voltages at the top and bottom terminals of Z_5 and across, with Z_5 disconnected, are

$$V_a = -2(1 + j1), \quad V_b = 0 \quad \text{and} \quad V_{ab} = V_a - V_b = -2(1 + j1).$$

In order to find Z_{eq}, we replace the two voltage sources by short-circuits and the current source by open-circuit. Then,

$$Z_{eq} = Z_6 = 1 + j1.$$

Due to the AC current source alone, the current through Z_5 is

$$I_L = \frac{V_{ab}}{Z_{eq} + Z_L} = \frac{-2(1 + j1)}{(1 - j1) + (1 + j1)} = -1 - j1.$$

Due to the rightmost DC source alone

$$I_L = \frac{2}{3} \, \text{A}.$$

The current source is open-circuited and the other voltage source is short-circuited. The sum of the two currents is the total current in R_L,

$$\frac{2}{3} + \sqrt{2}\cos(5t - \frac{3\pi}{4}).$$

No contribution from the leftmost voltage source, since the current source becomes open-circuited.

Fig. 3.35 A circuit with voltage and current sources with different frequencies with a bucking voltage inserted

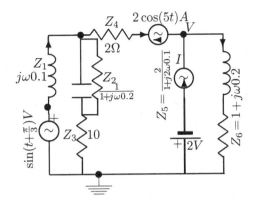

Bucking Voltage Method

Circuit analysis using Thévenin's theorem is a special case of nodal analysis, characterized by the equation

$$V = IZ_{eq} + V_{eq}.$$

Impedance Z_5 has been replaced by a current source with current I and voltage V at its top terminal, as shown in Fig. 3.35. Applying KCL at node with voltage V for the current source alone, we get

$$I - \frac{V}{1 + j1} = 2 \qquad (3.44)$$

$$V = I(1 + j1) - 2(1 + j1)$$

$$V_{ab} = V_{eq} = -2(1 + j1), \quad Z_{eq} = 1 + j1$$

Due to the AC current source alone

$$I_L = \frac{V_{ab}}{Z_{eq} + Z_L} = \frac{-2(1 + j1)}{(1 - j1) + (1 + j1)} = (-1 - j1)\,\text{A}$$

as found before.

Due to the DC source alone

$$I = \frac{V - 2}{1} \quad \text{or} \quad V = I + 2$$

$$V_{ab} = V_{eq} = 2, \quad Z_{eq} = 1$$

and

$$I_L = \frac{2}{2 + 1} = \frac{2}{3}\,\text{A}$$

Consider the circuit with a voltage-controlled voltage source, shown in Fig. 3.36a. The problem is to find the current through $Z_2 = 1\,\Omega$.

Nodal Analysis

With

$$Z_1 = j1, \quad Z_2 = 1, \quad Z_3 = 5 - j10, \quad V = \cos(t)$$

$$\frac{(V_2 - 1)}{Z_1} + \frac{V_2}{Z_2} + \frac{V_2 + 2(V_2 - 1)}{Z_3} = 0$$

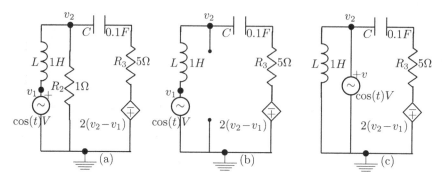

Fig. 3.36 (a) A circuit with a voltage-controlled voltage source; (b) with the load resistor disconnected; (c) the load resistor replaced by a voltage $\cos(t)$

$$V_2 = \frac{Z_2 Z_3 + 2Z_1 Z_2}{Z_2 Z_3 + Z_1 Z_3 + 3Z_1 Z_2} = 0.3974 - j0.4803$$

$$I_{Z2} = \frac{V_2}{Z_2} = 0.3974 - j0.4803.$$

Thévenin's Theorem Approach

The circuit is shown in Fig. 3.36b with the load resistor disconnected. Let us find the open-circuit voltage at the terminals shown by small discs. Applying KCL at node V_2, we get

$$\frac{(V_2 - 1)}{Z_1} + \frac{V_2 + 2(V_2 - 1)}{Z_3} = 0$$

$$V_2 = V_{eq} = \frac{Z_3 + 2Z_1}{Z_3 + 3Z_1} = 1.0946 - j0.0676.$$

In order to find the short-circuit current, we turn off all independent sources only. Then, a voltage source of $1\angle(0)$ V is applied at the open-circuited load terminals, as shown in Fig. 3.36c. The resultant current in that branch is found. Then, the inverse of that current is Z_{eq}. Applying $1\angle 0$ V,

$$i = \frac{1}{Z_1} + \frac{1+2}{Z_3} = 0.1200 - j0.7600$$

$$Z_{eq} = \frac{1}{i} = 0.2027 + j1.2838.$$

For linearity and Thévenin's theorem, leave the dependent sources, as they are controlled by circuit variables. Alternatively, applying a current of $1\angle 0$ A,

$$\frac{V_2}{Z_1} + \frac{V_2 + 2V_2}{Z_3} = 1$$

$$V_2(3Z_1 + Z_3) = Z_1 Z_3$$

$$V_2 = Z_{eq} = \frac{Z_1 Z_3}{(3Z_1 + Z_3)} = 0.2027 + j1.2838.$$

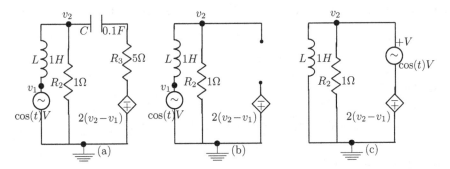

Fig. 3.37 (**a**) A circuit with a voltage-controlled voltage source; (**b**) with the load impedance disconnected; (**c**) the load impedance replaced by a voltage source $\cos(t)$

Bucking Voltage Method

$$\frac{(V-1)}{Z_1} + \frac{V + 2(V-1)}{Z_3} = I$$

$$V(3Z_1 + Z_3) = (Z_3 + 2Z_1) + IZ_1Z_3$$

$$V = \frac{(Z_3 + 2Z_1) + IZ_1Z_3}{(3Z_1 + Z_3)} = (1.0946 - j0.0676) + I(0.2027 + j1.2838)$$

$$I_L = \frac{(1.0946 - j0.0676)}{Z_2 + (0.2027 + j1.2838)} = 0.3974 - j0.4803.$$

Let us find the current through the resistor and capacitor in the circuit, shown in Fig. 3.37a, which is the same as the last one. The circuit is shown in Fig. 3.37b with the load impedance disconnected. By nodal analysis, in the last example, we found

$$V_2 = V_a = 0.3974 - j0.4803$$

$$V_b = -2(V_2 - 1) = 1.2052 + j0.9607$$

$$I_L = \frac{V_a - V_b}{Z_3} = 0.0830 - j0.1223.$$

Thévenin's Theorem Approach
With the load impedance disconnected, let us find the open-circuit voltage. Applying KCL at node V_2, we get

$$\frac{(V_2 - 1)}{Z_1} + \frac{V_2}{Z_2} = 0$$

$$V_2 = \frac{Z_2}{Z_1 + Z_2} = 0.5 - j0.5.$$

By voltage division also, we get the result.

$$V_{oc} = V_2 - (-2(V_2 - 1)) = 3V_2 - 2 = (-0.5 - j1.5).$$

Let us find Z_{eq}. The circuit is shown in Fig. 3.37c. Applying $1\angle 0$ V,

$$V_2 - 1 = -2V_2 \quad \text{or} \quad V_2 = \frac{1}{3}.$$

The current through the $1 - V$ source, which is the sum of the currents through Z_1 and Z_2, is

$$\frac{1}{3}(\frac{1}{j1} + 1) = \frac{1 + j1}{j3}.$$

The inverse of it is Z_{eq},

$$Z_{eq} = \frac{j3}{1 + j1} = j1.5(1 - j1) = 1.5 + j1.5.$$

Alternatively, applying $1\angle 0$ A,

$$V_2 = \frac{j1}{1 + j1}.$$

The voltage across the $1 - A$ source is

$$V_2 + 2V_2 = 3V_2 = Z_{eq} = \frac{j3}{1 + j1} = j1.5(1 - j1) = 1.5 + j1.5.$$

With $Z_3 = 5 - j10$, $Z_{eq} = 1.5 + j1.5$ and $V_{oc} = -0.5 - j1.5$,

$$I_L = \frac{-0.5 - j1.5}{Z_3 + (1.5 + j1.5)} = 0.0830 - j0.1223 \, \text{A}.$$

3.3.2   Norton's Theorem

Admittance is the reciprocal of the impedance.

$$E = ZI, \qquad I = YE$$

The first form is more convenient to find E, when I (the excitation) is given. The second form is more convenient to find I, when E (the excitation) is given. Any linear combination of voltage and current sources, independent or dependent, and impedances with two terminals can be replaced by a fixed current source I_{sc}, called Norton equivalent current, in parallel with an impedance Z_{eq}. The equivalent source is derived as follows:

1. Remove the load circuit, any combination of independent voltage and current sources and impedances. and find the short-circuit current, called I_{sc}, through the load circuit terminals.
2. Find the impedance across the load terminals, after short-circuiting all independent voltage sources and open-circuiting all independent current sources, called Z_{eq}. This step can also be carried out by finding the short-circuit current I_{sc} through the load circuit terminals and the open-circuit voltage V_{oc}. Then, $Z_{eq} = V_{oc}/I_{sc}$.

Figure 3.29b shows the Norton equivalent circuit. Norton equivalent circuit is characterized by the equation

$$I = \frac{V}{Z_{eq}} - I_{sc} = VY_{eq} - I_{sc}.$$

It is better use Thévenin's theorem for small Z_{eq} and Norton's theorem for large Z_{eq} compared with other impedances in the circuit. Once we find V_{oc} and Z_{eq}, as showed for Thévenin's equivalent circuit, then $I_{sc} = V_{oc}/Z_{eq}$.

3.3.3 Maximum Average Power Transfer Theorem

An AC circuit to find the condition for maximum average power transfer between a source and a load is shown in Fig. 3.38. A load impedance $Z_L = R_L + jX_L$ is connected to the AC source and its source impedance is $Z_s = R_s + jX_s$. The value of Z_L, for which the maximum average power transfer occurs from the source to the load, is to be found.

The current through the load impedance is

$$I = \frac{V}{Z_s + Z_L}.$$

The average power absorbed by the load, as shown in a later chapter, is

$$P_L = 0.5|I|^2 R_L = 0.5 \frac{|V|^2 R_L}{(R_s + R_L)^2 + (X_s + X_L)^2}.$$

Intuitively, as $(X_s + X_L)$ absorbs no power, it must be zero for maximum power transfer. That is, $X_L = -X_s$. Differentiating this expression with respect to X_L and equating to zero, we get

$$\frac{dP_L}{dX_L} = |V|^2 R_L \left(\frac{-(X_s + X_L)}{((R_s + R_L)^2 + (X_s + X_L)^2)^2} \right) = 0$$

For P_L to be maximum, $X_L = -X_s$. Substituting this condition in the expression for power, we get

$$P_L = 0.5 \frac{|V|^2 R_L}{(R_s + R_L)^2}.$$

This expression is the same as that obtained for DC circuits and yields the maximum power with $R_L = R_s$. The maximum average power delivered to the load is

$$P_L = \frac{|V|^2}{8R_L}.$$

Therefore, the condition for maximum average power transfer is that the source and load impedances are conjugate symmetric,

$$Z_L = Z_s^*.$$

When this condition is satisfied, the total impedance of the circuit becomes resistive. For a purely resistive load R_L, the condition for maximum power transfer is $R_L = |Z_s| = \sqrt{R_s^2 + X_s^2}$.

Fig. 3.38 An AC circuit to find the condition for maximum average power transfer

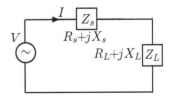

Example

Find the load impedance Z_2 in the bridge circuit, shown in Fig. 3.39a, for maximum power transfer from the source to the load.

The source voltage is $0.8660 + j0.5$. We have to find the Thévenin's equivalent voltage and impedance across the load impedance terminals. Replacing the bottom half of the Δ circuit by Y-circuit, we get, with

$$Z_{bc} = Z_4 = 0.9901 - j0.0990, \quad Z_{ac} = Z_3 = j1, \quad Z_{ab} = Z_5 = 1.$$

$$Z_a = \frac{j1 \times 1}{(0.9901 - j0.0990) + j1 + 1} = 0.1888 + j0.4170$$

$$Z_b = \frac{(0.9901 - j0.0990) \times 1}{(0.9901 - j0.0990) + j1 + 1} = 0.3942 - j0.2282$$

$$Z_c = \frac{(0.9901 - j0.0990) \times j1}{(0.9901 - j0.0990) + j1 + 1} = 0.2282 + j0.3942.$$

The transformed circuit, with the voltage source short-circuited, is shown in Fig. 3.39b. Now,

$$Z_{eq} = (0.3942 - j0.2282) + \frac{(-j10 + (0.1888 + j0.4170))(0.2282 + j0.3942)}{((-j10 + (0.1888 + j0.4170)) + (0.2282 + j0.3942))} = 0.6425 + j0.1763$$

The maximum average power transfer occurs with $Z_L = Z_{eq}^* = 0.6425 - j0.1763$. The V_{oc} across the load terminals of the transformed circuit, with the voltage source inserted, is

$$V_{oc} = \frac{((-j10 + (0.1888 + j0.4170)))(0.866 + j0.5)}{-j10 + (0.1888 + j0.4170) + (0.2282 + j0.3942)} = 0.9155 + j0.4977 = 1.0420\angle(0.4979)$$

Therefore, the maximum power transferred is

$$P_m = \frac{|V_{oc}|^2}{8R_L} = \frac{(1.0420^2)}{(8 \times 0.6425)} = 0.2113 \, \text{W}.$$

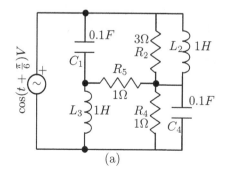
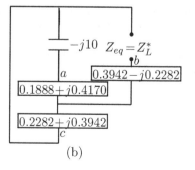

Fig. 3.39 (a) A bridge circuit; (b) circuit to find Z_{eq}

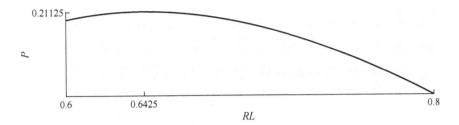

Fig. 3.40 Load resistor versus power transferred

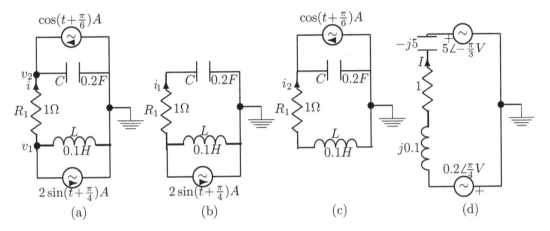

Fig. 3.41 (**a**) A circuit with current sources; (**b**) and (**c**) the circuits with one of the two current sources replaced by an open-circuit; (**d**) equivalent circuit with voltage sources in the frequency-domain

Figure 3.40 show the power transferred for a range of values of the load resistor. The peak value occurs when the load resistance is equal to the Thévenin equivalent resistance.

3.3.4 Source Transformation

Thévenin and Norton equivalent circuits provide the same volt–ampere relationship at the terminals. It is possible to substitute one source by another, called source transformation, to simplify the circuit analysis. Source transformation is most useful if the source is localized to some portion of the circuit.

Consider the circuit shown in Fig. 3.41a with current sources. The problem is to find the current through R_1. The sources are $2\cos(t - \frac{\pi}{4})A$ and $\cos(t + \frac{\pi}{6})A$. In the frequency-domain, they are $2\angle(-\frac{\pi}{4})$ and $1\angle(\frac{\pi}{6})$.

Nodal Method
Applying KCL at the right side nodes, we get, with $Z_L = j0.1$, $Z_C = -j5$ and $Z_{R1} = 1$,

$$\frac{V_1}{j0.1} + \frac{V_1 - V_2}{1} = -2\angle\left(-\frac{\pi}{4}\right)$$

$$\frac{V_2}{-j5} + \frac{V_2 - V_1}{1} = 1\angle\left(\frac{\pi}{6}.\right)$$

Simplifying, we get

$$V_1(1 + j0.1) - j0.1V_2 = -0.2\angle\left(\frac{\pi}{4}\right)$$

$$V_1(j5) + (1 - j5)V_2 = 5\angle\left(-\frac{\pi}{3}\right)$$

$$\begin{bmatrix} 1 + j0.1 & -j0.1 \\ j5 & 1 - j5 \end{bmatrix}\begin{bmatrix} V_1 \\ V_2 \end{bmatrix} = \begin{bmatrix} -0.1414 - j0.1414 \\ 2.5000 - j4.3301 \end{bmatrix}.$$

The determinant of the admittance matrix is $1 - j4.9$.
Solving the equilibrium equations, we get

$$\begin{bmatrix} V_1 \\ V_2 \end{bmatrix} = \begin{bmatrix} 1 + j0.1 & -j0.1 \\ j5 & 1 - j5 \end{bmatrix}^{-1}\begin{bmatrix} -0.1414 - j0.1414 \\ 2.5000 - j4.3301 \end{bmatrix}$$

$$= \begin{bmatrix} 1.0196 - j0.0040 & -0.0196 + j0.0040 \\ 0.9796 - j0.1999 & 0.0204 + j0.1999 \end{bmatrix}\begin{bmatrix} -0.1414 - j0.1414 \\ 2.5000 - j4.3301 \end{bmatrix}$$

$$= \begin{bmatrix} -0.1764 - j0.0488 \\ 0.7498 + j0.3012 \end{bmatrix}.$$

Since the resistance value is $1\,\Omega$, the current through it, from bottom to top, is $V_1 - V_2 = (-0.9263 - j0.35)$ A.

Let us find the current using the linearity property. The currents due to each of the two current sources, I_1 in Fig. 3.41b and I_2 in Fig. 3.41c, are found separately and added to find the total current. When one of the sources is open-circuited, the circuit becomes a current source feeding a parallel circuit. In Fig. 3.41b, the equivalent impedance is $(j0.1) \parallel (1 - j5)$. In Fig. 3.33c, the equivalent impedance is $(-j5) \parallel (1 + j0.1)$. The currents can be found by current division.

$$I_1 = -\frac{(j0.1)(2(\cos(-\frac{\pi}{4}) + j\sin(-\frac{\pi}{4})))}{(1 + j1 - j5)} = 0.0221 - j0.0334$$

$$I_2 = -\frac{(-j5)(\cos(\frac{\pi}{6}) + j\sin(\frac{\pi}{6}))}{(1 + j0.1 - j5)} = -0.9483 - j0.3167.$$

Now, $I = I_1 + I_2 = (-0.9263 - j0.35)$ A, as found by nodal analysis.

Consider the transformed circuit shown in Fig. 3.41d with voltage sources. It is a series circuit. The current sources are multiplied by the respective shunt impedances to get the equivalent voltage sources and the shunt impedances becomes the series impedances for the voltage sources. In the frequency-domain, the voltage sources are

$$2(0.1)\angle\left(-\frac{\pi}{4} + \frac{\pi}{2}\right) = 0.1414 + j0.1414 \quad \text{and} \quad 1(5)\angle\left(\frac{\pi}{6} - \frac{\pi}{2}\right) = 2.5 - j4.3301.$$

The current through the resistance is the sum of the voltages divided by the three impedances in series

$$\frac{2.6414 - j4.1887}{1 + j0.1 - j5} = (0.9263 + j0.35)\text{ A}.$$

This current flows in the opposite direction. The current is $I = (-0.9263 - j0.35)$ A, as obtained earlier. With voltage sources, we got the result by using the simpler Ohm's law.

3.4   Application

3.4.1   Filters

A practical signal is composed of different frequency components. At the time of generation or in transmission, desired signals get contaminated with undesirable signals, called noise signals. Typically, the desired signals are composed of low frequency components and noise signals are composed of high frequency components. Filters are commonly used in signal processing applications to filter out the unwanted noise component. Basic lowpass filters, which pass low frequency components and suppress high frequency components, are presented.

RL Filter
The RL lowpass filter is shown in Fig. 3.42a. The input signal to the filter is

$$v_1(t) = \cos(10t) + 0.1\cos(1000t).$$

It is desired that the output $v_2(t)$ contains the high frequency component of magnitude less than 10% of that at the input. By voltage division, the input–output relationship of the filter, in the frequency domain, is

$$\frac{V_2}{V_1} = \frac{R}{R + j\omega L}.$$

The output, for a given input is inversely proportional with the frequency due to the frequency ω appearing in the denominator. The magnitude has to meet the constraint

$$\left|\frac{R}{R + j\omega L}\right| = 0.1.$$

Substituting for the magnitude, we get

$$\frac{R}{\sqrt{R^2 + (\omega L)^2}} = 0.1 \quad \text{or} \quad 100R^2 = R^2 + (\omega L)^2.$$

With $\omega = 1000$ and $R = 1000\,\Omega$, solving for L, we get $L = 9.9499$ H. The magnitude of the transfer function, the input–output relationship in the frequency-domain, is shown in Fig. 3.43a. It

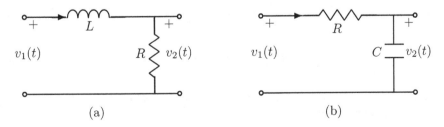

Fig. 3.42 (a) A RL lowpass filter circuit; (b) a RC lowpass filter circuit

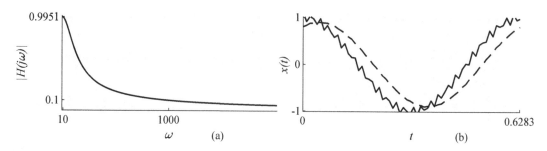

Fig. 3.43 (a) Magnitude of the transfer function of the filter; (b) excitation and the response of the RL circuit

is a plot of the magnitude and the phase (not shown) of the filter output for the input $e^{j\omega t}$ with the frequency varying. Since the filter is a lowpass one, it attenuates the high frequency components more. At $\omega = 10$ rad/s, the attenuation is 0.9951. At $\omega = 1000$ rad/s, the attenuation is 0.1. While the attenuation of the high frequency component is desirable, the low frequency components are also attenuated to a lesser extent. Filters are designed to suit the application requirements. The input and the output waveforms of the filter are shown in Fig. 3.43b. The period of the low frequency component is $2\pi/10 = 0.6283$ s. The output waveform, shown by the dashed line, has negligible undesirable high frequency component of the input. The filter circuit is a variable frequency voltage divider, since the impedance of the inductor increases with increasing frequency. Therefore, most of the voltage drop due to high frequency components occur across it and, thereby, the filter output, the voltage across the resistor, contains a low percent of the high frequency components. The output waveform is shifted due to the filtering operation.

RC Filter

The RC lowpass filter is shown in Fig. 3.42b. The input signal to the filter is

$$V_1(t) = \cos(10t) + 0.1\cos(1000t).$$

It is desired that the output $v_2(t)$ contains the high frequency component of magnitude less than 10% of that at the input. By voltage division, the input–output relationship of the filter, in the frequency domain, is

$$\frac{V_2}{V_1} = \frac{1}{1 + j\omega RC}.$$

The magnitude has to meet the constraint

$$\left|\frac{1}{1 + j\omega RC}\right| = 0.1.$$

Substituting for the magnitude, we get

$$\frac{1}{1 + (\omega RC)^2} = 0.1 \quad \text{or} \quad 100 = 1 + (\omega RC)^2.$$

With $\omega = 1000$ and $R = 1000\,\Omega$, solving for C, we get $C = 9.9499\,\mu\text{F}$. The transfer function and the output are similar to that in Fig. 3.43.

3.5   Summary

- Alternating current (AC) is also movement of electrical charge, but changes direction periodically.
- A sinusoidal waveform is a linear combination of sine and cosine waveforms.
- System models, such as differential equation and convolution, reduce to algebraic equations for a sinusoidal input for linear systems.
- The polar form of a sinusoid is $x(t) = A\cos(\omega t + \theta)$, $\quad -\infty < t < \infty$
- The rectangular form of a sinusoid is $A\cos(\omega t + \theta) = C\cos(\omega t) + D\sin(\omega t)$.
- The complex sinusoid, $v(t) = Ve^{j(\omega t + \theta)} = Ve^{j\theta}e^{j\omega t}$, $\quad -\infty < t < \infty$ is found to be indispensable for analysis due to its compact form and ease of manipulation of the exponential function. Due to Euler's identity,

$$v(t) = \frac{V}{2}\left(e^{j(\omega t + \theta)} + e^{-j(\omega t + \theta)}\right) = V\cos(\omega t + \theta)$$

- AC circuit analysis boils down to analysis with sources of the form $Ae^{j(\omega t + \theta)}$, which is a mathematical equivalent of the real sinusoid. Then, the output to the real sinusoid is deduced from the results.
- When the complex exponential source is used, the circuit elements, resistors, inductors, and capacitors, are to be represented appropriately.
- When the independent variable of signals, such as time t, is not frequency, then that representation is called the time-domain representation. On the other hand, when the independent variable of signals is frequency, such as f and ω, then that representation is called the frequency-domain representation.
- The first task in formulating the equilibrium equations for an electric circuit is to select a set of variables, voltages or currents. They must be independent and adequate to describe the state of the network at any instant. They must be uniquely and reversibly related to all the branch variables.
- After selecting an appropriate set of variables, we have to use Kirchhoff's laws.
- AC circuits must be analyzed at each frequency individually, if sources with more than one frequency are present in the circuit.
- As the sinusoidal waveform is everlasting, using a sinusoid as a source will result in the steady-state analysis.
- Any linear combination of voltage and current sources, independent or dependent, and impedances with two terminals can be replaced by a fixed voltage source V_{eq}, called Thévenin's equivalent voltage, in series with an impedance Z_{eq}, called Thévenin's equivalent impedance.
- Any linear combination of voltage and current sources, independent or dependent, and impedances with two terminals can be replaced by a fixed current source I_{sc}, called Nortan's equivalent current, in parallel with an impedance Z_{eq}.
- The condition for maximum power transfer is that the source and load impedances are conjugate symmetric,

$$Z_L = Z_s^*$$

When this condition is satisfied, the total impedance of the circuit becomes resistive.

- Source transformation is replacing a current source by a voltage source and vice versa for simpler circuit analysis. It is most useful if the source is localized to some portion of the circuit.

Exercises

3.1 Find the sinusoid in cosine form. Find the value of t closest to $t = 0$, where the first positive peak of the sinusoid occur.

    3.1.1 $x(t) = \sin(\frac{2\pi}{8}t + \frac{\pi}{4})$.

    3.1.2 $x(t) = 2\sin(\frac{2\pi}{6}t - \frac{\pi}{6})$.

    __* 3.1.3__ $x(t) = -5\sin(2\pi t + \frac{\pi}{3})$.

3.2 Find the rectangular form of the sinusoid. Find the value of t closest to $t = 0$, where the first positive peak of the sinusoid occur. Find the values of t at which the next two consecutive positive peaks occur. Get back the polar form from the rectangular form and verify that the given sinusoid is obtained.

    3.2.1 $x(t) = \sin(\frac{2\pi}{6}t + \frac{\pi}{6})$.

    3.2.2 $x(t) = 3\cos(\frac{2\pi}{3}t - \frac{\pi}{3})$.

    3.2.3 $x(t) = -2\sin(2\pi t + \frac{2\pi}{3})$.

    __* 3.2.4__ $x(t) = 4\sin(2\pi t - \frac{\pi}{4})$.

    3.2.5 $x(t) = 5\cos(\frac{2\pi}{4}t + \frac{2\pi}{6})$.

3.3 Express the signal in terms of complex exponentials.

    3.3.1 $x(t) = \sin(\frac{2\pi}{8}t)$

    3.3.2 $x(t) = \cos(2\pi t)$

    3.3.3 $x(t) = 2\cos(\frac{2\pi}{4}t - \frac{\pi}{3})$

    __*3.3.4__ $x(t) = 3\sin(\frac{2\pi}{6}t + \frac{\pi}{6})$

3.4 Find the current $i(t)$ in the series circuit in cosine sinusoidal form, using the frequency-domain representation.

    3.4.1

$$v(t) = 2\cos(\frac{2\pi}{8}t - \frac{\pi}{3}), R = 3, C = 0.2$$

    __* 3.4.2__

$$v(t) = 2\sin(\frac{2\pi}{4}t - \frac{\pi}{3}), R = 2, L = 0.1$$

3.5 Find the current I and the voltages across the impedances,

$$V_{Z1}, V_{Z2}, V_{Z3}, V_{Z4}, V_{Z5}$$

in the series circuit, shown in Fig. 3.44. Verify KVL.

    3.5.1

$$v(t) = 3\cos(2t)\,V, \quad Z_1 = \frac{1}{j\omega 0.3}, \quad Z_2 = 2,$$

$$Z_3 = \frac{(2)(j\omega 0.1)}{(2 + j\omega 0.1)}, \quad Z_4 = 1, \quad Z_5 = j\omega 0.2$$

Fig. 3.44 Series circuit with a voltage source

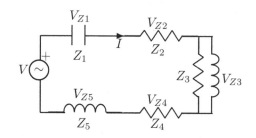

Fig. 3.45 Series circuit with a current source

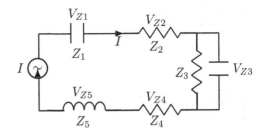

* 3.5.2

$$v(t) = 2\sin(3t)\,V, \quad Z_1 = \frac{1}{j\omega 0.1}, \quad Z_2 = 3,$$

$$Z_3 = \frac{(2)(j\omega 0.2)}{(2 + j\omega 0.2)}, \quad Z_4 = 2, \quad Z_5 = j\omega 0.3$$

3.5.3

$$v(t) = \cos(2t + \frac{\pi}{3})\,V, \quad Z_1 = \frac{1}{j\omega 0.2}, \quad Z_2 = 1,$$

$$Z_3 = \frac{(3)(j\omega 0.1)}{(3 + j\omega 0.1)}, \quad Z_4 = 1, \quad Z_5 = j\omega 0.4$$

3.6 Find the voltages across the impedances,

$$V_{Z1}, V_{Z2}, V_{Z3}, V_{Z4}, V_{Z5}$$

in the series circuit, shown in Fig. 3.45.

3.6.1

$$i(t) = \sin(2t + \frac{\pi}{3})\,A, \quad Z_1 = \frac{1}{j\omega 0.1}, \quad Z_2 = 2,$$

$$Z_3 = \frac{(4)(\frac{1}{j\omega 0.1})}{(4 + (\frac{1}{j\omega 0.1}))}, \quad Z_4 = 3, \quad Z_5 = j\omega 0.1$$

* 3.6.2

$$i(t) = 2\sin(2t)\,A, \quad Z_1 = \frac{1}{j\omega 0.2}, \quad Z_2 = 1,$$

$$Z_3 = \frac{(3)(\frac{1}{j\omega 0.1})}{(3 + (\frac{1}{j\omega 0.1}))}, \quad Z_4 = 4, \quad Z_5 = j\omega 0.1$$

3.6.3

$$i(t) = \cos(2t - \frac{\pi}{3})\,A, \quad Z_1 = \frac{1}{j\omega 0.1}, \quad Z_2 = 2,$$

$$Z_3 = \frac{(3)(\frac{1}{j\omega 0.1})}{(3 + (\frac{1}{j\omega 0.1}))}, \quad Z_4 = 3, \quad Z_5 = j\omega 0.4$$

Fig. 3.46 Parallel circuit
with a current source

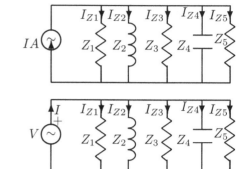

Fig. 3.47 Parallel circuit
with a voltage source

3.7 Find the voltage across the impedances,

$$V_{Z1}, V_{Z2}, V_{Z3}, V_{Z4}, V_{Z5}$$

in the parallel circuit, shown in Fig. 3.46. Find the currents through the impedances,

$$I_{Z1}, I_{Z2}, I_{Z3}, I_{Z4}, I_{Z5}$$

Verify that the sum of the currents through the impedances is equal to the source current.
3.7.1

$$i(t) = \sin(2t + \frac{\pi}{4})A, \quad Z_1 = 1, \quad Z_2 = j\omega 0.1,$$

$$Z_3 = 2, \quad Z_4 = \frac{1}{j\omega 0.2}, \quad Z_5 = 3$$

3.7.2

$$i(t) = 2\cos(t)A, \quad Z_1 = 3, \quad Z_2 = j\omega 0.2, \quad Z_3 = 2,$$

$$Z_4 = \frac{1}{j\omega 0.1}, \quad Z_5 = 1$$

* 3.7.3

$$i(t) = \cos(t + \frac{\pi}{6})A, \quad Z_1 = 1, \quad Z_2 = j\omega 0.1,$$

$$Z_3 = 2, \quad Z_4 = \frac{1}{j\omega 0.2}, \quad Z_5 = 3$$

3.8 Find the current I and the voltages across the impedances,

$$V_{Z1}, V_{Z2}, V_{Z3}, V_{Z4}, V_{Z5}$$

in the parallel circuit, shown in Fig. 3.47. Find the currents through the impedances,

$$I_{Z1}, I_{Z2}, I_{Z3}, I_{Z4}, I_{Z5}$$

Verify that the sum of the currents through the impedances is equal to the total current, I.

3.8.1

$$v(t) = 2\cos(t + \frac{\pi}{6})V, \quad Z_1 = 2, \quad Z_2 = j\omega 0.2, \quad Z_3 = 1,$$

$$Z_4 = \frac{1}{j\omega 0.3}, \quad Z_5 = 4$$

* 3.8.2

$$v(t) = 2\sin(2t + \frac{\pi}{4})V, \quad Z_1 = 3, \quad Z_2 = j\omega 0.3, \quad Z_3 = 2,$$

$$Z_4 = \frac{1}{j\omega 0.2}, \quad Z_5 = 3$$

3.8.3

$$v(t) = 3\cos(3t - \frac{\pi}{3})V, \quad Z_1 = 2, \quad Z_2 = j\omega 0.2, \quad Z_3 = 1,$$

$$Z_4 = \frac{1}{j\omega 0.3}, \quad Z_5 = 4$$

3.9 Find the current I and the voltages across the impedances,

$$V_{Z1}, V_{Z2}, V_{Z3}, V_{Z4}, V_{Z5}$$

and current through

$$I_{Z1}, I_{Z2}, I_{Z3}, I_{Z4}, I_{Z5}$$

in the series-parallel circuit, shown in Fig. 3.48. Verify KVL around the loops and KCL at node x.

3.9.1

$$v(t) = \cos(2t)V, \quad Z_1 = \frac{1}{j\omega 0.3}, \quad Z_2 = 3,$$

$$Z_3 = j\omega 0.3, \quad Z_4 = 2, \quad Z_5 = 1$$

* 3.9.2

$$v(t) = \sin(t)V, \quad Z_1 = \frac{1}{j\omega 0.1}, \quad Z_2 = 2,$$

$$Z_3 = j\omega 0.2, \quad Z_4 = 3, \quad Z_5 = 2$$

Fig. 3.48 Series-parallel circuit with a voltage source

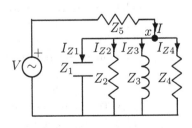

Fig. 3.49 Series-parallel
circuit with a current
source

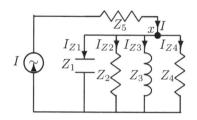

3.9.3

$$v(t) = \cos(2t + \frac{\pi}{6})V, \quad Z_1 = \frac{1}{j\omega 0.2}, \quad Z_2 = 4,$$

$$Z_3 = j\omega 0.1, \quad Z_4 = 3, \quad Z_5 = 3$$

3.10 Find the voltages across the impedances,

$$V_{Z1}, V_{Z2}, V_{Z3}, V_{Z4}, V_{Z5}$$

in the series-parallel circuit, shown in Fig. 3.49. Verify KCL at node x.

3.10.1

$$i(t) = 2\cos(2t + \frac{\pi}{3})A, \quad Z_1 = \frac{1}{j\omega 0.3}, \quad Z_2 = 2,$$

$$Z_3 = j\omega 0.2, \quad Z_4 = 1, \quad Z_5 = 3$$

* 3.10.2

$$i(t) = \cos(t - \frac{\pi}{6})A, \quad Z_1 = \frac{1}{j\omega 0.2}, \quad Z_2 = 3, \quad Z_3 = j\omega 0.3,$$

$$Z_4 = 4, \quad Z_5 = 1$$

3.10.3

$$i(t) = 3\sin(2t + \frac{\pi}{6})A, \quad Z_1 = \frac{1}{j\omega 0.4}, \quad Z_2 = 1, \quad Z_3 = j\omega 0.2,$$

$$Z_4 = 1, \quad Z_5 = 4$$

3.11 Given a circuit diagram with the independent currents, find the corresponding tree and the loops. Analyze the circuit by both nodal and loop methods to find the voltages and currents at all parts of the circuit. Verify that the results are the same by both the methods. Verify the results using KVL and KCL.

    3.11.1 The circuit is shown in Fig. 3.50.

    * 3.11.2 The circuit is shown in Fig. 3.51.

    3.11.3 The circuit is shown in Fig. 3.52.

3.12 Given a circuit diagram with the independent currents, analyze the circuit by both nodal and loop methods to find the voltages and currents at all parts of the circuit. Verify that the results are the same by both the methods. Verify the results using KVL and KCL.

    3.12.1 The circuit is shown in Fig. 3.53.

    * 3.12.2 The circuit is shown in Fig. 3.54.

Fig. 3.50 Circuit for
Exercise 3.11.1

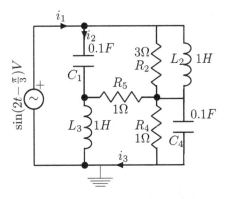

Fig. 3.51 Circuit for
Exercise 3.11.2

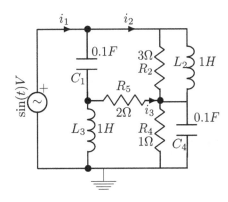

Fig. 3.52 Circuit for
Exercise 3.11.3

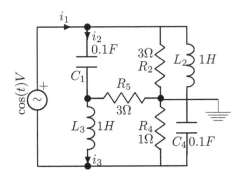

Fig. 3.53 Circuit for
Exercise 3.12.1

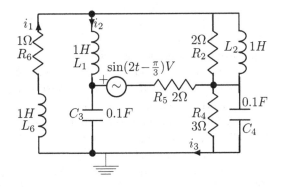

Fig. 3.54 Circuit for
Exercise 3.12.2

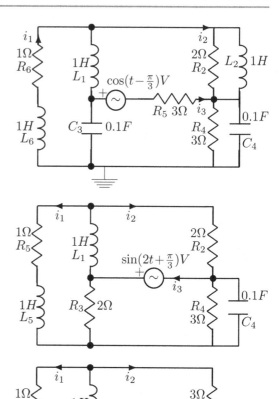

Fig. 3.55 Circuit for
Exercise 3.13.1

Fig. 3.56 Circuit for
Exercise 3.13.2

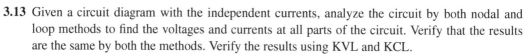

3.13 Given a circuit diagram with the independent currents, analyze the circuit by both nodal and
loop methods to find the voltages and currents at all parts of the circuit. Verify that the results
are the same by both the methods. Verify the results using KVL and KCL.

   3.13.1 The circuit is shown in Fig. 3.55.

   * **3.13.2** The circuit is shown in Fig. 3.56.

3.14 Analyze the circuit using the linearity property.

   3.14.1 The circuit is shown in Fig. 3.57.

   * **3.14.2** The circuit is shown in Fig. 3.58.

3.15 Given a circuit diagram with the independent currents, analyze the circuit by both nodal and
loop methods to find the voltages and currents at all parts of the circuit. Verify that the results
are the same by both the methods. Verify the results using KVL and KCL.

   3.15.1 The circuit is shown in Fig. 3.59.

   * **3.15.2** The circuit is shown in Fig. 3.60.

3.16 Given a circuit diagram with the independent currents, analyze the circuit by both nodal and
loop methods to find the voltages and currents at all parts of the circuit. Verify that the results
are the same by both the methods. Verify the results using KVL and KCL.

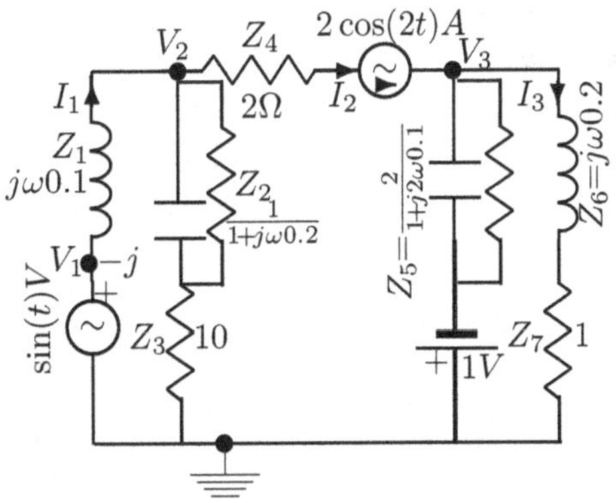

Fig. 3.57 Circuit for
Exercise 3.14.1

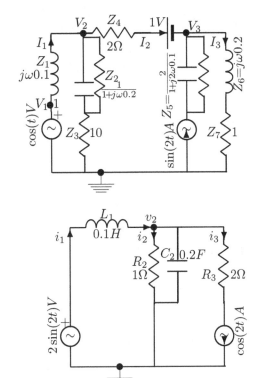

Fig. 3.58 Circuit for
Exercise 3.14.2

Fig. 3.59 Circuit for
Exercise 3.15.1

3.16.1 The circuit is shown in Fig. 3.61.

* **3.16.2** The circuit is shown in Fig. 3.62.

3.17 Given a circuit diagram with the independent currents, analyze the circuit by both nodal and loop methods to find the voltages and currents at all parts of the circuit. Verify that the results are the same by both the methods. Verify the results using KVL and KCL.

3.17.1 The circuit is shown in Fig. 3.63.

* **3.17.2** The circuit is shown in Fig. 3.64.

3.18 Find the Thévenin and Norton equivalent circuits and determine the current through the load impedance.

3.18.1 The load impedance is Z_1 in Fig. 3.13a.

3.18.2 The load impedance is Z_3 in Fig. 3.14a.

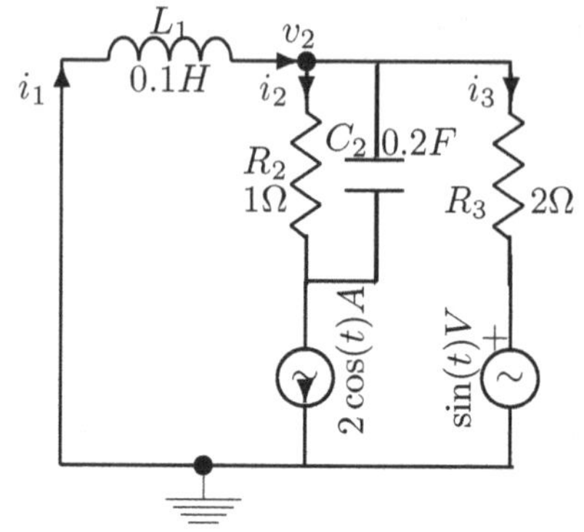

Fig. 3.60 Circuit for
Exercise 3.15.2

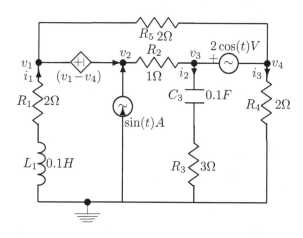

Fig. 3.61 Circuit for
Exercise 3.16.1

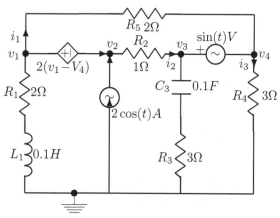

Fig. 3.62 Circuit for
Exercise 3.16.2

3.18.3 The load impedance is Z_3 in Fig. 3.15a.

3.18.4 The load impedance is Z_1 in Fig. 3.16a.

3.18.5 The load impedance is R_3 in Fig. 3.17a.

3.18.6 The load impedance is Z_1 in Fig. 3.19.

3.38.7 The load impedance is Z_3 in Fig. 3.20.

3.19 Find the value of the impedance Z for maximum power transfer from the source to the impedance Z.

Fig. 3.63 Circuit for
Exercise 3.17.1

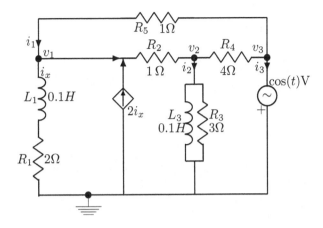

Fig. 3.64 Circuit for
Exercise 3.17.2

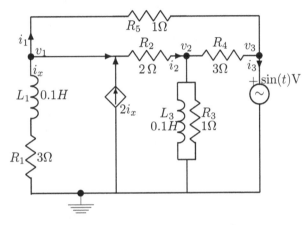

Fig. 3.65 Circuit for
Exercise 3.19.1

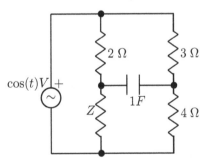

3.19.1 The circuit is shown in Fig. 3.65.

* **3.19.2** The circuit is shown in Fig. 3.66.

3.20 Find the current through the impedance Z_3 by analyzing the circuit as shown and also after source transformation.

* **3.20.1** The circuit is shown in Fig. 3.67.

3.20.2 The circuit is shown in Fig. 3.68.

Fig. 3.66 Circuit for
Exercise 3.19.2

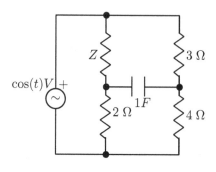

Fig. 3.67 A circuit with
current sources

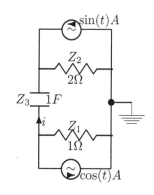

Fig. 3.68 A circuit with
voltage sources

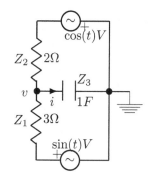

Steady-State Power

4

Electric power is the rate at which electric energy is transferred from an electric power source to a sink that absorbs power. For example, a bulb emits light by absorbing electric power. All the three entities, voltage, current, and power, are important in the generation, transmission, distribution, and utilization of electric power. All the devices, such as a generator, a motor, a heater, and an amplifier, are characterized by their ability to generate power or consume power to do some work.

Circuit element resistor dissipates energy. However, capacitors and inductors store energy, respectively, in their electric and magnetic fields, and, hence, are called storage elements. The stored energy in inductors and capacitors can sustain current in the circuit for some time after the power is switched off. That is why we get light, although diminishing, from the bulb for some time after the power is switched off. While the expression for the power is still of the form VI, the power in AC circuits has two components to take into account, the active power and reactive power. The stored energy is returned to the source. Therefore, the average power consumed by the storage devices, in the sinusoidal steady state, is zero. The kinetic energy is stored in the magnetic field generated by the flow of current in the inductor. Similar to the kinetic energy, $\frac{1}{2}Mv^2$ in a mechanical system, the energy stored in an inductor of L henries is $\frac{1}{2}Li^2(t)$. The current $i(t)$ flowing through an inductor generates a voltage $L\frac{di(t)}{dt}$. Therefore, power has to be delivered by the source deriving a current through the inductor to oppose the generated voltage. This power is used to store energy in the inductor. Therefore, energy delivered, $dT(t)$, during an infinitesimal interval dt is

$$dT(t) = v(t)i(t)dt = i(t)L\frac{di(t)}{dt}dt.$$

Integrating this expression, we get the expression for the instantaneous energy stored in the inductor as

$$T(t) = L\int_{-\infty}^{t} i(x)\frac{di(x)}{dx}\,dx = \frac{1}{2}Li^2(t),$$

since $i(-\infty) = 0$.

A capacitor is made of two conducting surfaces separated by a dielectric (nonconductor or insulator) material. When we apply a voltage $v(t)$ across a capacitor, an electric field is set up. Therefore, power has to be delivered by the source. This power is used to store energy in the capacitor. Therefore, energy delivered, $dV(t)$, during an interval dt is

© Springer Nature Switzerland AG 2020
D. Sundararajan, *Introductory Circuit Theory*,
https://doi.org/10.1007/978-3-030-31985-4_4

$$dV(t) = v(t)i(t)dt = v(t)C\frac{dv(t)}{dt}dt.$$

Integrating this expression, we get the expression for the instantaneous energy stored in the capacitor as

$$V(t) = C \int_{-\infty}^{t} v(x)\frac{dv(x)}{dx}\,dx = \frac{1}{2}Cv^2(t)$$

The current $i(t)$ and the voltage $v(t)$ may be any functions of time and $T(t)$ and $V(t)$ are the corresponding time functions representing the stored energies at instants of time.

Work has to be done to move a charge in a certain direction in a conductor. The work is carried out by the voltage between the two points of interest. Therefore,

$$v = \frac{dw}{dq},$$

where w is the work in joules, v is the voltage in volts, and q is the charge in coulombs. That is, one volt is one joule per coulomb. Now, power p is the rate of doing work measured in watts.

$$p = \frac{dw}{dt} = \frac{dw}{dq}\frac{dq}{dt} = vi.$$

If p is positive, power is absorbed by the element. Otherwise, power is generated by the element. Energy measured in joules, is the integral of power over an interval. A resistor always absorbs power. The average power absorbed by inductors and capacitors is zero. The power supplied by the sources is equal to the power dissipated in the resistors and that stored in the reactive elements.

4.1   Energy in Reactive Elements with Sinusoidal Sources

Let the source to a capacitor C be $v(t) = V \cos(\omega t)$. The instantaneous energy stored in its electric field associated with the capacitor is

$$V(t) = \frac{1}{2}CV^2 \cos^2(\omega t) = \frac{1}{4}CV^2(1 + \cos(2\omega t)).$$

We used the trigonometric identity

$$\cos^2(\omega t) = \frac{1}{2}(1 + \cos(2\omega t)).$$

This expression represents a shifted double-frequency sinusoid with its amplitude varying from 0 to $0.5CV^2$. The average energy is

$$V_{av} = \frac{1}{4}CV^2.$$

Differentiating the expression for $V(t)$ with respect to t, we get

$$\frac{dV(t)}{dt} = -\frac{\omega}{2}CV^2 \sin(2\omega t).$$

Therefore,

$$\frac{dV(t)}{dt}\bigg|_{av} = 0,$$

since the integral of a sinusoid over an integral number of its cycles is zero.

Let the source to an inductor L be $i(t) = I\cos(\omega t)$. The instantaneous energy stored in its magnetic field associated with the inductor is

$$T(t) = \frac{1}{2}LI^2\cos^2(\omega t) = \frac{1}{4}LI^2(1 + \cos(2\omega t)).$$

The average energy is

$$T_{av} = \frac{1}{4}LI^2.$$

Differentiating the expression for $T(t)$ with respect to t, we get

$$\frac{dT(t)}{dt} = -\frac{\omega}{2}LI^2\cos(2\omega t).$$

Therefore,

$$\frac{dT(t)}{dt}\bigg|_{av} = 0.$$

A reactive element cannot absorb energy indefinitely.

4.2 Power Relations in a Circuit

Let the voltage applied to a circuit be $v(t) = V\cos(\omega t + \theta)$. Let the current through the circuit be $i(t) = I\cos(\omega t + \phi)$. The instantaneous power absorbed by the circuit is

$$s(t) = v(t)i(t) = VI\cos(\omega t + \theta)\cos(\omega t + \phi).$$

Using the trigonometric identity

$$2\cos(A)\cos(B) = \cos(A - B) + \cos(A + B),$$

we get

$$s(t) = 0.5VI\cos(\theta - \phi) + 0.5VI\cos(2\omega t + \theta + \phi).$$

The first of the two components of $s(t)$ is a constant and its value depends on the difference between the phases of the voltage and current. The second component is a sinusoid of double the frequency of that of the source. Since $s(t)$ is time varying, the average value S of $s(t)$ over one period, with unit watts, is measured by the wattmeter, the instrument to measure the average power. Let the period of the source be $2T$. Since the period of $s(t)$ is T, the apparent average power, with unit volt amperes, is given by

$$S = \frac{1}{T}\int_0^T 0.5VI\cos(\theta - \phi)dt + \frac{1}{T}\int_0^T 0.5VI\cos(2\omega t + \theta + \phi)dt = 0.5VI\cos(\theta - \phi).$$

The second integral involves a sinusoid and the integral of a sinusoid over its period is zero.

In the frequency-domain representation of circuits, the voltages and currents are, respectively, represented as

$$V\angle\theta \quad \text{and} \quad I\angle\phi.$$

Then,

$$S = 0.5(V\angle\theta)(I\angle(-\phi)) = 0.5VI\angle(\theta - \phi) = 0.5VI(\cos(\theta - \phi) + j\sin(\theta - \phi)) = P_{av} + jQ_{av}.$$

For passive circuits, S will lie in the first or fourth quadrants of the complex plane. If S lies in the first quadrant, the load is inductive. If S lies in the fourth quadrant, the load is capacitive. Taking the magnitude, we get

$$|S| = \sqrt{P_{av}^2 + Q_{av}^2},$$

S is the apparent power. The first component, which is the power actually consumed, is called the active power P_{av}. The power that is swapped back and forth from the source and the circuit is called the reactive or wattless power Q_{av}. Therefore,

$$P_{av} = 0.5\text{Re}((V\angle\theta)(I\angle(-\phi))) = 0.5VI\cos(\theta - \phi)$$

$$Q_{av} = 0.5\text{Im}((V\angle\theta)(I\angle(-\phi))) = 0.5VI\sin(\theta - \phi).$$

That is the real part of the 0.5 times the product of the complex amplitude of the voltage and the conjugated complex amplitude of the current is the active power. The imaginary part is the reactive power. The active power is usually denoted by the symbol P_{av} and the reactive power by the symbol Q_{av}. Therefore, the power in complex form is

$$S = 0.5VI^* = 0.5II^*Z = 0.5|I|^2Z = 0.5|V|^2Y^* = P_{av} + jQ_{av},$$

where Z is the impedance, $Y = 1/Z$ is the admittance, and

$$\mathbf{V} = V\angle\theta \quad \text{and} \quad \mathbf{I} = I\angle\phi.$$

The active power in a linear passive circuit is always positive, while the reactive power is negative in capacitive circuits and positive in inductive circuits. The unit for reactive power is vars (volt–amperes reactive). The total power consumed by a circuit is the sum of the powers consumed by the constituent branches of the circuit. Since the reactive power may be positive or negative, it is possible to change its value by adding suitable passive circuits, a process called power-factor correction. It should be mentioned that the meters used to measure currents and voltages measure the RMS values, which is equal to $1/\sqrt{2}$ of the peak value of a sinusoid. If we use RMS values, then the expressions for the power become $V_{RMS}I_{RMS}$, as in the case of DC circuits without the term $1/2$.

If $\theta = \phi$, as in resistive circuits with the phase difference zero, the average power is

$$P = 0.5\,VI.$$

That is, the effective impedance has only resistive component. If $\theta - \phi = \pm0.5\pi$, the average power is zero. That is, the effective impedance has only reactive component. In the term $\cos(\theta - \phi)$, the phase difference is usually represented by θ. As $\cos(\theta)$ indicates the extent the circuit absorbs power from the source, it is called the power factor of the circuit, denoted by pf. The cosine of the phase difference between the voltage and current is the power factor. It is also the cosine of the angle of the impedance.

A capacitive impedance has a negative phase angle. Consequently, the reactive power Q_{av}, which is a function of the angle, is negative and vice versa for an inductive impedance. The pf is an indication of the relative magnitudes of the resistive and reactive components of a circuit. A higher pf indicates that the resistive component is greater. The total average power supplied is the sum of P_{av} for each element of the circuit.

$$P_{av} = P_{av1} + P_{av2} + P_{av3} + \cdots$$

Similarly, the total reactive power supplied is the algebraic sum of Q_{av} for each element of the circuit.

$$Q_{av} = Q_{av1} + Q_{av2} + Q_{av3} + \cdots$$

Therefore,

$$S_{av} = S_{av1} + S_{av2} + S_{av3} + \cdots$$

The equivalent definitions of the power factor are

$$\text{pf} = \frac{\text{active power}}{|\text{apparent power}|} = \cos(\theta) = \frac{R}{|Z|}.$$

Angle θ is the phase angle of the current with respect to the voltage. If the current leads the voltage, the power factor is said to be leading and is lagging if current lags voltage. A pf of 0.8 lagging implies that the current lags the voltage by $\cos^{-1}(0.8) = 0.6435$ radians or $36.8699°$.

1. For resistive loads, $Q = 0$ and pf $= 1$.
2. For inductive loads, $Q > 0$ with a lagging pf.
3. For capacitive loads, $Q < 0$ with a leading pf.

Example 4.1 Find the average active and reactive power consumed by the circuit, shown in Fig. 4.1a. What is the power factor of the circuit? Verify that the power supplied is equal to the sum of the power consumed by the components. Relate the stored energy with the reactive power.

Solution The source voltage is $0.5 - j0.866$ V. The frequency of the source is $\omega = 2$ radians. The values of the impedances are

$$Z_1 = 1, \quad Z_2 = j4.$$

The source current is $I = \frac{0.5 - j0.866}{1 + j4} = -0.1744 - j0.1686$ A.

The values of the currents through the impedances are

$$\{I_{Z_1} = -0.1744 - j0.1686, \quad I_{Z_2} = -0.1744 - j0.1686\}.$$

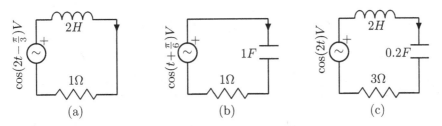

Fig. 4.1 (a) A RL series circuit; (b) a RC series circuit; (c) a RLC series circuit

The apparent power consumed by the impedances, the product of the current, its conjugate and the impedance divided by 2 are

$$0.0294, \quad j0.1176.$$

The total apparent power consumed by the circuit is the sum of those of all the impedances,

$$S = P_{av} + jQ_{av} = 0.0294 + j0.1176 \text{ va}.$$

The power supplied by the source $VI^*/2$ is also the same. The average power consumed by the resistor is

$$0.0294 \text{ W}$$

and is equal to P_{av}. The impedance of the circuit is $Z = V/I = 1 + j4$.

$$\text{pf} = \frac{\text{Re}(Z)}{|Z|} = \frac{0.0294}{|S|} = \cos(\angle(V) - \angle(I)) = 0.2425 \text{ lagging}.$$

The instantaneous energy dissipated by the resistor is given, in the time-domain, by

$$p(t) = i^2(t)R.$$

For this example, $R = 1\Omega$ and

$$i(t) = |I| \cos(2t + \angle(I))A = 0.2425 \cos(2t - 2.373).$$

Using the trigonometric double-angle formula

$$\cos^2(x) = \frac{1 + \cos(2x)}{2},$$

we can find $p(t) = i^2(t)$ to be

$$0.2425^2 \cos^2(2t - 2.373) = 0.0294 + 0.0294 \cos(4t + 1.5372).$$

The period of this waveform is $2\pi/(4) = \pi/2$ seconds. Alternatively, $p(t)$ can also be computed numerically at discrete points with sufficiently small interval to approximate the instantaneous power. The result is shown in Fig. 4.2a. It oscillates about the average power 0.0294. The phase is about 90°.

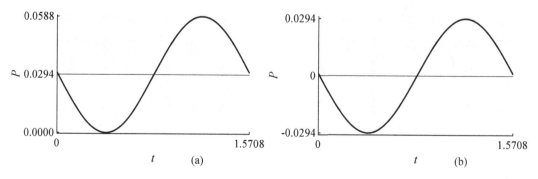

Fig. 4.2 (a) Instantaneous energy dissipated by the resistor; (b) the oscillating component of the energy

That is, it is an inverted sine wave. The oscillating part alone is shown in Fig. 4.2b. The oscillating part is a double-frequency sinusoid. Manually, one can do the computation at few points and find that the power waveform has two peaks and two valleys, in a period of the current through the resistor, with all the values positive.

Let the time-domain sinusoidal current be expressed in complex form. Then,

$$i_p(t) = \frac{1}{2}\left(I_p e^{j\omega t} + I_p^* e^{-j\omega t}\right).$$

The product of two currents $i_p(t)$ and $i_q(t)$ is

$$
\begin{aligned}
i_p(t)i_q(t) &= \frac{1}{4}\left(I_p e^{j\omega t} + I_p^* e^{-j\omega t}\right)\left(I_q e^{j\omega t} + I_q^* e^{-j\omega t}\right) \\
&= \frac{1}{4}\left(I_p I_q e^{j2\omega t} + I_p^* I_q^* e^{-j2\omega t} + I_p I_q^* + I_p^* I_q\right) \\
&= \frac{1}{2}\left(I_p I_q^*\right) + \frac{1}{2}\mathrm{Re}\left(I_p I_q e^{j2\omega t}\right).
\end{aligned}
$$

Since $i_p(t) = i_q(t) = i(t)$ and the energy dissipated in a resistor is

$$Ri^2(t),$$

we get the instantaneous power dissipated as

$$P(t) = \frac{R}{2}|I|^2 + \frac{R}{2}\mathrm{Re}\left(I^2 e^{j2\omega t}\right).$$

The formula $i^2 R$ is for a DC current. Since the amplitude profile of a sinusoidal waveform is time varying, it is found that

$$I_{eff} = I_{RMS} = 0.7071|I|$$

serves the same purpose in power calculations, where $|I|$ is the peak value. The constant value I_{eff} dissipates the same power in a resistor as that by a sinusoidal current with a peak value of $|I|$ on the average. That is the average power dissipated in a resistor due to a sinusoidal current flowing through it is

$$I_{eff}^2 R = \frac{|I|^2}{2}R = \frac{|I|}{\sqrt{2}}\frac{|I|}{\sqrt{2}}R$$

which is the first part of the expression for power derived above. The values $\frac{|I|}{\sqrt{2}}$ and $\frac{|V|}{\sqrt{2}}$ are referred as the effective values of sinusoidal current or voltage. They are also known as RMS (root-mean-square) values by their definition.

A similar expression for the instantaneous energy stored in a magnetic field is

$$T(t) = \frac{L}{4}|I|^2 + \frac{L}{4}\mathrm{Re}\left(I^2 e^{j2\omega t}\right).$$

The first part is the average energy stored. As the energy stored is $0.5Li^2(t)$ rather than $Ri^2(t)$, L replaces R and the denominator is multiplied by 2. For this example, as $L = 2$ and $R = 1$ and with the same current, we get the same figure, Fig. 4.3, for the stored energy.

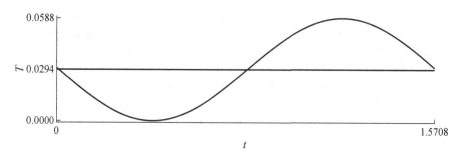

Fig. 4.3 Instantaneous energy stored in the inductor

As expected, the stored energy is related to the reactive power by a constant factor 2ω. The equilibrium equation for the circuit, in the frequency-domain with

$$v(t) \rightarrow Ve^{j\omega t} \quad \text{and} \quad i(t) \rightarrow Ie^{j\omega t},$$

is

$$IR + j\omega LI = V.$$

Multiplying both sides by $0.5I^*$, we get

$$0.5II^*R + j2\omega L0.25(II^*) = 0.5VI^* = P_{av} + jQ_{av}.$$

Therefore, for this example, the average reactive power is

$$2(2)0.0294 = 0.1176$$

as obtained above.   ∎

Example 4.2 Find the average active and reactive power consumed by the circuit, shown in Fig. 4.1b. What is the power factor of the circuit? Verify that the power supplied is equal to the sum of the power consumed by the components. Relate the stored energy with the reactive power.

Solution The source voltage is $0.8660 + j0.5$ V. The source frequency is $\omega = 1$ radian. The values of the impedances are

$$Z_1 = 1, \; Z_2 = -j1.$$

The source current is $I = \frac{0.8660 + j0.5}{1 - j1} = 0.1830 + j0.6830\,A = 0.7071\angle1.3090$. The values of the currents through the impedances are

$$\{I_{Z_1} = 0.1830 + j0.6830, \; I_{Z_2} = 0.1830 + j0.6830\}.$$

The apparent power consumed by the impedances, the product of the current, its conjugate and the impedance divided by 2 are

$$0.25, \; -j0.25.$$

The total apparent power consumed by the circuit is the sum of those of all the impedances,

$$S = P_{av} + jQ_{av} = 0.25 - j0.25 \text{ va.}$$

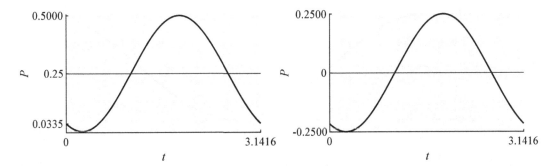

Fig. 4.4 (**a**) Instantaneous energy dissipated by the resistor; (**b**) the oscillating component of the energy

The power supplied by the source $VI^*/2$ is also the same. The average power consumed by the resistor is

$$0.25\,\text{W}$$

and is equal to P_{av}. The impedance of the circuit is $Z = V/I = 1 - j$.

$$\text{pf} = \frac{\text{Re}(Z)}{|Z|} = \frac{0.25}{|S|} = \cos(\angle(V) - \angle(I)) = 0.7071 \text{ leading.}$$

The instantaneous energy dissipated by the resistor is

$$0.7071^2 \cos^2(t + 1.3090) = 0.25 + 0.25\cos(2t + 2.618).$$

The period of this waveform is $2\pi/(2) = \pi$ seconds. The result is shown in Fig. 4.4a. It oscillates about the average power 0.25. The phase is about $150°$. That is, it is an inverted and shifted sine wave. The oscillating part alone is shown in Fig. 4.4b.

An expression, similar to that for magnetic energy, for the instantaneous energy stored in an electric field is

$$V(t) = \frac{1}{4C\omega^2}|I|^2 - \frac{1}{4C\omega^2}\text{Re}\left(I^2 e^{j2\omega t}\right),$$

where I is the current through the capacitor and C is the value of the capacitor. Note that

$$V = \frac{I}{j\omega C}.$$

The first part is the average energy stored.

The equilibrium equation for the circuit, in the frequency-domain with

$$v(t) \to V e^{j\omega t} \quad \text{and} \quad i(t) \to I e^{j\omega t},$$

is

$$IR - j\frac{I}{\omega C} = V.$$

Multiplying both sides by $0.5I^*$, we get

$$0.5II^*R - j2\omega\left(0.25\frac{(II^*)}{C\omega^2}\right) = 0.5VI^* = P_{av} + jQ_{av}.$$

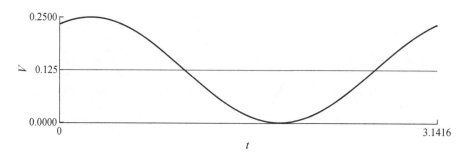

Fig. 4.5 Instantaneous energy stored in the capacitor

The reactive power is -2ω times the energy stored in the capacitor. Therefore, for this example, the average reactive power, with the squared magnitude of the current being 0.5, is

$$-2(1)0.125 = -0.25$$

as obtained above. The instantaneous energy stored in the electric field is shown in Fig. 4.5. ∎

Example 4.3 Find the average active and reactive power consumed by the circuit, shown in Fig. 4.1c. What is the power factor of the circuit? Verify that the power supplied is equal to the sum of the power consumed by the components. Relate the stored energy with the reactive power.

Solution The source voltage is 1 V. The frequency of the source is $\omega = 2$ radians. The values of the impedances are

$$Z_1 = 3, \ Z_2 = j4, \ Z_3 = -j2.5.$$

The source current is

$$I = \frac{1}{3 + j1.5} = (0.2667 - j0.1333) \text{ A}.$$

The values of the currents through the impedances are

$$\{I_{Z_1} = 0.2667 - j0.1333, \ I_{Z_2} = 0.2667 - j0.1333, \ I_{Z_3} = 0.2667 - j0.1333\}.$$

The apparent power consumed by the impedances, the product of the current, its conjugate and the impedance divided by 2 are

$$0.1333, \ j0.1778, \ -j0.1111.$$

The total apparent power consumed by the circuit is the sum of those of all the impedances,

$$S = P_{av} + jQ_{av} = 0.1333 + j0.0667 \text{ va}.$$

The power supplied by the source $VI^*/2$ is also the same. The average power consumed by the resistor is

$$0.1333 \text{ W}$$

and is equal to P_{av}.

The impedance of the circuit is $Z = V/I = 3 + j1.5$.

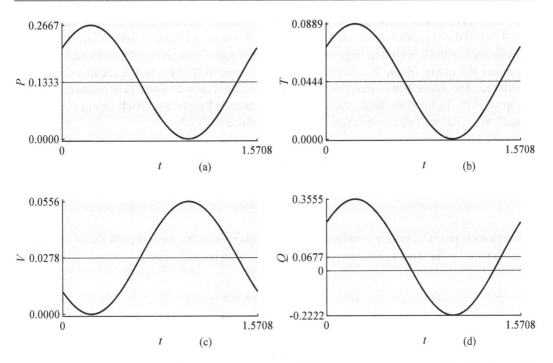

Fig. 4.6 (a) Instantaneous energy dissipated by the resistor; (b) instantaneous energy stored in the inductor; (c) instantaneous energy stored in the capacitor; (d) reactive energy component of the circuit

$$\text{pf} = \frac{\text{Re}(Z)}{|Z|} = \frac{0.1333}{|S|} = \cos(\angle(V) - \angle(I)) = 0.8944.$$

The instantaneous energy dissipated by the resistor is

$$(3)0.2981^2 \cos^2(2t - 0.4636) = 0.1333 + 0.1333 \cos(4t - 0.9273).$$

The period of this waveform is $2\pi/(4) = \pi/2$ seconds. The result is shown in Fig. 4.6a. It oscillates about the average energy 0.1333. The phase is about $-53.1301°$.

The instantaneous energy stored in the magnetic field is shown in Fig. 4.6b. The instantaneous energy stored in the electric field is shown in Fig. 4.6c. Figure 4.6d shows the reactive energy consumed by the circuit, which is obtained by multiplying the difference in the stored energies by the inductor and capacitor multiplied by 2ω. The lower limit is negative indicating that the source receives part of the reactive energy from the storage elements.

∎

4.3 Power-Factor Correction

Most of the loads, such as induction motors and furnaces, are inductive. The power has two components: (1) active power to do the work such as heating, producing light, motion, etc.; and (2) the reactive power to sustain the magnetic field associated with such loads. With a pf of 0.8, 100 va is required to get 80 watts of power. As the pf decreases, a higher current is required from the supply line to do the same work. If the used power is charged on va basis, obviously, pf correction saves money. If the used power is charged on watts basis, then no additional charges are made for pf between some limits, say 0.9 to 1. If the pf goes lower than 0.9, then additional charges are made as per the agreed contract.

For inductive loads with a voltage source, a suitable parallel capacitor C has to be connected to improve the power factor. With a fixed terminal voltage, the current to the net impedance will be reduced. The active power remains the same, while the reactive power gets reduced resulting in improved pf. To improve the pf, capacitors can be installed at the load itself or at the feeder or substation. Let the reactive power be Q_1 and the desired one be Q_2. Since,

$$Q_1 - Q_2 = 0.5\frac{V^2}{X_C} = 0.5\omega C V^2,$$

$$C = 2\frac{Q_1 - Q_2}{\omega V^2}.$$

Although most practical loads are inductive, it is possible to improve the power factor for capacitive loads also in a similar manner by connecting an inductor L in parallel. Since,

$$Q_1 - Q_2 = 0.5\frac{V^2}{X_L} = 0.5\frac{V^2}{\omega L},$$

$$L = \frac{V^2}{2\omega (Q_1 - Q_2)}.$$

With a current source, an appropriate impedance has to be connected in series with the load impedance.

Consider a circuit, shown in Fig. 4.7a, with the voltage applied being

$$v(t) = \cos\left(2t + \frac{\pi}{6}\right).$$

The impedance of the circuit is composed of a resistor 1Ω connected in series with a capacitor 0.5 F. Find the apparent, active, and reactive powers and the pf of the circuit. If the pf is to be 0.8 and 1, what is the value of the inductor to be connected in parallel with the impedance in each case?

The frequency-domain representation of the voltage and the impedance is

$$V = 1\angle\frac{\pi}{6} = 1\angle 30° = 0.8660 + j0.5 \quad \text{and} \quad Z = 1 - j1$$

$$pf = \frac{R}{|Z|} = 0.7071$$

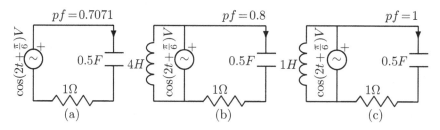

Fig. 4.7 (**a**) A RC series circuit; (**b**) the circuit with a pf correction inductor of 4 H; (**c**) the circuit with a pf correction inductor of 1 H

leading. The current through the circuit is

$$I = \frac{V}{Z} = 0.1830 + j0.6830 = 0.7071\angle 75°$$

$$S = 0.5VI^* = 0.5(0.8660 + j0.5)(0.1830 - j0.6830) = 0.25 - j0.25 = P_{av} + jQ_{av}$$

$$pf = \frac{P_{av}}{|S|} = 0.7071$$

as obtained above. The power consumed by the resistor is

$$0.5I^*I(1) = 0.25W.$$

Figure 4.8a shows the voltage and the current. The current is leading the voltage by 45°. Figure 4.8d shows the active and reactive power components.

The desired pf is 0.8. As the circuit is capacitive, an inductor has to be connected in parallel with the impedance to improve the pf from 0.7071 to 0.8. The active power remains the same, but the reactive power drawn from the source has to be reduced. The balance is supplied by the inductor, as shown in Fig. 4.7b. In the steady state, part of the reactive power is supplied by the inductor. That is, energy stored in the magnetic field associated with the inductor is swapped back and forth with the electric field associated with the capacitor, reducing the amount of reactive power to be supplied by the voltage source. The new apparent power is

$$\frac{P_{av}}{0.8} = 0.3125 \text{ va}.$$

Therefore, the new reactive power is

$$0.3125 \sin\left(\cos^{-1}(0.8)\right) = 0.1875 vars.$$

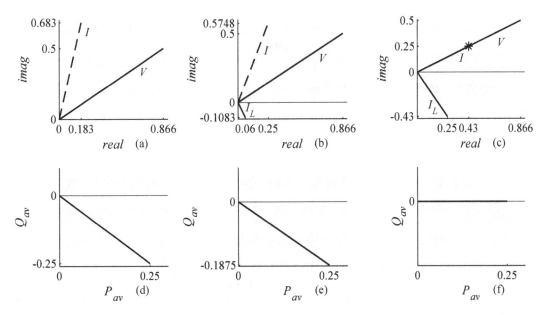

Fig. 4.8 (a) V and I in Fig. 4.7a; (b) V and I in Fig. 4.7b; (c) V and I in Fig. 4.7c; (d) apparent power in Fig. 4.7a; (e) apparent power in Fig. 4.7b; (f) apparent power in Fig. 4.7c

The difference between the reactive powers is $0.25 - 0.1875 = 0.0625$. Now, with $\omega = 2$ and $|V| = 1$,

$$L = \frac{1^2}{(2)(2)0.0625} = 4H.$$

The current through the inductor is

$$\frac{V}{j2(4)} = 0.0625 - j0.1083.$$

The new current supplied by the source is

$$In = (0.1830 + j0.6830) + (0.0625 - j0.1083) = 0.2455 + j0.5748.$$

The power supplied is

$$S = 0.5V In^* = 0.5(0.8660 + j0.5)(0.2455 - j0.5748) = 0.2500 - j0.1875 = P_{av} + jQ_{av}.$$

The new pf is

$$pf = \frac{P_{av}}{|S|} = 0.8.$$

Figure 4.8b shows the voltage, currents from the source and the inductor. The phase difference between the voltage and the current has been reduced, improving the pf (which is the cosine of this phase angle). Figure 4.8e shows the active and reactive power components.

The desired pf is 1. The whole reactive power 0.25 has to be supplied by the inductor. Obviously, the inductor has to supply more current and it requires a smaller inductor, as shown in Fig. 4.7c.

$$L = \frac{1^2}{(2)(2)0.25} = 1H.$$

The current through the inductor is

$$\frac{V}{j2(1)} = 0.2500 - j0.4330.$$

The new current supplied by the source is

$$In = (0.1830 + j0.6830) + (0.2500 - j0.4330) = 0.4330 + j0.25.$$

The power supplied is

$$S = 0.5V In^* = 0.5(0.8660 + j0.5)(0.4330 - j0.25) = 0.2500 - j0 = P_{av} + jQ_{av}.$$

The new pf is

$$pf = \frac{P_{av}}{|S|} = 1.$$

Figure 4.8c shows the voltage, currents from the source and the inductor. The phase difference between the voltage and the current is zero, making the pf equal to 1. Figure 4.8f shows the active and reactive power components. Figure 4.9 shows the instantaneous power consumed in the circuit with different pf.

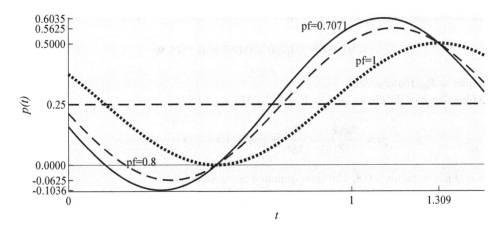

Fig. 4.9 The instantaneous power consumed by the circuit is the sum of a constant and a double-frequency sinusoidal component

Example 4.4 Find the average active and reactive power consumed by the circuit, shown in Fig. 3.13a. What is the power factor of the circuit? Verify that the power supplied is equal to the sum of the power consumed by the components. Find the value of the capacitor for power-factor improvement, for pf = 0.8 lagging and pf = 1.

Solution The source voltage is $0.866 + j0.5$ V. The source current is $I = I_1 = 0.6550 - j0.1845$ A. The values of the impedances are

$$Z_1 = -j10, \ Z_2 = \frac{j3}{3+j1} = 0.3 + j0.9, \ Z_3 = j1, \ Z_4 = \frac{-j10}{1-j10} = 0.9901 - j0.0990, \ Z_5 = 1.$$

The values of the currents through the impedances are

$$\{I_{Z_1} = -0.0283 + j0.0677, \ I_{Z_2} = 0.6833 - j0.2523, \ I_{Z_3} = 0.2171 - j0.1886,$$

$$I_{Z_4} = 0.4379 + j0.0041, \ I_{Z_5} = -0.2454 + j0.2564\}.$$

The apparent power consumed by the impedances, the product of the current, its conjugate and the impedance divided by 2 are

$$-j0.0269, \ 0.0796 + j0.2387, \ j0.0414, \ 0.0950 - j0.0095, \ 0.0630.$$

The total apparent power consumed by the circuit is the sum of those of all the impedances,

$$S = P_{av} + jQ_{av} = 0.2375 + j0.2437 \text{ va}.$$

The power supplied by the source $VI^*/2$ is also the same.

Now, let us compute the power consumed by the resistors alone, which is the active average power P_{av}. The resistors are

$$3, 1, 1$$

and the corresponding currents are found from the branch currents, respectively, as

$$0.1440 + j0.1798, \quad 0.4340 - j0.0393 \quad -0.2454 + j0.2564.$$

The total power consumed by the resistors is

$$0.0796 + 0.0950 + 0.063 = 0.2375 \text{ W}$$

and is equal to P_{av} found earlier.

The impedance of the circuit is $Z = V/I = 1.0257 + j1.0523$.

$$\text{pf} = \frac{\text{Re}(Z)}{|Z|} = \frac{P_{av}}{|S|} = \cos(\angle(V) - \angle(I)) = 0.698.$$

The desired power factor is 0.8. The corresponding apparent power is

$$\frac{P_{av}}{0.8} = 0.2969.$$

The corresponding reactive power is

$$0.2969 \sin(\cos^{-1}(0.8)) = 0.1781.$$

The difference between the reactive powers is $0.2437 - 0.1781 = 0.0655$. Now, with $\omega = 1$ and $|V| = 1$,

$$C = 2\frac{0.0655}{(1)(1^2)} = 0.1311 F.$$

The current through the capacitor is

$$jV(1)(0.1311) = -0.0655 + j0.1135.$$

The new current supplied by the source is

$$(0.6550 - j0.1845) + (-0.0655 + j0.1135) = 0.5895 - j0.0710.$$

The new pf is

$$\cos\left(\frac{\pi}{6} - \angle(0.5895 - j0.0710)\right) = 0.8.$$

The desired power factor is 1. The corresponding apparent power is

$$\frac{P_{av}}{1} = 0.2375.$$

The corresponding reactive power is

$$0.2375 \sin\left(\cos^{-1}(1)\right) = 0.$$

The difference between the reactive powers is $0.2437 - 0 = 0.2437$. Now, with $\omega = 1$ and $|V| = 1$,

$$C = 2\frac{0.2437}{(1)(1^2)} = 0.4873 F.$$

The current through the capacitor is

$$jV(1)(0.4873) = -0.2437 + j0.4220.$$

The new current supplied by the source is

$$(0.6550 - j0.1845) + (-0.2437 + j0.4220) = 0.4114 + j0.2375.$$

The new pf is

$$\cos\left(\frac{\pi}{6} - \angle(0.4114 + j0.2375)\right) = 1.$$

∎

Example 4.5 Find the average active and reactive power consumed by the circuit, shown in Fig. 3.14a. What is the power factor of the circuit? Verify that the power supplied is equal to the sum of the power consumed by the components. Find the value of the capacitor for power-factor improvement, for pf $= 0.8$ lagging.

Solution The source voltage is $0.5 + j0.866$ V. The source current is $I = I_{z5} = -0.4192 - j0.0678$ A. The values of the impedances are

$$Z_1 = j1, \quad Z_2 = \frac{j2}{2+j1} = 0.4 + j0.8, \quad Z_3 = -j10,$$

$$Z_4 = \frac{-j30}{3 - j10} = 2.7523 - j0.8257, \quad Z_5 = 1, \quad Z_6 = 1 + j1.$$

The values of the corresponding currents are

$$\{I_{Z_1} = -0.4837 - j0.0727, \ I_{Z_2} = 0.3912 + j0.0038, \ I_{Z_3} = -0.0645 - j0.0049,$$

$$I_{Z_4} = -0.0280 - j0.0640, \ I_{Z_5} = -0.4192 - j0.0678, \ I_{Z_6} = -0.0925 - j0.0689\}.$$

The apparent power consumed by the impedances, the product of the current, its conjugate and the impedance divided by 2 are

$$j0.1196, 0.0306 + j0.0612, -j0.0209, 0.0067 - j0.0020, 0.0902, 0.0067 + j0.0067.$$

The total apparent power consumed by the circuit is the sum of those of all the impedances,

$$S = P_{av} + jQ_{av} = 0.1342 + j0.1646 \text{ va}.$$

The power supplied by the source $VI^*/2$ is also the same.

Now, let us compute the power consumed by the resistors alone, which is the active average power P_{av}. The resistors are

$$2, 3, 1, 1$$

and the corresponding currents are found from the branch currents, respectively, as

$$0.0767 - j0.1572, \quad -0.0433 + j0.0510 \quad -0.4192 + j0.0678, \quad -0.0925 + j0.0689.$$

The total power consumed by the resistors is

$$0.0306 + 0.0067 + 0.0902 + 0.0067 = 0.1342 \text{ W}$$

and is equal to P_{av} found earlier.

The impedance of the circuit is $Z = V/I = 1.4878 + j1.8252$.

$$\text{pf} = \frac{\text{Re}(Z)}{|Z|} = \frac{P_{av}}{|S|} = \cos(\angle(V) - \angle(I)) = 0.6318.$$

The desired power factor is 0.8. The corresponding apparent power is

$$\frac{P_{av}}{0.8} = 0.1677.$$

The corresponding reactive power is

$$0.2969 \sin\left(\cos^{-1}(0.8)\right) = 0.1006.$$

The difference between the reactive powers is $0.1646 - 0.1006 = 0.0640$. Now, with $\omega = 1$ and $|V| = 1$,

$$C = 2\frac{0.0640}{(1)\left(1^2\right)} = 0.1279F.$$

The current through the capacitor is

$$jV(1)(0.1279) = -0.1108 + j0.0640.$$

The new current supplied by the source is

$$(0.4192 + j0.0678) + (-0.1108 + j0.0640) = 0.3084 + j0.1318.$$

The new pf is

$$\cos\left(\frac{\pi}{3} - \angle(0.3084 + j0.1318)\right) = 0.8.$$

■

Example 4.6 Find the average active and reactive power consumed by the circuit, shown in Fig. 3.15a. What is the power factor of the circuit? Verify that the power supplied is equal to the sum of the power consumed by the components. Find the value of the capacitor for power-factor improvement, for pf= 0.9 lagging.

Solution The source voltage is $0.5 + j0.8660$ V. The source current is $I = -0.5112 + j0.1373$ A. The values of the impedances are

$$Z_1 = j1, \quad Z_2 = \frac{j2}{2 + j1} = 0.4 + j0.8, \quad Z_3 = -j10,$$

$$Z_4 = \frac{-j30}{3 - j10} = 2.7523 - j0.8257, \quad Z_5 = 1 + j1.$$

The values of the corresponding currents are

$$\{I_{Z_1} = -0.4837 - j0.0727, \quad I_{Z_2} = 0.3912 + j0.0038, \quad I_{Z_3} = -0.0645 - j0.0049,$$

$$I_{Z_4} = -0.0280 - j0.0640, \quad I_{Z_5} = -0.4192 - j0.0678, \quad I_{Z_6} = -0.0925 - j0.0689\}.$$

The apparent power consumed by the impedances, the product of the current, its conjugate and the impedance divided by 2 are

$$j0.1859, 0.0476 + j0.0951, -j0.0325, 0.0104 - j0.0031, 0.0103 + j0.0103.$$

The total apparent power consumed by the circuit is the sum of those of all the impedances,

$$S = P_{av} + jQ_{av} = 0.0683 + j0.2557 \text{ va}.$$

The power supplied by the source $VI^*/2$ is also the same.

Now, let us compute the power consumed by the resistors alone, which is the active average power P_{av}. The resistors are

$$2, 3, 1$$

and the corresponding currents are found from the branch currents, respectively, as

$$0.1676 + j0.1395, \quad -0.0753 - j0.0359 \quad -0.1404 - j0.0310.$$

The total power consumed by the resistors is

$$0.0476 + 0.0104 + 0.0103 = 0.0683 \text{ W}$$

and is equal to P_{av} found earlier.

The impedance of the circuit is $Z = V/I = 0.4878 + j1.8252$.

$$\text{pf} = \frac{\text{Re}(Z)}{|Z|} = \frac{P_{av}}{|S|} = \cos(\angle(V) - \angle(I)) = 0.2582.$$

The desired power factor is 0.9. The corresponding apparent power is

$$\frac{P_{av}}{0.9} = 0.0759.$$

The corresponding reactive power is

$$0.0759 \sin\left(\cos^{-1}(0.9)\right) = 0.0331.$$

The difference between the reactive powers is $0.2557 - 0.0331 = 0.2226$. Now, with $\omega = 1$ and $|V| = 1$,

$$C = 2\frac{0.2226}{(1)(1^2)} = 0.4452F.$$

The current through the capacitor is

$$jV(0.4452) = -0.3855 + j0.2226.$$

The new current supplied by the source is

$$(0.5112 - j0.1373) + (-0.3855 + 0.2226) = 0.1257 + j0.0853.$$

The new pf is

$$\cos\left(\frac{\pi}{3} - \angle(0.1257 + j0.0853)\right) = 0.9.$$

∎

Example 4.7 Find the average active and reactive power consumed by the circuit, shown in Fig. 3.16a. What is the power factor of the circuit?

Solution The source voltages are $-j$ and $2\,\mathrm{V}$. The source current is $-j\,\mathrm{A}$. The values of the impedances are

$$Z_1 = j0.1, \quad Z_2 = 2 - j5, \quad Z_3 = 2, \quad Z_4 = -j10, \quad Z_5 = 1 + j0.2.$$

The values of the corresponding currents are

$$\{I_{Z_1} = 0.1821 + j0.9461, \; I_{Z_2} = 0.1821 - j0.0539, \; I_{Z_3} = j1,$$
$$I_{Z_4} = -0.0824 + j0.1921, \; I_{Z_5} = 0.0824 + j0.8079\}.$$

The apparent power consumed by the impedances, the product of the current, its conjugate and the impedance divided by 2, are

$$j0.0464, 0.0361 - j0.0901, 1, -j0.2185, 0.3298 + j0.0660.$$

The total apparent power consumed by the circuit is the sum of those of all the impedances

$$S = P_{av} + jQ_{av} = 1.3658 - j0.1962 \text{ va.}$$

The power supplied by the sources is

$$(-0.4730 - j0.0910) + (1.9213 + j0.0869) + (-0.0824 - j0.1921)$$

also the same.

Now, let us compute the power consumed by the resistors alone, which is the active average power P_{av}. The resistors are

$$2, 2, 1$$

and the corresponding currents are found from the branch currents, respectively, as

$$0.1821 - j0.0539, \quad 1, \quad 0.0824 + j0.8079.$$

The total power consumed by the resistors is

$$0.0361 + 1 + 0.3298 = 1.3658 \text{ W}$$

and is equal to P_{av} found earlier.

The power factor of the circuit is

$$\text{pf} = \frac{P_{av}}{|S|} = 0.9898$$

leading. As it is close to 1, power-factor correction may not be needed.

∎

Example 4.8 Find the average active and reactive power consumed by the circuit, shown in Fig. 3.17a. What is the power factor of the circuit?

Solution The source voltage is $-j$. The source current is 2 A. The values of the impedances are

$$Z_1 = j0.1, \quad Z_2 = 0.9615 - j0.1923, \quad Z_3 = 2.$$

The values of the corresponding currents are

$$\{I_{Z_1} = 1.9701 - j0.5985, \ I_{Z_2} = 2, \ I_{Z_3} = -0.0299 - j0.5985\}.$$

The apparent power consumed by the impedances, the product of the current, its conjugate and the impedance divided by 2, are

$$j0.2120, 1.9231 - j0.3846, 0.3591.$$

The total apparent power consumed by the circuit is the sum of those of all the impedances,

$$S = P_{av} + jQ_{av} = 2.2822 - j0.1726 \text{ va}.$$

The power supplied by the sources is

$$(0.2993 - j0.9850) + (1.9830 + j0.8124)$$

also the same.

Now, let us compute the power consumed by the resistors alone, which is the active average power P_{av}. The resistors are

$$1, 2$$

and the corresponding currents are found from the branch currents, respectively, as

$$1.9231 - j0.3846, \quad -0.0299 - j0.5985.$$

The total power consumed by the resistors is

$$1.9231 + 0.3591 = 2.2822 \text{ W}$$

and is equal to P_{av} found earlier.

The power factor of the circuit is

$$\text{pf} = \frac{P_{av}}{|S|} = 0.9972$$

leading. As it is close to one, power-factor correction may not be needed.

∎

If the current enters the positive terminal of the source, the source is absorbing power. Otherwise, the source is supplying power. The total average power is equal to the sum of the powers supplied by each source acting alone, when the frequencies are harmonic. That is, the higher frequencies are integral multiple of the lowest nonzero frequency.

Example 4.9 Find the average active and reactive power consumed by the circuit, shown in Fig. 3.18. What is the power factor of the circuit?

Solution The DC source is 2 V. The AC source voltage is $0.8660 - j0.5$ V at $\omega = 1$ radian. The source current is 2 A at $\omega = 5$ radians. The impedances are

$$Z_1 = j\omega 0.1, Z_2 = \frac{1}{(1 + j\omega 0.2)}, Z_3 = 10, Z_4 = 2, Z_5 = \frac{(1/(j\omega 0.1))2}{(2 + (1/(j\omega 0.1)))}, Z6 = j\omega 0.2, Z7 = 1.$$

With the DC source alone, the resistor values are 1 and 2 in series. The current is 2/3 A. Therefore, the power consumed is

$$\frac{2}{3}\frac{2}{3}3 = \frac{4}{3} = 2\frac{2}{3}\,\text{W}.$$

The pf is equal to 1.

With $\omega = 1$, the values of the impedances are

$$Z_1 = j0.1, \quad Z_2 = 0.9615 - j0.1923, \quad Z_3 = 10.$$

The value of the current is

$$I = 0.0794 - j0.0449.$$

The apparent power consumed by the impedances, the product of the current, its conjugate and the impedance divided by 2 are

$$j0.0004, 0.0040 - j0.0008, 0.0416.$$

The total apparent power consumed by the circuit is the sum of those of all the impedances,

$$S = P_{av} + jQ_{av} = 0.0456 - j0.0004 \text{ va}.$$

The power supplied by the source is

$$0.0456 - j0.0004$$

also the same.

Now, let us compute the power consumed by the resistors alone, which is the active average power P_{av}. The resistors are

$$10, 1$$

and the corresponding currents are found from the branch currents, respectively, as

$$0.0794 - j0.0449, \quad 0.0677 - j0.0585.$$

The total power consumed by the resistors is

$$0.0416 + 0.0040 = 0.0456 \text{ W}$$

and is equal to P_{av} found earlier.

The power factor of the circuit is

$$\text{pf} = \frac{P_{av}}{|S|} = 0.999965$$

leading. As it is close to 1, power-factor correction may not be needed.

With $\omega = 5$, the values of the impedances are

$$Z_1 = j0.5, \quad Z_2 = 0.5 - j0.5, \quad Z_3 = 10$$

$$Z_4 = 2, \quad Z_5 = 1 - j1, \quad Z_6 = j1, \quad Z_7 = 1.$$

The value of the currents, respectively, are

$$2 - j0.0952, \, j0.0952, \, j0.0952, \, 2, \, 1 + j1, \, 1 - j1, \, 1 - j1.$$

The apparent power consumed by the impedances, the product of the current, its conjugate and the impedance divided by 2 are

$$j1.0023, \, 0.0023 - j0.0023, \, 0.0454, \, 4, \, 1 - j1, \, j1, \, 1.$$

The total apparent power consumed by the circuit is the sum of those of all the impedances

$$S = P_{av} + j Q_{av} = 6.0476 + j1 \text{ va.}$$

The power supplied by the source is also the same.

Now, let us compute the power consumed by the resistors alone, which is the active average power P_{av}. The resistors are

$$1, 10, 2, 2, 1$$

and the corresponding currents are found from the branch currents, respectively, as

$$0.0476 + j0.0476, \, j0.0952, \, 2, \, 1, \, 1 - j1.$$

The total power consumed by the resistors is

$$0.0023 + 0.0454 + 4 + 1 + 1 = 6.0476 \text{ W}$$

and is equal to P_{av} found earlier.

The power factor of the circuit is

$$\text{pf} = \frac{P_{av}}{|S|} = 0.9866$$

lagging.   ∎

Example 4.10 Find the average active and reactive power consumed by the circuit, shown in Fig. 3.19. What is the power factor of the circuit? Verify that the power supplied is equal to the sum of the power consumed by the components.

Solution The source voltage is $V = 2$ V. The source current is $I = -j$ A. The values of the impedances are

$$Z_1 = 2 + j0.1, \quad Z_2 = 1, \quad Z_3 = 3 - j10, \quad Z_4 = 3, \quad Z_5 = 2.$$

The values of the corresponding currents are

$$\{I_{Z_1} = 1.7257 + j0.7736, \ I_{Z_2} = -5.4514 + j0.4529, \ I_{Z_3} = 0.3666 + j0.4978,$$
$$I_{Z_4} = 1.3591 - j0.7242, \ I_{Z_5} = -3.7257 + j0.2264\}.$$

The apparent power consumed by the impedances, the product of the current, its conjugate and the impedance divided by 2 are

$$3.5765 + j0.1788, 14.9615, 0.5732 - j1.9107, 3.5575, 13.9321.$$

The total apparent power consumed by the circuit is the sum of those of all the impedances,

$$S = P_{av} + j Q_{av} = 36.6008 - j1.7319 \text{ va}.$$

The power supplied by the sources is also the same.

$$(1.3127 + j5.7644) + (-5.0848 - j0.9506) + (40.3729 - j6.5457) = 36.6008 - j1.7319.$$

Now, let us compute the power consumed by the resistors alone, which is the active average power P_{av}. The resistors are

$$2, 1, 3, 3, 2$$

and the corresponding currents are the same as the branch currents. The total power consumed by the resistors is

$$3.5765 + 14.9615 + 0.5732 + 3.5575 + 13.9321 = 36.6008 \text{ W}$$

and is equal to P_{av} found earlier.

$$\text{pf} = \frac{P_{av}}{|S|} = 0.9989$$

leading.   ∎

Example 4.11 Find the average active and reactive power consumed by the circuit, shown in Fig. 3.20. What is the power factor of the circuit? Verify that the power supplied is equal to the sum of the power consumed by the components.

Solution The source voltage is j V. The values of the impedances are

$$Z_1 = 1 + j0.1, \quad Z_2 = 3, \quad Z_3 = 0.0099 + j0.0990, \quad Z_4 = 3, \quad Z_5 = 1.$$

The values of the corresponding currents, respectively, are

$$0.9472 + j2.6635, -0.2664 - j0.9053, 0.3058 + j1.2245, 0.0394 + j0.3192, 0.6808 + j1.7582.$$

The apparent power consumed by the impedances, the product of the current, its conjugate and the impedance divided by 2 are

$$3.9958 + j0.3996, 1.3357, 0.0079 + j0.0789, 0.1552, 1.7775.$$

The total apparent power consumed by the circuit is the sum of those of all the impedances,

$$S = P_{av} + jQ_{av} = 7.2720 + j0.4784 \text{ va.}$$

The power supplied by the sources is also the same.

$$(-0.7195 - j0.3207) + (7.9915 + j0.7992) = 7.2720 + j0.4784.$$

The first component is the power supplied by the voltage source.

Now, let us compute the power consumed by the resistors alone, which is the active average power P_{av}. The resistors are

$$1, 3, 1, 3, 1$$

and the corresponding currents are the same as the branch currents. The total power consumed by the resistors is

$$3.9958 + 1.3357 + 0.0079 + 0.1552 + 1.7775 = 7.2720 \text{ W}$$

and is equal to P_{av} found earlier.

$$\text{pf} = \frac{P_{av}}{|S|} = 0.9978$$

lagging.   ■

4.4   Application

While electric power provides comfort in home, office, and industry, of course, we have to pay for it. Periodically, the reading of the kilowatthour meter is taken and we are billed for our use of power by the utility. A kilowatthour meter is used to measure the consumption of electric power. It is located between the power lines and the distribution panel of the building. The energy is the product of rate of power and the time period over which the power is used and its unit is wattseconds or joules. Since wattseconds is too small, watthour and kilowatthour are often used. One kilowatthour (kWh) is the energy dissipated by a 200 W bulb in 5 h. Typical wattage ratings of a laptop computer, washing machine, and a smoothing iron are, respectively, 70, 500, and 1000. For example, using a laptop for 7 h a day requires $(70 \times 7)/1000 = 0.49 \text{ kWh}$ of energy.

4.5   Summary

- Electric power is the rate at which electric energy is transferred by an electric source to a sink that absorbs power.
- Circuit element resistor dissipates energy. Capacitors and inductors store energy, respectively, in their electric and magnetic fields and, hence, are called storage elements.
- The stored energy is returned to the source. Therefore, the average power consumed by the storage devices, in the sinusoidal steady state, is zero.

- The power in AC circuits has two components to take into account, as the sinusoid and the impedance are characterized, at a frequency, by two parameters, the magnitude and the phase.
- The instantaneous energy stored in an inductor is $\frac{1}{2}Li^2(t)$. The instantaneous energy stored in a capacitor is $\frac{1}{2}Cv^2(t)$.
- Energy measured in joules is the integral of power over an interval. A resistor always absorbs power. The average power absorbed by inductors and capacitors is zero.
- The instantaneous power absorbed by the circuit is

$$s(t) = 0.5VI\cos(\theta - \phi) + 0.5VI\cos(2\omega t + \theta + \phi).$$

- In the frequency-domain representation of circuits, the voltages and currents are, respectively, represented as

$$V\angle\theta \quad \text{and} \quad I\angle\phi.$$

Then,

$$0.5(V\angle\theta)(I\angle(-\phi)) = 0.5VI\angle(\theta - \phi) = 0.5VI(\cos(\theta - \phi) + j\sin(\theta - \phi)).$$

The first component, which is the power actually consumed, is called the active power P_{av}. The power that is swapped back and forth from the source and the circuit is called the reactive or wattless power Q_{av}.
- Since the reactive power may be positive or negative, it is possible to change its value by adding suitable passive circuits, a process called power-factor correction.
- The equivalent definitions of the power factor are

$$\text{pf} = \frac{\text{active power}}{|\text{apparent power}|} = \cos(\theta) = \frac{R}{|Z|}.$$

Angle θ is the phase angle of the current with respect to the voltage.
- For inductive loads, with a voltage source, a suitable parallel capacitor C has to be connected to improve the power factor. It is possible to improve the power factor for capacitive loads by connecting an inductor L in parallel.

Exercises

4.1 Consider the circuit, shown in Fig. 4.10, with the voltage applied being

$$v(t) = \sin(2t)\ V.$$

The impedance of the circuit is composed of a resistor 2Ω connected in series with an inductor 1 H. Find the apparent, active, and reactive powers and the pf of the circuit. If the pf is to be 0.8

Fig. 4.10 A RL series circuit

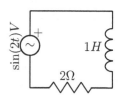

Fig. 4.11 A RC series
circuit

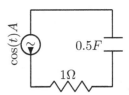

and 1, what is the value of the capacitor to be connected in parallel with the impedance in each
case?

* **4.2** Consider the circuit, shown in Fig. 4.11, with the current source

$$i(t) = \cos(t)\, A.$$

The impedance of the circuit is composed of a resistor 1Ω connected in series with a capacitor
0.5 F. Find the apparent, active, and reactive powers and the pf of the circuit. If the pf is to be
0.8 and 1, what is the value of the inductor to be connected in series with the impedance in each
case?

4.3 Find the average active and reactive power consumed by the circuit, shown in Fig. 3.50. What is
the power factor of the circuit? Verify that the power supplied is equal to the sum of the power
consumed by the components. Find the value of the capacitor for power-factor improvement,
for pf = 0.8 lagging and pf = 1.

* **4.4** Find the average active and reactive power consumed by the circuit, shown in Fig. 3.53. What is
the power factor of the circuit? Verify that the power supplied is equal to the sum of the power
consumed by the components. Find the value of the capacitor for power-factor improvement,
for pf = 0.8 lagging.

4.5 Find the average active and reactive power consumed by the circuit, shown in Fig. 3.55. What is
the power factor of the circuit? Verify that the power supplied is equal to the sum of the power
consumed by the components. Find the value of the capacitor for power-factor improvement,
for pf = 1.

Magnetically Coupled Circuits

5

5.1 Mutual Inductance

The circuit elements in circuits, we analyzed thus far, are conductively coupled. Conductors are used to connect elements. Circuit elements can also be connected through magnetic fields, called magnetically coupled. An inductor is basically a conducting coil. The induction of a voltage in another inductance by the current flowing in an inductance is mutual induction. A flywheel is a mechanical device that stores rotational energy and returns it to the system. Similarly, an inductor stores magnetic energy and returns it to the system. It does not dissipate energy like a resistor. While a resistor is required to limit the current in the circuit, an inductor smooths the flow of current. A time-varying current $i(t)$ flowing through an inductor of value L_{11} henries induces a voltage, which tends to oppose the current increase,

$$v(t) = L_{11} \frac{di(t)}{dt}.$$

The positive value L_{11} of the inductance is its self-inductance, or inductance because the voltage induced is due to the current flowing through itself, as we have seen earlier. A voltage

$$v(t) = L_{12} \frac{dx(t)}{dt}$$

is also induced in the inductor, which is affected by the magnetic field of another inductance due to current $x(t)$. The mutually induced voltage may be positive or negative depending upon the relative directions of the induced voltages. The value L_{12} is called the mutual inductance between the two inductances and also measured in henries. A current in an inductance may induce voltages in a number of other inductances in its proximity, in addition to its self-induced voltage, and the open-circuit mutual voltages are designated as

$$v_1(t) = L_{11} \frac{di_1(t)}{dt}, \quad v_2(t) = L_{21} \frac{di_1(t)}{dt}, \quad v_3(t) = L_{31} \frac{di_1(t)}{dt}, \ldots$$

In the determination of the mutual inductances, only a pair of coils is considered at a time, assuming that no other induced currents in the other inductances (by open-circuiting others). The study of mutual inductance is mandatory in analyzing circuits with often used components such as induction motors, transformers, induction heaters, etc.

© Springer Nature Switzerland AG 2020
D. Sundararajan, *Introductory Circuit Theory*,
https://doi.org/10.1007/978-3-030-31985-4_5

The currents and voltages induced in various inductances (for example with three inductances) are related as

$$v_1(t) = L_{11}\frac{di_1(t)}{dt} + L_{12}\frac{di_2(t)}{dt} + L_{13}\frac{di_3(t)}{dt}$$

$$v_2(t) = L_{21}\frac{di_1(t)}{dt} + L_{22}\frac{di_2(t)}{dt} + L_{23}\frac{di_3(t)}{dt}$$

$$v_3(t) = L_{31}\frac{di_1(t)}{dt} + L_{32}\frac{di_2(t)}{dt} + L_{33}\frac{di_3(t)}{dt}.$$

The self-inductances and the mutual inductances are, respectively,

$$\{L_{11}, L_{22}, L_{33}\} \quad \text{and} \quad \{L_{12}, L_{13}, L_{21}, L_{23}, L_{31}, L_{32}\}$$

with $L_{ij} = L_{ji}$. In these expressions, i_1, i_2, i_3 are the respective currents and v_1, v_2, v_3 are the respective voltage drops. As $L_{ij} = L_{ji}$, the mutual inductance between two coils is also denoted by M.

5.2   Stored Energy

Consider the case of two mutually coupled inductors with their coefficients being

$$\{L_{11}, L_{12}, L_{21}, L_{22}\}$$

with $L_{12} = L_{21}$. Then, equations relating their voltages and currents are

$$v_1(t) = L_{11}\frac{di_1(t)}{dt} + L_{12}\frac{di_2(t)}{dt} \tag{5.1}$$

$$v_2(t) = L_{21}\frac{di_1(t)}{dt} + L_{22}\frac{di_2(t)}{dt}. \tag{5.2}$$

Multiplying the two equations, respectively, by $i_1(t)$ and by $i_2(t)$ and adding, we get

$$v_1(t)i_1(t) + v_2(t)i_2(t) = L_{11}i_1(t)\frac{di_1(t)}{dt} + L_{12}i_1(t)\frac{di_2(t)}{dt} + L_{21}i_2(t)\frac{di_1(t)}{dt} + L_{22}i_2(t)\frac{di_2(t)}{dt}$$

Let

$$T(t) = \frac{1}{2}\left(L_{11}i_1^2(t) + L_{12}i_1(t)i_2(t) + L_{21}i_1(t)i_2(t) + L_{22}i_2^2(t)\right) = \frac{1}{2}\sum_{p,q=1}^{n} L_{pq}i_p(t)i_q(t)$$

where n is the number of coupled inductors. Then, with $n = 2$, we get

$$\frac{dT(t)}{dt} = L_{11}i_1(t)\frac{di_1(t)}{dt} + L_{12}i_1(t)\frac{di_2(t)}{dt} + L_{21}i_2(t)\frac{di_1(t)}{dt} + L_{22}i_2(t)\frac{di_2(t)}{dt} = v_1(t)i_1(t) + v_2(t)i_2(t)$$

The instantaneous power absorbed by the two coils is equal to the derivative of the energy T stored in the associated magnetic fields.

The equation for $T(t)$ is in quadratic form. That is,

$$2T(t) = \begin{bmatrix} i_1(t) & i_2(t) \end{bmatrix} \begin{bmatrix} L_{11} & L_{12} \\ L_{21} & L_{22} \end{bmatrix} \begin{bmatrix} i_1(t) \\ i_2(t) \end{bmatrix} = L_{11}i_1^2(t) \pm 2L_{12}i_1(t)i_2(t) + L_{22}i_2^2(t).$$

The energy stored in the magnetic fields must be positive regardless of the signs of the currents. The sign is positive for the mutual inductance term, if both the currents leave or enter the dotted terminals of the coils (as shown in Fig. 5.1). Otherwise, the energy contribution is negative.

A quadratic form is positive definite if all the principle minors in the top-left corner of its corresponding matrix are all zero. That is,

$$|L_{11} > 0, \begin{vmatrix} L_{11} & L_{12} \\ L_{21} & L_{22} \end{vmatrix} > 0.$$

Then,

$$L_{11} > 0, \quad L_{22} > 0, \quad \text{and} \quad L_{11}L_{22} - L_{12}^2 > 0 \quad \text{or} \quad \frac{L_{12}^2}{L_{11}L_{22}} < 1$$

The coupling coefficient is defined as

$$k = \frac{|L_{12}|}{\sqrt{L_{11}L_{22}}},$$

the magnitude of which must be less than 1 for practical inductances. The mutual inductance increases or decreases the stored energy of the two isolated inductors. Obviously, the closer the coils are placed the larger is the value of k. The linkage between the magnetic fields is stronger. If they play music at a place, the audibility decreases directly proportional to the distance of the point of the listener and we cannot hear it after a certain distance.

We are particularly interested in the values of the associated energies when the sources and variables are steady sinusoids, as is the case in the analysis of circuits. Let the time-domain currents and voltages be sinusoids expressed in complex form. Then,

$$i_p(t) = \frac{1}{2}(I_p e^{j\omega t} + I_p^* e^{-j\omega t}).$$

The product of two currents $i_p(t)$ and $i_q(t)$ is given by

$$i_p(t)i_q(t) = \frac{1}{4}(I_p e^{j\omega t} + I_p^* e^{-j\omega t})(I_q e^{j\omega t} + I_q^* e^{-j\omega t})$$

$$= \frac{1}{4}(I_p I_q e^{j2\omega t} + I_p^* I_q^* e^{-j2\omega t} + I_p I_q^* + I_p^* I_q)$$

$$= \frac{1}{2}\text{Re}(I_p I_q^*) + \frac{1}{2}\text{Re}(I_p I_q e^{j2\omega t}).$$

Since the stored energy in an inductor is

$$\frac{1}{2}L_{pq}i_p(t)i_q(t),$$

we get

$$T(t) = \frac{1}{4} \sum_{p,q=1}^{n} (Lpq I_p I_q^*) + \frac{1}{4} \mathrm{Re} \left(e^{j2\omega t} \sum_{p,q=1}^{n} (L_{pq} I_p I_q) \right).$$

The real part of an expression is denoted by the symbol "Re." The Re sign is not needed in the first term.

Similarly, energy stored in the electric field associated with a capacitor C is

$$\frac{1}{4\omega^2} \sum_{p,q=1}^{n} \left(\frac{I_p I_q^*}{Cpq} \right) - \frac{1}{4\omega^2} \mathrm{Re} \left(e^{j2\omega t} \sum_{p,q=1}^{n} \left(\frac{I_p I_q}{Cpq} \right) \right).$$

Note that

$$v(t) = \frac{1}{C} \int i(t) dt \quad \text{and} \quad V = \frac{1}{C} \frac{1}{j2\omega} (I e^{j\omega t} - I^* e^{-j\omega t})$$

Similar expressions can also be derived using voltage variables.

Figure 5.1 shows the voltages induced by mutual inductance. The polarity of the voltage depends on the way both the coils are physically wound. Any conductor has inductive effect. In order to enhance the inductive effect, a practical inductor is essentially a coil of current conducting wire. To determine the polarity, dots are placed in the circuit. Current entering the dotted end of a coil induces a voltage in the other coil such that the polarity of the mutually induced voltage is positive at the dotted end, as shown in Fig. 5.1a. That is, the polarity of the voltage is the same as that of the self-induced voltage due to current entering the inductor. Current entering the undotted end of a coil induces a voltage in the other coil such that the polarity of the mutually induced voltage is positive at the undotted end, as shown in Fig. 5.1c. In the other two cases, shown in Fig. 5.1b and d, one current enters at the dotted end while the other current leaves the dotted end and the polarity of the mutually induced voltage is negative.

Assume that the circuit is excited by a sinusoidal source and the circuit is represented in the frequency-domain. Then, two magnetically coupled inductors, with self-inductances L_1 and L_2 and currents I_1 and I_2 through them, respectively, and mutual inductance M, can be replaced, using Eqs. (5.1) and (5.2), by two isolated inductors, in series with current-controlled voltage sources.

$$V_1 = j\omega L_1 I_1 \pm j\omega M I_2 \quad \text{and} \quad V_2 = j\omega L_2 I_2 \pm j\omega M I_1$$

Remember that any source waveform, in practice, can be decomposed into sinusoidal components by Fourier analysis. As already mentioned earlier also, this is the reason for the importance of analyzing circuits assuming sinusoidal sources.

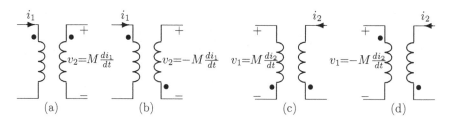

Fig. 5.1 Determining the polarity of the mutually induced voltage

5.3   Examples

Example 5.1 Analyze the circuit shown in Fig. 5.2a. The circuit includes two coupled inductors, with the coupling coefficient 0.5.

Solution Source is $\cos(t)V$. With

$$V = 1\angle 0, \ \omega = 1, \ Z_1 = R_1 = 1, \ Z_2 = j\omega L_1 = j1, \ Z_3 = j\omega L_2 = j4, \ k = 0.5,$$

$$Z_M = j\omega k\sqrt{L_1 L_2} = j1, \ Z_4 = R_2 = 2.$$

The frequency-domain version of the equivalent circuit, with two isolated inductors in series with current-controlled voltage sources, is shown in Fig. 5.2b. The current flowing into the dotted terminal of L_1 induces a voltage that is positive at the dotted terminal of L_2. Therefore, the controlled voltage source is jI_1. The current flowing out of the dotted terminal of L_2 induces a voltage that is negative at the dotted terminal of L_1. Therefore, the controlled voltage source is $-jI_2$. The equilibrium equations, using mesh analysis, are

$$\begin{bmatrix} (Z_1 + Z_2) & -Z_M \\ -Z_M & (Z_3 + Z_4) \end{bmatrix} \begin{bmatrix} I_1 \\ I_2 \end{bmatrix} = \begin{bmatrix} 1\angle 0 \\ 0 \end{bmatrix}.$$

Substituting the numerical values, we get

$$\begin{bmatrix} 1 + j1 & -j1 \\ -j1 & 2 + j4 \end{bmatrix} \begin{bmatrix} I_1 \\ I_2 \end{bmatrix} = \begin{bmatrix} 1 \\ 0 \end{bmatrix}.$$

Determinant of the impedance matrix is $-1 + j6$. Solving for I_1 and I_2, we get

$$\begin{bmatrix} I_1 \\ I_2 \end{bmatrix} = \begin{bmatrix} 1 + j1 & -j1 \\ -j1 & 2 + j4 \end{bmatrix}^{-1} \begin{bmatrix} 1 \\ 0 \end{bmatrix}$$

$$= \begin{bmatrix} 0.5946 - j0.4324 & 0.1622 - j0.0270 \\ 0.1622 - j0.0270 & 0.1351 - j0.1892 \end{bmatrix} \begin{bmatrix} 1 \\ 0 \end{bmatrix}$$

$$= \begin{bmatrix} 0.5946 - j0.4324 \\ 0.1622 - j0.0270 \end{bmatrix}.$$

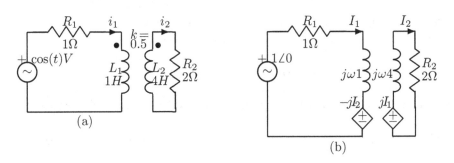

Fig. 5.2 (a) Circuit with a mutual inductance; (b) its equivalent frequency-domain representation

Let us find the power stored and consumed by the circuit. The values of the impedances are

$$Z_1 = 1, \ Z_2 = j1, \ Z_3 = j4, \ Z_4 = 2.$$

The values of the corresponding currents are

$$\{0.5946 - j0.4324, \ 0.5946 - j0.4324, \ 0.1622 - j0.0270, \ 0.1622 - 0.0270\}.$$

The average apparent power consumed by the impedances, the product of the current, its conjugate and the impedance divided by 2 are

$$0.2703, \ j0.2703, \ j0.0541, \ 0.0270.$$

The sum of these powers is $0.2973 + j0.3243$. The mutual coupling between the inductances is destructive. Therefore, the power due to mutual impedance,

$$0.5 Z_m (I_1 I_2^* + I_2 I_1^*) = j0.1081$$

has to be subtracted. The total average apparent power consumed by the circuit is the sum of those of all the impedances,

$$S = P_{av} + j Q_{av} = 0.2973 + j0.2162 \text{ va.}$$

The power supplied by the source $V I_1^*/2$ is also the same. The pf is 0.8087 lagging.

Now, let us compute the power consumed by the resistors alone, which is the active average power P_{av}. The resistors are

$$1, 2$$

and the corresponding currents are, respectively,

$$0.5946 - j0.4324, \ 0.1622 - j0.0270.$$

The total power consumed by the resistors is

$$0.2703 + 0.0270 = 0.2973 \text{ W}$$

and is equal to P_{av} found earlier. The time-domain representation of the currents are

$$i_1(t) = 0.7352 \cos(t - 0.6288), \ i_2(t) = 0.1644 \cos(t - 0.1651).$$

The instantaneous energy dissipated by the resistors is $i_1^2(t) + 2i_2^2(t)$, shown in Fig. 5.3.

Let us find the energy stored in the circuit. The energy stored at $t = 1$ is

$$0.5(1)0.6851^2 + 0.5(4)(0.1104)^2 - 1(0.6851)(0.1104) = 0.1835 J.$$

The energy stored at $t = 1$ is, using the frequency-domain representation of the currents,

$$0.25(I_1 I_1^*(1) + I_2^* I_2(4) + 2I_1 I_2^*(-1) + \text{Re}(e^{j2(1)}(I_1 I_1(1) + I_2 I_2(4) + 2I_1 I_2(-1))) = 0.1835,$$

as found earlier.

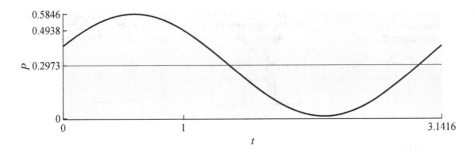

Fig. 5.3 The instantaneous energy dissipated in the resistors of the circuit with a sinusoidal current of frequency 1 rad/s

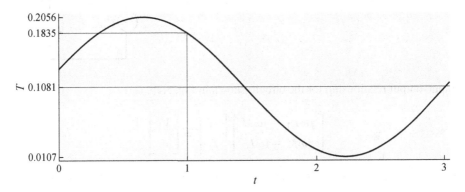

Fig. 5.4 The instantaneous energy stored in the magnetic field of the circuit with a sinusoidal current of frequency 1 rad/s

Figure 5.4 shows the instantaneous energy stored in the magnetic field of the circuit with a sinusoidal current of frequency 1 rad/s. The stored energy, which is sinusoidal, varies about the average value. As the frequency is double that of the current, it completes a cycle in one-half period of that of the current. That is, $2\pi/\omega = \pi/1 = 3.1416$ s. The average value is 0.1081. As the frequency is 1, the average reactive power is twice this value 0.2162, as found earlier.

A magnetically coupled circuit can also be replaced by equivalent T or Π circuits with no magnetic coupling. Figure 5.5a and b shows, respectively, a magnetically coupled circuit and its equivalent T circuit. The input–output relationship for the circuit is

$$\begin{bmatrix} j\omega L_1 & j\omega M \\ j\omega M & j\omega L_2 \end{bmatrix} \begin{bmatrix} I_1 \\ I_2 \end{bmatrix} = \begin{bmatrix} V_1 \\ V_2 \end{bmatrix}.$$

The input–output relationship for the equivalent T circuit with three isolated inductors is

$$\begin{bmatrix} j\omega(L_a + L_c) & j\omega L_c \\ j\omega L_c & j\omega(L_b + L_c) \end{bmatrix} \begin{bmatrix} I_1 \\ I_2 \end{bmatrix} = \begin{bmatrix} V_1 \\ V_2 \end{bmatrix}.$$

Therefore,

$$L_a = L_1 - M, \quad L_b = L_2 - M, \quad L_c = M.$$

This model can be adapted to suit different current directions.

Fig. 5.5 (a) A
magnetically coupled
circuit; (b) its equivalent T
circuit

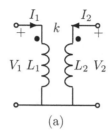 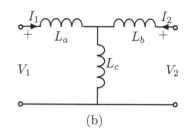

(a)   (b)

Fig. 5.6 An equivalent
frequency-domain version
of the circuit in Fig. 5.2a

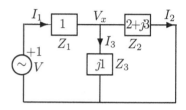

The input–output relationship for the circuit, shown in Fig. 5.2b, is

$$\begin{bmatrix} j\omega L_1 & -j\omega M \\ j\omega M & -j\omega L_2 \end{bmatrix} \begin{bmatrix} I_1 \\ I_2 \end{bmatrix} = \begin{bmatrix} V_1 \\ V_2 \end{bmatrix}.$$

The input–output relationship for the equivalent T circuit with three isolated inductors is

$$\begin{bmatrix} j\omega(L_a + L_c) & -j\omega L_c \\ j\omega L_c & -j\omega(L_b + L_c) \end{bmatrix} \begin{bmatrix} I_1 \\ I_2 \end{bmatrix} = \begin{bmatrix} V_1 \\ V_2 \end{bmatrix}.$$

Therefore,
$$L_a = L_1 - M, \quad L_b = L_2 - M, \quad L_c = M.$$

Let us redo the problem using the T circuit model, shown in Fig. 5.6. It consists of three isolated inductors and two resistors.

$$V = 1\angle 0, \quad Z_1 = 1, \quad Z_2 = 2 + j3, \quad Z_3 = j1$$

Using nodal analysis, we get

$$\frac{V_x - 1}{Z_1} + \frac{V_x}{Z_2} + \frac{V_x}{Z_3} = 0 \quad \text{and} \quad V_x = \frac{Z_2 Z_3}{Z_1 Z_2 + Z_2 Z_3 + Z_1 Z_3}.$$

This result could also be obtained using voltage division formula. Substituting numerical values and solving, we get $V_x = 0.4054 + j0.4324$. Now,

$$I_1 = -\frac{V_x - 1}{Z_1} = 0.5946 - j0.4324, \quad I_2 = \frac{V_x}{Z_2} = 0.1622 - j0.0270, \quad I_3 = \frac{V_x}{Z_3} = 0.4324 - 0.4054$$

Note that $I_1 = I_2 + I_3$.

Let us find the power stored and consumed by the circuit. The values of the impedances are

$$V = 1\angle 0, \quad Z_1 = 1, \quad Z_2 = 2 + j3, \quad Z_3 = j1.$$

The values of the corresponding currents are

$$\{0.5946 - j0.4324, \ 0.1622 - j0.0270, \ 0.4324 - 0.4054\}.$$

The average apparent power consumed by the impedances, the product of the current, its conjugate and the impedance divided by 2 are

$$0.2703, \ 0.0270 + 0.0405, \ j0.1757.$$

Therefore, the total average apparent power consumed by the circuit is the sum of those of all the impedances,

$$S = P_{av} + jQ_{av} = 0.2973 + j0.2162 \text{ va.}$$

The power supplied by the source $VI_1^*/2$ is also the same. The pf is 0.8087.
    The time-domain representation of the currents are

$$i_1(t) = 0.7352\cos(t - 0.6288), \ i_2(t) = 0.1644\cos(t - 0.1651), \ i_3(t) = 0.5927\cos(t - -0.7532)$$

The instantaneous power dissipated by the resistors is $i_1^2(t) + 2i_2^2(t)$, shown in Fig. 5.3.
    Let us find the energy stored in the circuit. The energy stored at $t = 1$ is

$$0.5(0)(0.6851)^2 + 0.5(3)(0.1104)^2 + (0.5)1(0.5748)^2 = 0.1835\,\text{J.}$$

The instantaneous energy stored, which can also be computed in the frequency-domain, is shown in Fig. 5.4. ∎

Example 5.2 Analyze the circuit shown in Fig. 5.7a. The circuit includes two coupled inductors, with the coupling coefficient 0.5.

Solution Source is $\sin(2t)V$. With

$$V = -j, \ \omega = 2, \ Z_1 = R_1 = 2, \ Z_2 = j\omega L_1 = j8, \ Z_3 = j\omega L_2 = j2, \ k = 0.5,$$

$$Z_M = j\omega k\sqrt{L_1 L_2} = j2, \ Z_4 = R_2 = 1, \ Z_5 = 1/(j\omega C_2) = -j2.5.$$

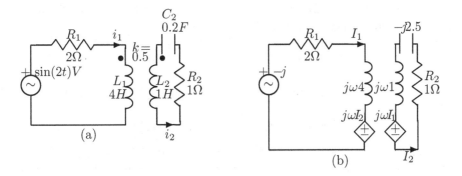

Fig. 5.7 (a) Circuit with a mutual inductance; (b) its equivalent frequency-domain representation

The frequency-domain version of the equivalent circuit, with two isolated inductors in series with current-controlled voltage sources, is shown in Fig. 5.7b. The current flowing into the dotted terminal of L_1 induces a voltage that is positive at the dotted terminal of L_2. Therefore, the controlled voltage source is $j2I_1$. Similarly, the other controlled voltage source is $j2I_2$. The equilibrium equations, using mesh analysis, are

$$\begin{bmatrix} (Z_1 + Z_2) & Z_M \\ Z_M & (Z_3 + Z_4 + Z_5) \end{bmatrix} \begin{bmatrix} I_1 \\ I_2 \end{bmatrix} = \begin{bmatrix} -j \\ 0 \end{bmatrix}.$$

Substituting the numerical values, we get

$$\begin{bmatrix} 2 + j8 & j2 \\ j2 & 1 - j0.5 \end{bmatrix} \begin{bmatrix} I_1 \\ I_2 \end{bmatrix} = \begin{bmatrix} -j \\ 0 \end{bmatrix}.$$

Determinant of the impedance matrix is $10 + j7$. Solving for I_1 and I_2, we get

$$\begin{bmatrix} I_1 \\ I_2 \end{bmatrix} = \begin{bmatrix} 2 + j8 & j2 \\ j2 & 1 - j0.5 \end{bmatrix}^{-1} \begin{bmatrix} -j \\ 0 \end{bmatrix}$$

$$= \begin{bmatrix} 0.0436 - j0.0805 & -0.0940 - 0.1342 \\ -0.0940 - j0.1342 & 0.5101 + j0.4430 \end{bmatrix} \begin{bmatrix} -j \\ 0 \end{bmatrix}$$

$$= \begin{bmatrix} -0.0805 - j0.0436 \\ -0.1342 + j0.0940 \end{bmatrix}.$$

Let us find the power stored and consumed by the circuit. The values of the impedances are

$$Z_1 = 2, \; Z_2 = j8, \; Z_3 = j2, \; Z_4 = 1, \; Z_5 = -j2.5.$$

The values of the corresponding currents are

$$\{-0.0805 - j0.0436, \; -0.0805 - j0.0436, \; -0.1342 + j0.0940, \; -0.1342 + j0.0940, \; -0.1342 + j0.0940\}$$

The average apparent power consumed by the impedances, the product of the current, its conjugate and the impedance divided by 2 are

$$0.0084, \; j0.0336, \; j0.0268, \; 0.0134, \; -j0.0336.$$

The sum of these powers is $0.0218 + j0.0268$. The mutual coupling between the inductances is constructive. Therefore, the power due to mutual impedance,

$$0.5Z_m(I_1 I_2^* + I_2 I_1^*) = j0.0134$$

has to be added. The total average apparent power consumed by the circuit is the sum of those of all the impedances,

$$S = P_{av} + jQ_{av} = 0.0218 + j0.0403 \text{ va}.$$

The power supplied by the source $VI_1^*/2$ is also the same. The pf is 0.4763 lagging.

Now, let us compute the power consumed by the resistors alone, which is the active average power P_{av}. The resistors are

$$2, 1$$

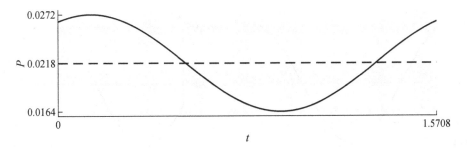

Fig. 5.8 The instantaneous energy dissipated in the resistors of the circuit with a sinusoidal current of frequency 2 rad/s

and the corresponding currents are, respectively,

$$-0.0805 - j0.0436, \quad -0.1342 + j0.0940.$$

The total power consumed by the resistors is

$$0.0084 + 0.0134 = 0.0218 \text{ W}$$

and is equal to P_{av} found earlier. The time-domain representation of the currents are

$$i_1(t) = 0.0916\cos(2t - 2.6452), \quad i_2(t) = 0.1638\cos(2t + 2.5309).$$

The instantaneous power dissipated by the resistors is $2i_1^2(t) + i_2^2(t)$, is shown in Fig. 5.8.
    Let us find the energy stored in the circuit. The energy stored in the magnetic fields at $t = 1$ is

$$0.5(4)0.0732^2 + 0.5(1)(-0.0296)^2 + 1(0.0732)(-0.0296) = 0.0090 \text{ J}.$$

The voltage across the capacitor is $0.2349 + j0.3356$. Its time-domain representation is

$$v_1(t) = 0.4096\cos(2t + 0.9601).$$

The energy stored in the electric field at $t = 1$ is

$$0.5(0.2)(-0.4029)^2 = 0.0162 \text{ J}.$$

The energy stored in the inductors at $t = 1$ is, using the frequency-domain representation of the currents,

$$0.25(I_1 I_1^*(4) + I_2^* I_2(1) + 2I_1 I_2^*(1) + \text{Re}(e^{j4(1)}(I_1 I_1(4) + I_2 I_2(1) + 2I_1 I_2(1)))) = 0.0090,$$

as found earlier.
    The energy stored at $t = 1$ in the capacitor is, using the frequency-domain representation of the currents,

$$\frac{1}{4\omega^2 C}(I_2 I_2^*) - \frac{1}{4\omega^2 C}(I_2 I_2 e^{j4t}) = 0.0162$$

as found earlier.

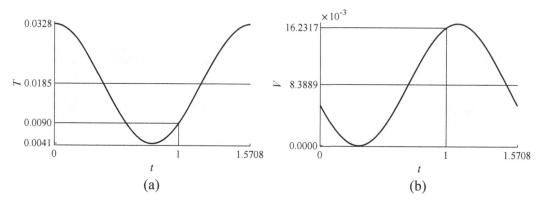

Fig. 5.9 (a) The instantaneous energy stored in the magnetic field of the circuit with a sinusoidal current of frequency 2 rad/s; (b) the instantaneous energy stored in the electric field

Fig. 5.10 An equivalent frequency-domain version of the circuit in Fig. 5.7a

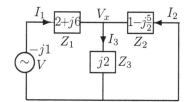

Figure 5.9a shows the instantaneous energy stored in the magnetic fields of the circuit with a sinusoidal current of frequency 2 rad/s. The stored energy, which is sinusoidal, varies about the average value. As the frequency is double that of the current, it completes a cycle in one-half period of that of the current. That is, $\pi/\omega = \pi/2 = 1.5708$ s. The average value is 0.0185.

Figure 5.9b shows the instantaneous energy stored in the electric field of the circuit with a sinusoidal current of frequency 2 rad/s. The average value is 0.0084.

As the frequency is 2, the average reactive power in the circuit is the difference between the average energy stored in the magnetic and electric fields multiplied by 2ω, $4(0.0185 - 0.0084) = 0.0403$, as found earlier.

Let us redo the problem using the T circuit model, shown in Fig. 5.10. It consists of two isolated inductor, two resistors, and a capacitor. Since $M = 1$ and $\omega = 2$,

$$V = -j1, \quad Z_1 = 2 + j6, \quad Z_2 = 1 - j2.5, \quad Z_3 = j2.$$

Using nodal analysis, we get

$$\frac{V_x - j1}{Z_1} + \frac{V_x}{Z_2} + \frac{V_x}{Z_3} = 0 \quad \text{and} \quad V_x = \frac{(-j1)Z_2 Z_3}{Z_1 Z_2 + Z_2 Z_3 + Z_1 Z_3}.$$

This result could also be obtained using voltage division formula. Substituting numerical values and solving, we get $V_x = -0.1007 - j0.4295$. Now,

$$I_1 = -\frac{V_x - j1}{Z_1} = -0.0805 - j0.0436, \quad I_2 = -\frac{V_x}{Z_2} = -0.1342 + j0.0940, \quad I_3 = \frac{V_x}{Z_3} = -0.2148 + j0.0503$$

Note that $I_1 + I_2 = I_3$.

Let us find the power stored and consumed by the circuit. The values of the impedances are

$$Z_1 = 2 + j6, \quad Z_2 = 1 - j2.5, \quad Z_3 = j2.$$

The values of the corresponding currents are

$$\{-0.0805 - j0.0436, \ -0.1342 + j0.0940, \ -0.2148 + j0.0503\}.$$

The average apparent power consumed by the impedances, the product of the current, its conjugate and the impedance divided by 2 are

$$0.0084 + j0.0252, \ 0.0134 - j0.0336, \ j0.0487.$$

Therefore, the total average apparent power consumed by the circuit is the sum of those of all the impedances,

$$S = P_{av} + jQ_{av} = 0.0218 + j0.0403 \text{ va.}$$

The power supplied by the source $VI_1^*/2$ is also the same. The pf is 0.4763.

The time-domain representation of the currents are

$$i_1(t) = 0.0916\cos(2t - 2.6452), \ i_2(t) = 0.1638\cos(2t + 2.5309), \ i_3(t) = 0.2206\cos(2t + 2.9114)$$

The instantaneous power dissipated by the resistors is $2i_1^2(t) + i_2^2(t)$, is shown in Fig. 5.8.

Let us find the energy stored in the circuit. The energy stored at $t = 1$ is

$$0.5(3)0.0732^2 + 0.5(0)(-0.0296)^2 + 0.5(1)(0.0436)(0.0436) = 0.0090 \text{ J},$$

as found earlier. The instantaneous energy stored, which can also be computed in the frequency-domain, is shown in Fig. 5.9.

Example 5.3 Analyze the bridge circuit, with mutually coupled inductors, shown in Fig. 5.11a. The coupling coefficient is 0.5.

Solution An equivalent frequency-domain circuit is shown in Fig. 5.11b.

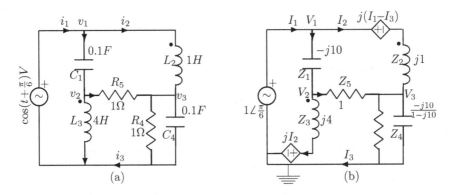

Fig. 5.11 (a) A bridge circuit with mutually coupled inductors; (b) an equivalent frequency-domain circuit

Mesh Analysis

The equilibrium equations are

$$Z_1(I_1 - I_2) + Z_3(I_1 - I_3) + Z_m I_2 = 1 \angle \frac{\pi}{6}$$

$$Z_1(I_1 - I_2) + Z_5(I_3 - I_2) - I_2 Z_2 - Z_m(I_1 - I_3) = 0$$

$$-Z_3(I_1 - I_3) + Z_5(I_3 - I_2) + Z_4 I_3 - Z_m I_2 = 0.$$

As expected, these equations are the same as those derived in Chaps. 2 and 3 for the bridge circuit with additional terms added to take into account of the mutual coupling of inductors.

Simplifying, we get

$$(Z_1 + Z_3)I_1 + (Z_m - Z_1)I_2 - Z_3 I_3 = \left(\frac{\sqrt{3}}{2} + j\frac{1}{2} \right)$$

$$(Z_1 - Z_m)I_1 - (Z_1 + Z_2 + Z_5)I_2 + (Z_m + Z_5)I_3 = 0$$

$$-Z_3 I_1 - (Z_m + Z_5)I_2 + (Z_3 + Z_4 + Z_5)I_3 = 0$$

The equations in matrix form is

$$\begin{bmatrix} (Z_1 + Z_3) & (Z_m - Z_1) & -Z_3 \\ (Z_1 - Z_m) & -(Z_1 + Z_2 + Z_5) & (Z_m + Z_5) \\ -Z_3 & -(Z_m + Z_5) & (Z_3 + Z_4 + Z_5) \end{bmatrix} \begin{bmatrix} I_1 \\ I_2 \\ I_3 \end{bmatrix} = \begin{bmatrix} \left(\frac{\sqrt{3}}{2} + j\frac{1}{2} \right) \\ 0 \\ 0 \end{bmatrix}.$$

With

$$Z_1 = -j10, \quad Z_2 = j1, \quad Z_3 = j4, \quad Z_4 = \frac{-j10}{1 - j10}, \quad Z_5 = 1, \quad Z_m = j1$$

$$\begin{bmatrix} I_1 \\ I_2 \\ I_3 \end{bmatrix} = \begin{bmatrix} -j6.0000 & 0.0000 + j11.0000 & 0.0000 - j4.0000 \\ -j11.0000 & -1.0000 + j9.0000 & 1.0000 + j1.0000 \\ -j4.0000 & -1.0000 - j1.0000 & 1.9901 + j3.9010 \end{bmatrix}^{-1} \begin{bmatrix} \left(\frac{\sqrt{3}}{2} + j\frac{1}{2} \right) \\ 0 \\ 0 \end{bmatrix}$$

$$= \begin{bmatrix} 0.4325 - j0.3237 & -0.4502 + j0.3538 & 0.5892 - j0.2374 \\ 0.4502 - j0.3538 & -0.4751 + j0.2780 & 0.6314 - j0.2337 \\ 0.5892 - j0.2374 & -0.6314 + j0.2337 & 0.8524 - j0.2867 \end{bmatrix} \begin{bmatrix} \left(\frac{\sqrt{3}}{2} + j\frac{1}{2} \right) \\ 0 \\ 0 \end{bmatrix}$$

$$= \begin{bmatrix} 0.5364 - j0.0641 \\ 0.5667 - j0.0813 \\ 0.6290 + j0.0890 \end{bmatrix}$$

$$\{I_{Z_1} = -0.0304 + j0.0172, \quad I_{Z_3} = -0.0926 - j0.1531, \quad I_{Z_5} = 0.0623 + j0.1704\}$$

Nodal Analysis

We have to express the current variables in terms of voltage variables and we get

$$I_2 = -\frac{(V_2 - V_3)}{Z_5} + \frac{V_3}{Z_4}, \quad (I_1 - I_3) = \frac{(V_1 - V_2)}{Z_1} - \frac{(V_2 - V_3)}{Z_5}.$$

The voltage source is

$$V_1 = 1\angle\frac{\pi}{6} = \left(\frac{\sqrt{3}}{2} + j\frac{1}{2}\right).$$

The equilibrium equations are

$$\frac{(V_2 - V_1)}{Z_1} + \frac{(V_2 - V_3)}{Z_5} + \frac{(V_2 - jI_2)}{Z_3} = 0$$

$$\frac{(V_3 - V_1 + j(I_1 - I_3))}{Z_2} + \frac{(V_3 - V_2)}{Z_5} + \frac{V_3}{Z_4} = 0.$$

As expected, these equations are the same as those derived in Chaps. 2 and 3 for the bridge circuit with additional terms added to take into account of the mutual coupling of inductors. The first equation is the application of KCL at the left middle node. The second equation is the application of KCL at the right middle node. Substituting for I_2 and $I_1 - I_3$, we get

$$\frac{(V_2 - V_1)}{Z_1} + \frac{(V_2 - V_3)}{Z_5} + \frac{\left(V_2 + j\frac{(V_2-V_3)}{Z_5} - j\frac{V_3}{Z_4}\right)}{Z_3} = 0$$

$$\frac{\left(V_3 - V_1 + j\frac{(V_1-V_2)}{Z_1} - j\frac{(V_2-V_3)}{Z_5}\right)}{Z_2} + \frac{(V_3 - V_2)}{Z_5} + \frac{V_3}{Z_4} = 0.$$

Simplifying, we get

$$(Z_3 Z_5 + jZ_1 + Z_1 Z_3 + Z_1 Z_5)V_2 - \left(Z_1 Z_3 + (jZ_1) + \left(\frac{jZ_1 Z_5}{Z_4}\right)\right)V_3 = Z_3 Z_5 V_1$$

$$-\left(Z_2 Z_4 + jZ_4 + \left(\frac{jZ_4 Z_5}{Z_1}\right)\right)V_2 + (Z_4 Z_5 + Z_2 Z_4 + jZ_4 + Z_2 Z_5)V_3 = \left(Z_4 Z_5 + \left(-j\frac{Z_4 Z_5}{Z_1}\right)\right)V_1$$

With

$$Z_1 = -j10, \quad Z_2 = j1, \quad Z_3 = j4, \quad Z_4 = \frac{-j10}{1 - j10}, \quad Z_5 = 1, \quad Z_m = j1$$

$$\begin{bmatrix} 50.0000 - j6.0000 & -60.0000 - j1.0000 \\ -0.0990 - j1.9901 & 1.1881 + j2.8812 \end{bmatrix} \begin{bmatrix} V_2 \\ V_3 \end{bmatrix} = \begin{bmatrix} -2.0000 + j3.4641 \\ 0.9977 + j0.4502 \end{bmatrix}.$$

The determinant of the admittance matrix is $72.7426 + j17.4257$.
Solving the equilibrium equations, we get

$$\begin{bmatrix} V_2 \\ V_3 \end{bmatrix} = \begin{bmatrix} 50.0000 - j6.0000 & -60.0000 - j1.0000 \\ -0.0990 - j1.9901 & 1.1881 + j2.8812 \end{bmatrix}^{-1} \begin{bmatrix} -2.0000 + j3.4641 \\ 0.9977 + j0.4502 \end{bmatrix}$$

$$= \begin{bmatrix} 0.0244 + j0.0338 & 0.7832 - j0.1739 \\ 0.0075 + j0.0256 & 0.6314 - j0.2337 \end{bmatrix} \begin{bmatrix} -2.0000 + j3.4641 \\ 0.9977 + j0.4502 \end{bmatrix}$$

$$= \begin{bmatrix} 0.6938 + j0.1962 \\ 0.6316 + j0.0259 \end{bmatrix}$$

$$\{V_{Z_1} = 0.1722 + j0.3038, \quad V_{Z_2} = 0.2344 + j0.4741, \quad V_{Z_5} = 0.0623 + j0.1704\}.$$

Let us get the currents in the various branches of the circuit from the voltages found by the nodal analysis.

$$V_1 = 0.8660 + j0.5, \quad V_2 = 0.6938 + j0.1962, \quad V_3 = 0.6316 + j0.0259$$

The voltage induced across a magnetically coupled inductor is the sum of the self- and mutually induced voltages. Therefore, in order to find the current through a coupled inductor, we subtract the mutually induced voltage from the voltage across it and then divide the result by its self-impedance.

$$I_{Z_1} = \frac{V_1 - V_2}{Z_1} = -0.0304 + j0.0172$$

$$I_{Z_5} = \frac{V_2 - V_3}{Z_5} = 0.0622 + j0.1703$$

$$I_{Z_3} = \frac{V_2 - jI_2}{Z_3} = -0.0926 - j0.1531 = I_{Z_1} - I_{Z_5}$$

$$I_3 = \frac{V_3}{Z_4} = 0.6290 + j0.0891$$

$$I_1 = I_{Z_3} + I_3 = 0.5364 - j0.0641$$

$$I_2 = \frac{V_1 - V_3 - j(I_1 - I_3)}{Z_2} = 0.5667 - j0.0813 = I_1 - I_{Z_1}.$$

Let us find the power stored and consumed by the circuit. The values of the impedances are

$$Z_1 = -j10, \quad Z_2 = j1, \quad Z_3 = j4, \quad Z_4 = 0.9901 - j0.0990, \quad Z_5 = 1.$$

The values of the corresponding currents are

$$\{-0.0304 + j0.0172, \ 0.5667 - j0.0813, \ -0.0926 - j0.1531, \ 0.6290 + j0.0890, \ 0.0623 + j0.1704\}$$

The average apparent power consumed by the impedances, the product of the current, its conjugate and the impedance divided by 2 are

$$-j0.0061, \quad j0.1639, \quad j0.0641, \quad 0.1998 - j0.0200, \quad 0.0164.$$

The sum of these powers is $0.2162 + j0.2019$. The mutual coupling between the inductances is constructive. Therefore, the power due to mutual impedance,

$$0.5Z_m(I_2 I_3^* + I_3 I_2^*) = -j0.04$$

has to be added. Therefore, The total average apparent power consumed by the circuit is the sum of those of all the impedances,

$$S = P_{av} + jQ_{av} = 0.2162 + j0.1618 \text{ va.}$$

The power supplied by the source $VI_1^*/2$ is also the same. The pf is 0.8006 lagging.

Now, let us compute the power consumed by the resistors alone, which is the active average power P_{av}. The resistors are

1, 1

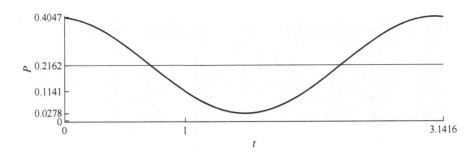

Fig. 5.12 The instantaneous energy dissipated in the resistors of the circuit with a sinusoidal current of frequency 1 rad/s

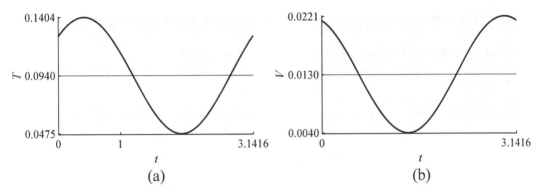

(a)   (b)

Fig. 5.13 (a) The instantaneous energy stored in the magnetic field of the circuit with a sinusoidal current of frequency 1 rad/s; (b) the instantaneous energy stored in the electric field

and the corresponding currents are, respectively,

$$0.0623 + j0.1704, \ 0.6316 + j0.0259$$

The total power consumed by the resistors is

$$0.0164 + 0.1998 = 0.2162 \text{ W}$$

and is equal to P_{av} found earlier.

The instantaneous energy dissipated in the resistors of the circuit is shown in Fig. 5.12. The average value is 0.2162, as found earlier.

Figure 5.13a shows the instantaneous energy stored in the magnetic fields of the circuit with a sinusoidal current of frequency 1 rad/s. The stored energy, which is sinusoidal, varies about the average value. As the frequency is double that of the current, it completes a cycle in one-half period of that of the current. That is, $2\pi/\omega = 2\pi/2 = 3.1416$ s. The average value is 0.0940. Figure 5.13b shows the instantaneous energy stored in the electric field of the circuit with a sinusoidal current of frequency 1 rad/s. The average value is 0.0130. As the frequency is 1, the average reactive power in the circuit is the difference between the average energy stored in the magnetic and electric fields multiplied by 2ω, $2(0.0940 - 0.0130) = 0.1618$, as found earlier.

Example 5.4 Analyze the circuit, with mutually coupled inductors, shown in Fig. 5.14a. The coupling coefficient is 0.5.

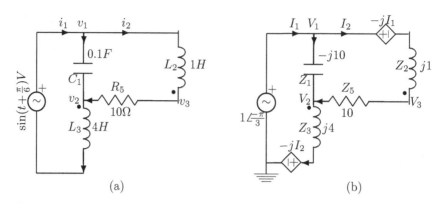

Fig. 5.14 (a) A circuit with mutually coupled inductors; (b) an equivalent frequency-domain circuit

Solution An equivalent frequency-domain circuit is shown in Fig. 5.14b.

Mesh Analysis
The equilibrium equations are

$$Z_1(I_1 - I_2) + Z_3 I_1 - Z_m I_2 = 1\angle\frac{-\pi}{3}$$

$$Z_1(I_1 - I_2) - Z_5 I_2 - I_2 Z_2 + Z_m I_1 = 0.$$

Simplifying, we get

$$(Z_1 + Z_3)I_1 - (Z_m + Z_1)I_2 = \left(\frac{1}{2} - j\frac{\sqrt{3}}{2}\right)$$

$$(Z_1 + Z_m)I_1 - (Z_1 + Z_2 + Z_5)I_2 = 0.$$

The equations in matrix form is

$$\begin{bmatrix} (Z_1 + Z_3) & -(Z_m + Z_1) \\ (Z_1 + Z_m) & -(Z_1 + Z_2 + Z_5) \end{bmatrix}\begin{bmatrix} I_1 \\ I_2 \end{bmatrix} = \begin{bmatrix} \left(\frac{1}{2} - j\frac{\sqrt{3}}{2}\right) \\ 0 \end{bmatrix}.$$

With

$$Z_1 = -j10, \quad Z_2 = j1, \quad Z_3 = j4, \quad Z_5 = 10, \quad Z_m = j1$$

$$\begin{bmatrix} I_1 \\ I_2 \end{bmatrix} = \begin{bmatrix} 0.0000 - j6.0000 & 0.0000 + j9.0000 \\ 0.0000 - j9.0000 & -10.0000 + j9.0000 \end{bmatrix}^{-1} \begin{bmatrix} \left(\frac{1}{2} - j\frac{\sqrt{3}}{2}\right) \\ 0 \end{bmatrix}$$

$$= \begin{bmatrix} 0.1871 + j0.0825 & -0.1247 + j0.0561 \\ 0.1247 - j0.0561 & -0.0832 + j0.0374 \end{bmatrix}\begin{bmatrix} \left(\frac{1}{2} - j\frac{\sqrt{3}}{2}\right) \\ 0 \end{bmatrix}$$

$$= \begin{bmatrix} 0.1650 - j0.1208 \\ 0.0138 - j0.1361 \end{bmatrix}$$

$$\{I_1 - I_2 = I_{Z_1} = 0.1512 + j0.0153\}.$$

Nodal Analysis

We have to express the current variables in terms of voltage variables and we get

$$I_2 = \frac{(V_3 - V_2)}{Z_5}, \quad I_1 = \frac{(V_1 - V_2)}{Z_1} + \frac{(V_3 - V_2)}{Z_5}.$$

The voltage source is

$$V_1 = 1\angle\frac{-\pi}{3} = \left(\frac{1}{2} - j\frac{\sqrt{3}}{2}\right).$$

The equilibrium equations are

$$\frac{(V_2 - V_1)}{Z_1} + \frac{(V_2 - V_3)}{Z_5} + \frac{(V_2 + jI_2)}{Z_3} = 0$$

$$\frac{(V_3 - V_1 - jI_1)}{Z_2} + \frac{(V_3 - V_2)}{Z_5} = 0.$$

The first equation is the application of KCL at the left middle node. The second equation is the application of KCL at the right middle node. Substituting for I_2 and I_1, we get

$$\frac{(V_2 - V_1)}{Z_1} + \frac{(V_2 - V_3)}{Z_5} + \frac{\left(V_2 + j\frac{(V_3-V_2)}{Z_5}\right)}{Z_3} = 0$$

$$\frac{\left(V_3 - V_1 - j\frac{(V_1-V_2)}{Z_1} - j\frac{(V_3-V_2)}{Z_5}\right)}{Z_2} + \frac{(V_3 - V_2)}{Z_5} = 0.$$

Simplifying, we get

$$(Z_3Z_5 - jZ_1 + Z_1Z_3 + Z_1Z_5)V_2 - (Z_1Z_3 - jZ_1)V_3 = Z_3Z_5V_1$$

$$\left(-Z_2 + j + \left(\frac{jZ_5}{Z_1}\right)\right)V_2 + (Z_5 + Z_2 - j)V_3 = \left(Z_5 + \left(j\frac{Z_5}{Z_1}\right)\right)V_1.$$

With

$$Z_1 = -j10, \quad Z_2 = j1, \quad Z_3 = j4, \quad Z_5 = 10, \quad Z_m = j1$$

$$\begin{bmatrix} 30.0000 - j60.0000 & -30.0000 + j0.0000 \\ -1.0000 + j0.0000 & 10.0000 + j0.0000 \end{bmatrix} \begin{bmatrix} V_2 \\ V_3 \end{bmatrix} = \begin{bmatrix} 34.6410 + j20.0000 \\ 4.5000 - j7.7942 \end{bmatrix}.$$

The determinant of the admittance matrix is $270 - j600$.
Solving the equilibrium equations, we get

$$\begin{bmatrix} V_2 \\ V_3 \end{bmatrix} = \begin{bmatrix} 30.0000 - j60.0000 & -30.0000 + j0.0000 \\ -1.0000 + j0.0000 & 10.0000 + j0.0000 \end{bmatrix}^{-1} \begin{bmatrix} 34.6410 + j20.0000 \\ 4.5000 - j7.7942 \end{bmatrix}$$

$$= \begin{bmatrix} 0.0062 + j0.0139 & 0.0187 + j0.0416 \\ 0.0006 + j0.0014 & 0.1019 + j0.0042 \end{bmatrix} \begin{bmatrix} 34.6410 + j20.0000 \\ 4.5000 - j7.7942 \end{bmatrix}$$

$$= \begin{bmatrix} 0.3471 + j0.6461 \\ 0.4847 - j0.7148 \end{bmatrix}.$$

Example 5.5 Analyze the circuit, with mutually coupled inductors, shown in Fig. 5.15a. The coupling coefficient is 0.5.

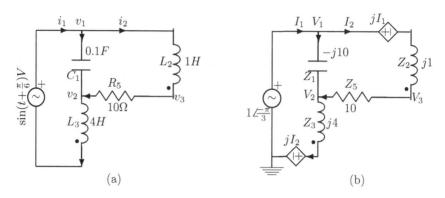

Fig. 5.15 (a) A circuit with mutually coupled inductors; (b) an equivalent frequency-domain circuit

Solution An equivalent frequency-domain circuit is shown in Fig. 5.15b.

Mesh Analysis
The equilibrium equations are

$$Z_1(I_1 - I_2) + Z_3 I_1 + Z_m I_2 = 1\angle \frac{-\pi}{3}$$

$$Z_1(I_1 - I_2) - Z_5 I_2 - I_2 Z_2 - Z_m I_1 = 0.$$

Simplifying, we get

$$(Z_1 + Z_3)I_1 + (Z_m - Z_1)I_2 = \left(\frac{1}{2} - j\frac{\sqrt{3}}{2}\right)$$

$$(Z_1 - Z_m)I_1 - (Z_1 + Z_2 + Z_5)I_2 = 0.$$

The equations in matrix form is

$$\begin{bmatrix} (Z_1 + Z_3) & (Z_m - Z_1) \\ (Z_1 - Z_m) & -(Z_1 + Z_2 + Z_5) \end{bmatrix} \begin{bmatrix} I_1 \\ I_2 \end{bmatrix} = \begin{bmatrix} \left(\frac{1}{2} - j\frac{\sqrt{3}}{2}\right) \\ 0 \end{bmatrix}.$$

With

$$Z_1 = -j10, \quad Z_2 = j1, \quad Z_3 = j4, \quad Z_5 = 10, \quad Z_m = j1$$

$$\begin{bmatrix} I_1 \\ I_2 \end{bmatrix} = \begin{bmatrix} 0.0000 - j6.0000 & 0.0000 + j11.0000 \\ 0.0000 - j11.0000 & -10.0000 + j9.0000 \end{bmatrix}^{-1} \begin{bmatrix} \left(\frac{1}{2} - j\frac{\sqrt{3}}{2}\right) \\ 0 \end{bmatrix}$$

$$= \begin{bmatrix} 0.1496 - j0.0004 & -0.0816 + j0.0911 \\ 0.0816 - j0.0911 & -0.0445 + j0.0497 \end{bmatrix} \begin{bmatrix} \left(\frac{1}{2} - j\frac{\sqrt{3}}{2}\right) \\ 0 \end{bmatrix}$$

$$= \begin{bmatrix} 0.0745 - j0.1297 \\ -0.0381 - j0.1162 \end{bmatrix}$$

$$\{I_1 - I_2 = I_{Z_1} = 0.1126 - j0.0135\}.$$

Nodal Analysis

We have to express the current variables in terms of voltage variables and we get

$$I_2 = \frac{(V_3 - V_2)}{Z_5}, \quad I_1 = \frac{(V_1 - V_2)}{Z_1} + \frac{(V_3 - V_2)}{Z_5}.$$

The voltage source is

$$V_1 = 1\angle\frac{-\pi}{3} = \left(\frac{1}{2} - j\frac{\sqrt{3}}{2}\right).$$

The equilibrium equations are

$$\frac{(V_2 - V_1)}{Z_1} + \frac{(V_2 - V_3)}{Z_5} + \frac{(V_2 - jI_2)}{Z_3} = 0$$

$$\frac{(V_3 - V_1 + jI_1)}{Z_2} + \frac{(V_3 - V_2)}{Z_5} = 0.$$

The first equation is the application of KCL at the left middle node. The second equation is the application of KCL at the right middle node. Substituting for I_2 and I_1, we get

$$\frac{(V_2 - V_1)}{Z_1} + \frac{(V_2 - V_3)}{Z_5} + \frac{\left(V_2 - j\frac{(V_3 - V_2)}{Z_5}\right)}{Z_3} = 0$$

$$\frac{\left(V_3 - V_1 + j\frac{(V_1 - V_2)}{Z_1} + j\frac{(V_3 - V_2)}{Z_5}\right)}{Z_2} + \frac{(V_3 - V_2)}{Z_5} = 0.$$

Simplifying, we get

$$(Z_3 Z_5 + jZ_1 + Z_1 Z_3 + Z_1 Z_5)V_2 - (Z_1 Z_3 + jZ_1)V_3 = Z_3 Z_5 V_1$$

$$\left(-Z_2 - j - \left(\frac{jZ_5}{Z_1}\right)\right)V_2 + (Z_5 + Z_2 + j)V_3 = \left(Z_5 - \left(j\frac{Z_5}{Z_1}\right)\right)V_1.$$

With

$$Z_1 = -j10, \quad Z_2 = j1, \quad Z_3 = j4, \quad Z_5 = 10, \quad Z_m = j1$$

$$\begin{bmatrix} 50.0000 - j60.0000 & -50.0000 + j0.0000 \\ 1.0000 - j2.0000 & 10.0000 + j2.0000 \end{bmatrix} \begin{bmatrix} V_2 \\ V_3 \end{bmatrix} = \begin{bmatrix} 34.6410 + j20.0000 \\ 5.5000 - j9.5263 \end{bmatrix}.$$

The determinant of the admittance matrix is $670 - j600$.

Solving the equilibrium equations, we get

$$\begin{bmatrix} V_2 \\ V_3 \end{bmatrix} = \begin{bmatrix} 50.0000 - j60.0000 & -50.0000 + j0.0000 \\ 1.0000 - j2.0000 & 10.0000 + j2.0000 \end{bmatrix}^{-1} \begin{bmatrix} 34.6410 + j20.0000 \\ 5.5000 - j9.5263 \end{bmatrix}$$

$$= \begin{bmatrix} 0.0068 + j0.0091 & 0.0414 + j0.0371 \\ -0.0023 + j0.0009 & 0.0859 - j0.0126 \end{bmatrix} \begin{bmatrix} 34.6410 + j20.0000 \\ 5.5000 - j9.5263 \end{bmatrix}$$

$$= \begin{bmatrix} 0.6351 + j0.2598 \\ 0.2541 - j0.9024 \end{bmatrix}.$$

5.4   Application

Two of the most commonly used devices in electrical systems are transformers and induction motors. Induction motors are most often used to convert electrical power to mechanical power. Currents in the secondary winding of the rotor are induced by the magnetic field of the primary winding of the stator.

5.4.1   Transformers

Transformers are used for changing voltage levels, isolating circuits and impedance matching. It transfers energy through coupled magnetic fields from one circuit to another. The operation of transformers is based on mutual inductance. The transformer consists of a core, which makes a path for the coupling of the magnetic field. Typically, iron is used to make the core and it improves the coupling coefficient. The primary winding on the core receives the energy from a AC source. The secondary winding receives the energy from the primary winding and delivers it to the load. The transformer has an enclosure to protect it due to mechanical damage, dirt, and moisture. Transformers provide electrical isolation, since there is no electrical connection between the two windings. Transformers provide DC isolation, since it does not pass DC.

Ideal Transformers
In introducing practical devices, their ideal form is usually first presented. It is easier to analyze and they set an ideal bound in terms of performance for their practical approximations. For example, the ideal filter, which is practically unrealizable, is introduced first in the study of filters. In practice, ideal devices are approximated to satisfy the required specifications. Similarly, the ideal transformer is presented here as an example of the applications of mutual inductance. This implies that coils making the inductances have zero resistance and infinite inductance.

Consider the frequency-domain relations for two coupled inductors L_1 and L_2, presented earlier.

$$V_1 = j\omega L_1 I_1 + j\omega M I_2 \quad \text{and} \quad V_2 = j\omega L_2 I_2 + j\omega M I_1.$$

From the first equation, we get

$$I_1 = \frac{V_1 - j\omega M I_2}{j\omega L_1}.$$

Substituting for I_1 in the second equation, we get

$$V_2 = j\omega L_2 I_2 + \frac{M V_1}{L_1} - \frac{j\omega M^2 I_2}{L_1}.$$

With perfect coupling, $k = 1$ and $M = \sqrt{L_1 L_2}$. Therefore,

$$V_2 = j\omega L_2 I_2 + \frac{\sqrt{L_1 L_2} V_1}{L_1} - \frac{j\omega L_1 L_2 I_2}{L_1} = \sqrt{\frac{L_2}{L_1}} V_1 = n V_1.$$

As the values of the inductances tend to infinity, n, called the turns ratio, remains the same. The inductance L of the windings is proportional to the square of the number of turns N_2 and N_1 of the wire in the windings and, therefore,

$$\frac{V_2}{V_1} = \frac{N_2}{N_1} = n.$$

Further, the input and output powers are equal in an ideal transformer, since there are no losses. That is,

$$V_1 I_1 = V_2 I_2.$$

Consequently,

$$\frac{I_2}{I_1} = \frac{N_1}{N_2} = \frac{1}{n}.$$

For an isolation transformer, $n = 1$. With $n > 1$, the transformer is called a step-up transformer, since $V_2 > V_1$. For a step-down transformer $n < 1$, since $V_2 < V_1$.

5.5   Summary

- Circuit elements can also be connected through magnetic fields, called magnetically coupled.
- The induction of a voltage in another inductance by the current flowing in an inductance is mutual induction.
- A voltage

$$v(t) = L_{12}\frac{dx(t)}{dt}$$

  is also induced in the inductor, which is affected by the magnetic field of another inductance due to current $x(t)$. The mutually induced voltage may be positive or negative depending upon the relative directions of the induced voltages. The value L_{12} is called the mutual inductance between the two inductances and also measured in henries.
- In the determination of the mutual inductances, only a pair of coils is considered at a time, assuming that no other induced currents in the other inductances (by open-circuiting others).
- The coupling coefficient is defined as

$$k = \frac{|L_{12}|}{\sqrt{L_{11}L_{22}}},$$

  the magnitude of which must be less than 1 for practical inductances. The mutual inductance increases or decreases the stored energy of the two isolated inductors.
- Two magnetically coupled inductors, with self-inductances L_1 and L_2 and currents I_1 and I_2 through them, respectively, and mutual inductance M, can be replaced, in the frequency-domain, by two isolated inductors, in series with current-controlled voltage sources.

$$V_1 = j\omega L_1 I_1 \pm j\omega M I_2 \quad \text{and} \quad V_2 = j\omega L_2 I_2 \pm j\omega M I_1$$

- A magnetically coupled circuit can also be replaced by equivalent T or Π circuits with no magnetic coupling.
- The average reactive power in the circuit is the difference between the average energy stored in the magnetic and electric fields multiplied by 2ω, where ω is the frequency of sinusoidal excitation.
- Two of the most commonly used devices in electrical systems are transformers and induction motors. Induction motors are most often used to convert electrical power to mechanical power. Currents in the secondary winding of the rotor are induced by the magnetic field of the primary winding of the stator.
- Transformers are used for changing voltage levels, isolating circuits and impedance matching. It transfers energy through coupled magnetic fields from one circuit to another.

Exercises

5.1 Analyze the circuit. Find the average energy consumed and stored in the circuit. Relate the reactive energy of the circuit to the energy stored in the magnetic fields. Find the energy stored in the magnetic fields at $t = 1$ s. Verify the results using two equivalent circuits.

    5.1.1 The circuit is shown in Fig. 5.16.

    * 5.1.2 The circuit is shown in Fig. 5.17.

5.2 Analyze the circuit. Find the average energy consumed and stored in the circuit. Relate the reactive energy of the circuit to the energy stored in the magnetic and electric fields. Find the energy stored in the magnetic and electric fields at $t = 1$ s. Verify the results using two equivalent circuits.

    5.2.1 The circuit is shown in Fig. 5.18.

    * 5.2.2 The circuit is shown in Fig. 5.19.

Fig. 5.16 Circuit for Exercise 5.1.1

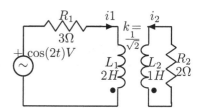

Fig. 5.17 Circuit for Exercise 5.1.2

Fig. 5.18 Circuit for Exercise 5.2.1

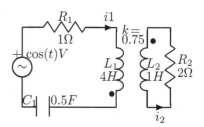

Fig. 5.19 Circuit for Exercise 5.2.2

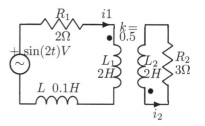

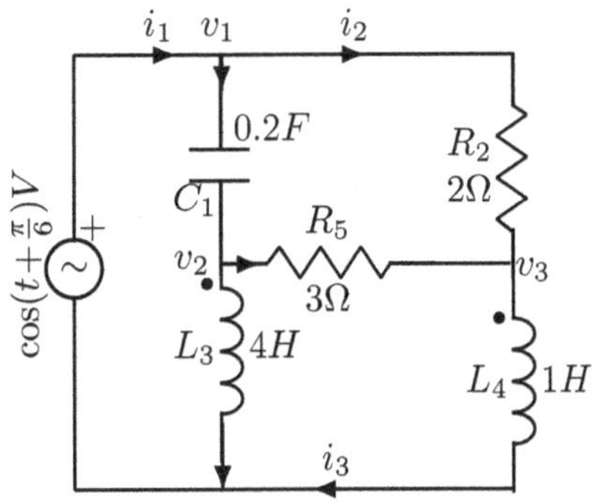

Fig. 5.20 Circuit for
Exercise 5.3.1

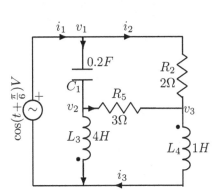

Fig. 5.21 Circuit for
Exercise 5.3.2

5.3 Analyze the circuit. Find the average energy consumed and stored in the circuit. Relate the reactive energy of the circuit to the energy stored in the magnetic and electric fields. Find the energy stored in the magnetic and electric fields at $t = 1$ s. Verify the results using both the mesh and loop methods of analysis.

    5.3.1 The circuit is shown in Fig. 5.20.

    ***5.3.2** The circuit is shown in Fig. 5.21.

Three-Phase Circuits

6

6.1 Three-Phase Voltages

The sinusoidal waveform, at a given frequency, is characterized by its amplitude and phase. In a single-phase source we studied so far, it is assumed that a sinusoidal wave is the input. However, keeping the amplitude the same, a source with three sinusoidal waveforms differing in equal amount of phase turns out to be better for electric-power generation, transmission and distribution systems, and in various other practical applications. The resulting circuits exhibit symmetry both with respect to geometrical structure and element values. Circuits with three sources differing in phase by $120°$ and of the same amplitude, called three-phase circuits, are mostly used in electrical power systems. The two major advantages of three-phase systems are reduced losses in transmission and the net instantaneous power is uniform, not pulsating. A single-phase source can be obtained just by taking from one of the three sources. The three sinusoidal voltage sources, V_a, V_b, and V_c, forming the three-phase source are related as

$$V_a = V_c \angle \pm \frac{2\pi}{3}, \quad V_b = V_a \angle \pm \frac{2\pi}{3}, \quad V_c = V_b \angle \pm \frac{2\pi}{3}.$$

If the algebraic signs of the phase displacements are consistently negative,

$$V_a = V_c \angle - \frac{2\pi}{3}, \quad V_b = V_a \angle - \frac{2\pi}{3}, \quad V_c = V_b \angle - \frac{2\pi}{3}$$

the voltages form a positive or (abc) sequence as shown in Fig. 6.1a. That is, V_b lags V_a by $2\pi/3$ radians. Further, it is normally assumed that $\angle V_a = 0$. A set of example waveforms are

$$v_a(t) = 1 \sin\left(\frac{2\pi}{3}t\right), \quad v_b(t) = 1 \sin\left(\frac{2\pi}{3}t - \frac{2\pi}{3}\right), \quad v_c(t) = 1 \sin\left(\frac{2\pi}{3}t - \frac{4\pi}{3}\right).$$

The peak value of the waveforms is 1 and the angular frequency is $2\pi/3$ radians. Therefore, the period is 3. The peak (zero) of $v_b(t)$ occurs after 1/3 of the cycle of that of $v_a(t)$. The peak of $v_c(t)$ occurs after 1/3 of the cycle of that of $v_b(t)$.

If the algebraic signs of the phase displacements are consistently positive,

$$V_a = V_c \angle + \frac{2\pi}{3}, \quad V_b = V_a \angle + \frac{2\pi}{3}, \quad V_c = V_b \angle + \frac{2\pi}{3}$$

© Springer Nature Switzerland AG 2020
D. Sundararajan, *Introductory Circuit Theory*,
https://doi.org/10.1007/978-3-030-31985-4_6

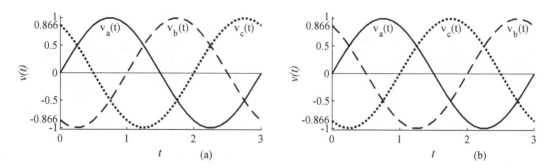

Fig. 6.1 (a) Three-phase positive sequence voltages; (b) three-phase negative sequence voltages

the voltages form a negative or (acb) sequence as shown in Fig. 6.1b. A set of example waveforms are

$$v_a(t) = 1 \sin\left(\frac{2\pi}{3}t\right), \quad v_b(t) = 1 \sin\left(\frac{2\pi}{3}t + \frac{2\pi}{3}\right), \quad v_c(t) = 1 \sin\left(\frac{2\pi}{3}t + \frac{4\pi}{3}\right).$$

The peak (zero) of $v_c(t)$ occurs after 1/3 of the cycle of that of $v_a(t)$. The peak of $v_b(t)$ occurs after 1/3 of the cycle of that of $v_c(t)$. One effect of applying the two sequences to a motor results in the change of direction of the rotation of the motor.

Given $v_c(t) = 1 \cos\left(\frac{\pi}{8}t + \frac{\pi}{4}\right)$, find $v_a(t)$ and $v_b(t)$ for positive and negative sequence three-phase voltages.

Positive Sequence
Since

$$V_a = V_c\angle -\frac{2\pi}{3}, \quad V_b = V_a\angle -\frac{2\pi}{3}, \quad V_c = V_b\angle -\frac{2\pi}{3},$$

the voltages are

$$v_a(t) = 1\cos\left(\frac{\pi}{8}t + \frac{\pi}{4} + \frac{4\pi}{3}\right), \quad v_b(t) = 1\cos\left(\frac{\pi}{8}t + \frac{\pi}{4} + \frac{2\pi}{3}\right), \quad v_c(t) = 1\cos\left(\frac{\pi}{8}t + \frac{\pi}{4}\right)$$

Negative Sequence
Since

$$V_a = V_c\angle +\frac{2\pi}{3}, \quad V_b = V_a\angle +\frac{2\pi}{3}, \quad V_c = V_b\angle +\frac{2\pi}{3},$$

the voltages are

$$v_a(t) = 1\cos\left(\frac{\pi}{8}t + \frac{\pi}{4} - \frac{4\pi}{3}\right), \quad v_b(t) = 1\cos\left(\frac{\pi}{8}t + \frac{\pi}{4} - \frac{2\pi}{3}\right), \quad v_c(t) = 1\cos\left(\frac{\pi}{8}t + \frac{\pi}{4}\right)$$

The sum of the set of such voltages or currents is zero. For example,

$$1\sin\left(\frac{2\pi}{3}t\right) + 1\sin\left(\frac{2\pi}{3}t + \frac{2\pi}{3}\right) + 1\sin\left(\frac{2\pi}{3}t + \frac{4\pi}{3}\right)$$

$$= (1 - 0.5 - 0.5)\sin\left(\frac{2\pi}{3}t\right) + \left(0 + \frac{\sqrt{3}}{2} - \frac{\sqrt{3}}{2}\right)\cos\left(\frac{2\pi}{3}t\right) = 0.$$

6.1.1 The Instantaneous Power

Consider the instantaneous voltage and current in phase a of a balanced three-phase system,

$$v_a(t) = V\cos(\omega t) \quad \text{and} \quad i_a(t) = I\cos(\omega t + \theta).$$

The instantaneous power $p_a(t)$, which is the product of the voltage and current, is

$$p_a(t) = VI\cos(\omega t)\cos(\omega t + \theta)$$

$$= \frac{VI}{2}(\cos(2\omega t + \theta) + \cos(\theta)).$$

Depending on the sequence of the phases (positive or negative), the voltages and currents in the other two phases have additional phases of $\mp\frac{2\pi}{3}$ and $\mp\frac{4\pi}{3}$, respectively. Therefore, the total instantaneous power p in the system is

$$p(t) = p_a(t) + p_b(t) + p_c(t) = \frac{VI}{2}\left(\cos(2\omega t + \theta) + 3\cos(\theta) + \cos\left(2\omega t + \theta \mp \frac{4\pi}{3}\right) + \cos\left(2\omega t + \theta \mp \frac{8\pi}{3}\right)\right).$$

Since the terms, except $3\cos(\theta)$, add up to zero, the total instantaneous power in the system is

$$p(t) = 3\frac{VI}{2}\cos(\theta)\ W.$$

As it is constant with respect to time, power delivery is smooth.

In this chapter, it is assumed that the three-phase circuit is balanced. In an unbalanced circuit, the amplitudes of the source voltages are not equal and the phases differ by unequal angles or the load impedances are unequal. Unbalanced circuits can be decomposed into three balanced circuits by the symmetrical components method of analysis using complex linear transformation.

6.2 The Three-Phase Balanced $Y - Y$ Circuit

A balanced 3-phase $Y - Y$ circuit consists of three identical and symmetrically associated circuits with voltages of equal amplitude and phase differences equal to $2\pi/3$ radians. The points, called neutral, are at the same potential, since $I_a + I_b + I_c = 0$. Therefore, the current in the neutral line I_n is zero in a balanced system. However, the neutral line is usually provided in the system to reduce the extent of system unbalance in case of an accidental unbalance occurring in the system.

Under balanced conditions, we can compute the current in a single phase considering it as a single-phase system. Then, the other currents in the other phases can be obtained by adding suitable phases to the result. It is simpler to analyze Δ-connected circuits by converting them to equivalent Y-connected circuits, using the transformation formulas presented in Chap. 2. These transformations are equally applicable to the analysis of circuits with impedances.

Let the phase voltages, the voltages between the lines and the neutral, be

$$V_{an} = V_p\angle 0, \quad V_{bn} = V_p\angle -\frac{2\pi}{3}, \quad V_{cn} = V_p\angle\frac{2\pi}{3}.$$

Fig. 6.2 Balanced
three-phase $Y - Y$ circuit

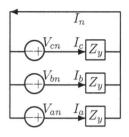

The line voltages, the voltage between the lines, can be found from the phase voltages.

$$V_{ab} = V_{an} - V_{bn} = V_p \left(1 + \frac{1}{2} + j\frac{\sqrt{3}}{2} \right) = V_p\sqrt{3} \left(\frac{\sqrt{3}}{2} + j\frac{1}{2} \right) = \sqrt{3}V_p \angle 30°$$

(6.1)

$$V_{bc} = V_{bn} - V_{cn} = \sqrt{3}V_p \angle - 90°$$ (6.2)

$$V_{ca} = V_{cn} - V_{an} = \sqrt{3}V_p \angle - 210°.$$ (6.3)

The last two expressions are obtained using the phase difference between the phases. Alternatively,

$$V_{bc} = V_{ab} \angle - 120°, \qquad V_{ca} = V_{ab} \angle 120°.$$

Therefore, the magnitude of the line voltage, V_L, in a Y-connected source is

$$V_L = \sqrt{3}V_p.$$

All the line voltages lead their corresponding phase voltages by 30°.

In balanced systems, only one phase of a $Y - Y$ circuit needs to be analyzed. For example,

$$I_a = \frac{V_{an}}{Z_Y}.$$

The other phase currents are deduced from this result by adding suitable phase angles. That is,

$$I_b = I_a \angle - 120°, \qquad I_c = I_a \angle 120°.$$

Example 6.1 Analyze the balanced three-phase $Y - Y$ circuit shown in Fig. 6.2. Let the phase voltages be

$$V_{an} = 1\angle 0, \quad V_{bn} = 1\angle - \frac{2\pi}{3}, \quad V_{cn} = 1\angle\frac{2\pi}{3}$$

at frequency $\omega = 1$ rad/s. The impedance is $Z_Y = 1 + j2$. Determine the currents I_a, I_b, I_c. Find also the power consumed and the pf of the circuit. Find the value of the capacitors required to improve the pf to 0.9.

Solution The current $I_a = 1/(1 + j2) = 0.2 - j0.4$ and

$$I_{bn} = I_a \angle - \frac{2\pi}{3} = -0.4464 + j0.0268, \quad I_{cn} = I_a \angle\frac{2\pi}{3} = 0.2464 + j0.3732$$

$$I_a + I_b + I_c = 0.$$

The power stored and consumed by the circuit in phase a is

$$0.5 I_a I_a^* Z_Y = S = P_{av} + j Q_{av} = 0.1 + j0.2 \text{ va}.$$

The power supplied by the source $V_{an} I_a^*/2$ is also the same. The total power consumed by the circuit is three times that of a phase, $3(0.1 + j0.2) = 0.3 + j0.6$. The pf is

$$pf = \frac{\text{real}(Z_Y)}{\text{abs}(Z_Y)} = \frac{\text{real}(S)}{\text{abs}(S)} = 0.4472.$$

The desired power factor is 0.9. The corresponding apparent power is

$$\frac{P_{av}}{0.9} = 0.1111.$$

The corresponding reactive power is

$$0.1111 \sin(\cos^{-1}(0.9)) = 0.0484.$$

The difference between the reactive powers is $0.2 - 0.0484 = 0.1516$. Now, with $\omega = 1$ and $|V| = 1$,

$$C = 2\frac{0.1516}{(1)(1^2)} = 0.3031 \text{ F}.$$

The current through the capacitor is

$$j V_{an}(0.3031) = j0.3031.$$

The new current supplied by the source is

$$I_a' = (0.2 - j0.4) + j0.3031 = 0.2000 - j0.0969.$$

The power supplied by the source $V_{an} I_a'^*/2$ is $0.1000 + j0.0484$ and the pf is 0.9. ∎

In each phase, a capacitor of value C is connected in parallel with the load impedance for power-factor improvement.

6.3 The Three-Phase Balanced $Y - \Delta$ Circuit

$\Delta - Y$ transformation derived in Chap. 2 is

$$Z_a = \frac{Z_{ab} Z_{ac}}{Z_{ac} + Z_{ab} + Z_{bc}}$$

$$Z_b = \frac{Z_{ba} Z_{bc}}{Z_{ac} + Z_{ab} + Z_{bc}}$$

$$Z_c = \frac{Z_{ca} Z_{cb}}{Z_{ac} + Z_{ab} + Z_{bc}}.$$

Fig. 6.3 (a) Balanced three-phase $Y - \Delta$ circuit; (b) its equivalent $Y - Y$ circuit

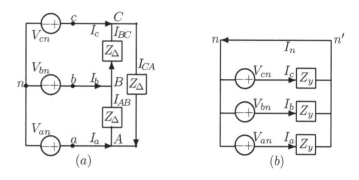

In a balanced circuit, the transformation reduces to

$$Z_Y = \frac{Z_\Delta}{3}.$$

For supplying the same reactive power, the Y-connected load circuit requires three times the value of the capacitor required for an equivalent Δ-connected load circuit.

Example 6.2 Analyze the balanced three-phase $Y - \Delta$ circuit shown in Fig. 6.3a. Let the phase voltages be

$$V_{an} = 1\angle 0, \quad V_{bn} = 1\angle -\frac{2\pi}{3}, \quad V_{cn} = 1\angle \frac{2\pi}{3}$$

at frequency $\omega = 1$ rad/s. The impedance is $Z_\Delta = 3 + j6$. Determine the currents I_a, I_b, I_c. Find also the power consumed and the pf of the circuit. Find the value of the capacitors required to improve the pf to 0.9.

Solution The equivalent $Y - Y$ circuit shown in Fig. 6.3b. The impedance is $Z_Y = Z_\Delta/3 = 1 + j2$. Now, the problem reduces to Example 6.1. The current $I_a = 1/(1 + j2) = 0.2 - j0.4$ and

$$I_{bn} = I_a \angle -\frac{2\pi}{3} = -0.4464 + j0.0268, \quad I_{cn} = I_a \angle \frac{2\pi}{3} = 0.2464 + j0.3732$$

$$I_a + I_b + I_c = 0.$$

The power stored and consumed by the circuit in phase a is

$$0.5 I_a I_a^* Z_Y = S = P_{av} + j Q_{av} = 0.1 + j0.2 \text{ va.}$$

The power supplied by the source $V_{an} I_a^*/2$ is also the same. The total power consumed by the circuit is three times that of a phase, $3(0.1 + j0.2) = 0.3 + j0.6$. The pf is

$$pf = \frac{\text{real}(Z_Y)}{\text{abs}(Z_Y)} = \frac{\text{real}(S)}{\text{abs}(S)} = 0.4472.$$

The desired power factor is 0.9. The corresponding apparent power is

$$\frac{P_{av}}{0.9} = 0.1111.$$

The corresponding reactive power is

$$0.1111 \sin(\cos^{-1}(0.9)) = 0.0484.$$

The difference between the reactive powers is $0.2 - 0.0484 = 0.1516$. Now, with $\omega = 1$ and $|V| = 1$,

$$C = 2\frac{0.1516}{(1)(1^2)} = 0.3031\,\text{F}.$$

The current through the capacitor is

$$j V_{an}(0.3031) = j0.3031.$$

The new current supplied by the source is

$$I'_a = (0.2 - j0.4) + j0.3031 = 0.2000 - j0.0969.$$

The power supplied by the source $V_{an} I'^*_a/2$ is $0.1000 + j0.0484$ and the pf is 0.9. A capacitor of value $0.3031/3\,\text{F}$ is to be connected across the lines of the Δ-connected load. That is $C = 0.1010\,\text{F}$ is connected in parallel with each Z_Δ. Then, the effective load impedance is

$$Z_\Delta \parallel Z_C = (3 + j6) \parallel (-j9.8966) = 12.1500 + j5.8845.$$

The improved pf is

$$pf = \frac{12.15}{|12.1500 + j5.8845|} = \frac{12.15}{13.5} = 0.9.$$

The line voltages are

$$V_{AB} = \sqrt{3}V_{an}\angle\frac{\pi}{6} = 1.5 + j0.866, \quad V_{BC} = V_{AB}\angle-\frac{2\pi}{3} = -j1.7321, \quad V_{CA} = V_{AB}\angle\frac{2\pi}{3} = -1.5 + j0.866$$

The currents through the impedances of the Δ-connected load are obtained by dividing the line voltages by Z_Δ.

$$I_{AB} = \frac{V_{AB}}{Z_\Delta} = 0.2155 - j0.1423, \quad I_{BC} = \frac{V_{BC}}{Z_\Delta} = -0.2309 - j0.1155, \quad I_{CA} = \frac{V_{CA}}{Z_\Delta} = 0.0155 + j0.2577$$

The power consumed can also be computed from these values to get the same values obtained earlier. For example,

$$0.5V_{AB}I^*_{AB} = 0.5(1.5 + j0.866)(0.2155 + j0.1423) = 0.1 + j0.2.$$

■

6.4   The Three-Phase Balanced $\Delta - Y$ Circuit

Let the Δ source be

$$V_{ab} = V_L \angle 0°, \quad V_{bc} \angle -120°, \quad V_{ca} \angle 120°.$$

Since, from Eq. (6.1), with V_p being the phase voltage

$$V_{ab} = \sqrt{3} V_p \angle 30° \quad \text{and} \quad V_p = \frac{V_{ab}}{\sqrt{3}} \angle -30°.$$

Then, an equivalent Y source, with V_L being the line voltage, is

$$V_{an} = \frac{V_L}{\sqrt{3}} \angle -30° = V_p \angle -30°, \quad V_{bn} = \frac{V_L}{\sqrt{3}} \angle -150° = V_p \angle -150°, \quad V_{cn} = \frac{V_L}{\sqrt{3}} \angle -270° = V_p \angle 90°$$

Therefore, in analyzing circuits with a Δ source, the source can be converted to an equivalent Y source. Then, the circuit can be analyzed using the Y source.

Example 6.3 Analyze the balanced three-phase $\Delta - Y$ circuit shown in Fig. 6.4a. Let the phase voltages be

$$V_{AB} = 1\sqrt{3} \angle \left(0 + \frac{\pi}{6}\right), \ V_{BC} = 1\sqrt{3} \angle \left(-\frac{2\pi}{3} + \frac{\pi}{6}\right), \ V_{CA} = 1\sqrt{3} \angle \left(\frac{2\pi}{3} + \frac{\pi}{6}\right)$$

at frequency $\omega = 1$ rad/s. The impedance is $Z_Y = 1 + j2$. Determine the currents I_a, I_b, I_c. Find also the power consumed and the pf of the circuit.

Solution The equivalent Y-connected source phase voltages are

$$V_{an} = 1\angle 0, \quad V_{bn} = 1\angle -\frac{2\pi}{3}, \quad V_{cn} = 1\angle \frac{2\pi}{3}.$$

Now, the problem reduces to Example 6.1. The equivalent $Y - Y$ circuit is shown in Fig. 6.4b. The current $I_a = 1/(1 + j2) = 0.2 - j0.4$ and

Fig. 6.4 (**a**) Balanced three-phase $\Delta - Y$ circuit; (**b**) its equivalent $Y - Y$ circuit

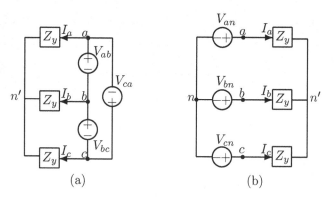

(a)   (b)

$$I_{bn} = I_a \angle -\frac{2\pi}{3} = -0.4464 + j0.0268, \quad I_{cn} = I_a \angle \frac{2\pi}{3} = 0.2464 + j0.3732$$

$$I_a + I_b + I_c = 0.$$

The power stored and consumed by the circuit in phase a is

$$0.5 I_a I_a^* Z_Y = S = P_{av} + j Q_{av} = 0.1 + j0.2 \text{ va}.$$

The power supplied by the source $V_{an} I_a^*/2$ is also the same. The total power consumed by the circuit is three times that of a phase, $3(0.1 + j0.2) = 0.3 + j0.6$. The pf is

$$pf = \frac{\text{real}(Z_Y)}{\text{abs}(Z_Y)} = \frac{\text{real}(S)}{\text{abs}(S)} = 0.4472.$$

∎

6.5 The Three-Phase Balanced Δ − Δ Circuit

Example 6.4 Analyze the balanced three-phase Δ − Δ circuit shown in Fig. 6.5a. Let the phase voltages be

$$V_{AB} = 1\sqrt{3} \angle \left(0 + \frac{\pi}{6}\right), \quad V_{BC} = 1\sqrt{3} \angle \left(-\frac{2\pi}{3} + \frac{\pi}{6}\right), \quad V_{CA} = 1\sqrt{3} \angle \left(\frac{2\pi}{3} + \frac{\pi}{6}\right)$$

at frequency $\omega = 1$ rad/s. The impedance is $Z_\Delta = 3 + j6$. Determine the currents I_a, I_b, I_c. Find also the power consumed and the pf of the circuit.

Solution The equivalent Y-connected source phase voltages are

$$V_{an} = 1\angle 0, \quad V_{bn} = 1\angle -\frac{2\pi}{3}, \quad V_{cn} = 1\angle \frac{2\pi}{3}.$$

The impedance is $Z_Y = Z_\Delta/3 = 1 + j2$. Now, the problem reduces to Example 6.1. The equivalent $Y − Y$ circuit is shown in Fig. 6.5b. The current $I_a = 1/(1 + j2) = 0.2 - j0.4$ and

Fig. 6.5 (**a**) Balanced three-phase Δ − Δ circuit; (**b**) its equivalent $Y − Y$ circuit

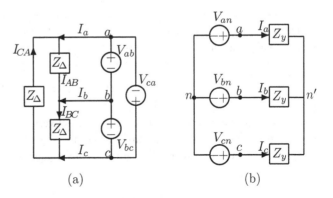

(a)   (b)

$$I_{bn} = I_a \angle -\frac{2\pi}{3} = -0.4464 + j0.0268, \quad I_{cn} = I_a \angle \frac{2\pi}{3} = 0.2464 + j0.3732$$

$$I_a + I_b + I_c = 0.$$

The power stored and consumed by the circuit in phase a is

$$0.5 I_a I_a^* Z_Y = S = P_{av} + j Q_{av} = 0.1 + j0.2 \text{ va}.$$

The power supplied by the source $V_{an} I_a^* / 2$ is also the same. The total power consumed by the circuit is three times that of a phase, $3(0.1 + j0.2) = 0.3 + j0.6$. The pf is

$$pf = \frac{\text{real}(Z_Y)}{\text{abs}(Z_Y)} = \frac{\text{real}(S)}{\text{abs}(S)} = 0.4472.$$

■

6.6   Application

The two major advantages of three-phase systems are that they supply uniform instantaneous power and provide efficient power transmission. The power loss in transmission of a certain amount of power by a three-phase system is one-fourth of that of a single-phase line.

The three-phase induction motor is the workhorse in practical applications for converting electric power to mechanical power. Such machines produce a steady torque.

6.7   Summary

- Systems with three sources differing in phase by 120° and of the same amplitude, called three-phase systems, are mostly used in electrical power systems.
- The two major advantages of three-phase systems are reduced losses in transmission and the net instantaneous power is uniform, not pulsating.
- A single-phase source can be obtained just by taking from one of the three sources.
- In the positive sequence, the peak (zero) of $v_b(t)$ occurs after 1/3 of the cycle of that of $v_a(t)$. The peak of $v_c(t)$ occurs after 1/3 of the cycle of that of $v_b(t)$.
- In the negative sequence, the peak (zero) of $v_c(t)$ occurs after 1/3 of the cycle of that of $v_a(t)$. The peak of $v_b(t)$ occurs after 1/3 of the cycle of that of $v_c(t)$.
- The sum of the set of three-phase voltages or currents is zero.
- The total power consumed by the circuit is three times that of a phase.
- The total instantaneous power in the system is

$$p(t) = 3\frac{VI}{2}\cos(\theta)\ W.$$

As it is constant with respect to time, power delivery is smooth.
- In an unbalanced system, the amplitudes of the source voltages are not equal and the phases differ by unequal angles or the load impedances are unequal.

- Unbalanced systems are decomposed into three balanced systems by the symmetrical components method of analysis using complex linear transformation.
- A balanced three-phase system consists of three identical and symmetrically associated circuits with voltages of equal amplitude and phase differences equal to $2\pi/3$ radians.
- Under balanced conditions, we can compute the current in a single phase of a Y-connected circuits, considering it as a single-phase system. Then, the other currents in the other phases can be obtained by adding suitable phases to the result.
- It is simpler to analyze Δ-connected circuits by converting them to equivalent Y-connected circuits, using transformation formulas.
- In each phase, a capacitor of value C is connected in parallel with the load impedance for power-factor improvement.
- For supplying the same reactive power, the Y-connected load circuit requires three times the value of the capacitor required for an equivalent Δ-connected load circuit.

Exercises

* 6.1.1 Given $v_a(t) = 2\cos\left(\frac{2\pi}{15}t\right)$, find $v_b(t)$ and $v_c(t)$ for positive and negative sequence three-phase voltages.

6.1.2 Given $v_b(t) = 3\cos\left(\frac{2\pi}{9}t - \frac{\pi}{3}\right)$, find $v_a(t)$ and $v_c(t)$ for positive and negative sequence three-phase voltages.

* 6.2.1 Analyze the balanced three-phase $Y - Y$ circuit. Let the phase voltages be

$$V_{an} = 2\angle 0, \quad V_{bn} = 2\angle -\frac{2\pi}{3}, \quad V_{cn} = 2\angle\frac{2\pi}{3}$$

at frequency $\omega = 2\,\text{rad/s}$. The impedance is $Z_Y = 1 + j3$. Determine the currents I_a, I_b, I_c. Find also the power consumed and the pf of the circuit. Find the value of the capacitors required to improve the pf to 0.8.

6.2.2 Analyze the balanced three-phase $Y - Y$ circuit. Let the phase voltages be

$$V_{an} = 3\angle 0, \quad V_{bn} = 3\angle -\frac{2\pi}{3}, \quad V_{cn} = 3\angle\frac{2\pi}{3}$$

at frequency $\omega = 2\,\text{rad/s}$. The impedance is $Z_Y = 1 + j4$. Determine the currents I_a, I_b, I_c. Find also the power consumed and the pf of the circuit. Find the value of the capacitors required to improve the pf to 0.7.

6.3.1 Analyze the balanced three-phase $Y - \Delta$ circuit. Let the phase voltages be

$$V_{an} = 2\angle 0, \quad V_{bn} = 2\angle -\frac{2\pi}{3}, \quad V_{cn} = 2\angle\frac{2\pi}{3}$$

at frequency $\omega = 2\,\text{rad/s}$. The impedance is $Z_\Delta = 3 + j9$. Determine the currents I_a, I_b, I_c. Find also the power consumed and the pf of the circuit. Find the value of the capacitors required to improve the pf to 0.8.

* 6.3.2 Analyze the balanced three-phase $Y - \Delta$ circuit. Let the phase voltages be

$$V_{an} = 3\angle 0, \quad V_{bn} = 3\angle -\frac{2\pi}{3}, \quad V_{cn} = 3\angle\frac{2\pi}{3}$$

at frequency $\omega = 2\,\text{rad/s}$. The impedance is $Z_\Delta = 3 + j12$. Determine the currents I_a, I_b, I_c. Find also the power consumed and the pf of the circuit. Find the value of the capacitors required to improve the pf to 0.7.

* **6.4.1** Analyze the balanced three-phase $\Delta - Y$ circuit. Let the phase voltages be

$$V_{AB} = 2\sqrt{3}\angle\left(0 + \frac{\pi}{6}\right), \quad V_{BC} = 2\sqrt{3}\angle\left(-\frac{2\pi}{3} + \frac{\pi}{6}\right), \quad V_{CA} = 2\sqrt{3}\angle\left(\frac{2\pi}{3} + \frac{\pi}{6}\right)$$

at frequency $\omega = 2\,\text{rad/s}$. The impedance is $Z_Y = 1 + j3$. Determine the currents I_a, I_b, I_c. Find also the power consumed and the pf of the circuit. Find the value of the capacitors required to improve the pf to 0.8.

6.4.2 Analyze the balanced three-phase $\Delta - Y$ circuit. Let the phase voltages be

$$V_{AB} = 3\sqrt{3}\angle\left(0 + \frac{\pi}{6}\right), \quad V_{BC} = 3\sqrt{3}\angle\left(-\frac{2\pi}{3} + \frac{\pi}{6}\right), \quad V_{CA} = 3\sqrt{3}\angle\left(\frac{2\pi}{3} + \frac{\pi}{6}\right)$$

at frequency $\omega = 2\,\text{rad/s}$. The impedance is $Z_Y = 1 + j4$. Determine the currents I_a, I_b, I_c. Find also the power consumed and the pf of the circuit. Find the value of the capacitors required to improve the pf to 0.7.

* **6.5.1** Analyze the balanced three-phase $\Delta - \Delta$ circuit. Let the phase voltages be

$$V_{AB} = 2\sqrt{3}\angle\left(0 + \frac{\pi}{6}\right), \quad V_{BC} = 2\sqrt{3}\angle\left(-\frac{2\pi}{3} + \frac{\pi}{6}\right), \quad V_{CA} = 2\sqrt{3}\angle\left(\frac{2\pi}{3} + \frac{\pi}{6}\right)$$

at frequency $\omega = 2\,\text{rad/s}$. The impedance is $Z_\Delta = 3 + j9$. Determine the currents I_a, I_b, I_c. Find also the power consumed and the pf of the circuit. Find the value of the capacitors required to improve the pf to 0.8.

6.5.2 Analyze the balanced three-phase $\Delta - \Delta$ circuit. Let the phase voltages be

$$V_{AB} = 3\sqrt{3}\angle\left(0 + \frac{\pi}{6}\right), \quad V_{BC} = 3\sqrt{3}\angle\left(-\frac{2\pi}{3} + \frac{\pi}{6}\right), \quad V_{CA} = 3\sqrt{3}\angle\left(\frac{2\pi}{3} + \frac{\pi}{6}\right)$$

at frequency $\omega = 2\,\text{rad/s}$. The impedance is $Z_\Delta = 3 + j12$. Determine the currents I_a, I_b, I_c. Find also the power consumed and the pf of the circuit. Find the value of the capacitors required to improve the pf to 0.7.

Two-Port Networks

7

A pair of terminals, which are external terminal points for currents to enter or leave a circuit, is called a port. Elements, such as a resistor with two terminals, are one-port circuit. A two-port circuit has two separate ports for input and output. Several currents flow through circuit elements in a circuit and, consequently, voltages appear at their terminals. But, not all of them are interest in applications. Modeling a circuit by a two-port circuit reduces the current variables to two. The two currents and the corresponding voltages are related by nodal and loop analysis. A two-port circuit may have dependent sources but no independent source. Although a circuit may have any number of ports, two-port models are widely used in the study of amplifier, communication and power system circuits. Depending on the choice of dependent and independent variables, one has to choose the most appropriate two-port model. This model is a generalization of Thévenin's theorem and Norton's theorem. Each port can be represented by Thévenin or Norton equivalent circuit. In Thévenin's theorem, a part of a circuit across a pair of terminals is replaced by a simple model. However, in modeling devices such as amplifiers and filters with input and output ports, a part of a circuit across pairs of terminals has to be replaced by a simple model.

7.1 Impedance Parameters

Figure 7.1 shows a two-port circuit with conventional voltage and current reference directions indicated. The top terminals are positive with respect to the lower terminals and currents enter the ports at the top terminals. The left port is the input port and the right is the output.

The relations between the voltage and current variables can be derived by nodal or loop analysis. With loop analysis, we get

$$Z_{11}I_1 + Z_{12}I_2 = V_1 \tag{7.1}$$

$$Z_{21}I_1 + Z_{22}I_2 = V_2, \tag{7.2}$$

where

$$Z_{11} = \frac{V_1}{I_1}\big|_{I_2=0} \quad \text{and} \quad Z_{22} = \frac{V_2}{I_2}\big|_{I_1=0}$$

are the driving-point impedances. Current zero implies an open-circuit at the corresponding port. That is, Z_{11} multiplied by the excitation current at terminal pair 1 equals the voltage across the terminal

© Springer Nature Switzerland AG 2020
D. Sundararajan, *Introductory Circuit Theory*,
https://doi.org/10.1007/978-3-030-31985-4_7

Fig. 7.1 A two-port circuit with conventional voltage and current reference directions indicated

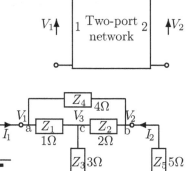

Fig. 7.2 A bridged-T circuit

pair 1 with $I_2 = 0$. Similarly, Z_{22} multiplied by the excitation current at terminal pair 2 equals the voltage across the terminal pair 2 with $I_1 = 0$.

$$Z_{12} = \frac{V_1}{I_2}\Big|_{I_1=0} \quad \text{and} \quad Z_{21} = \frac{V_2}{I_1}\Big|_{I_2=0}$$

are called transfer impedances. That is, Z_{12} multiplied by the excitation current at terminal pair 2 equals the voltage across the terminal pair 1 with $I_1 = 0$. Similarly, Z_{21} multiplied by the excitation current at terminal pair 1 equals the voltage across the terminal pair 2 with $I_2 = 0$. In matrix form, we get

$$\begin{bmatrix} Z_{11} & Z_{12} \\ Z_{21} & Z_{22} \end{bmatrix} \begin{bmatrix} I_1 \\ I_2 \end{bmatrix} = \begin{bmatrix} V_1 \\ V_2 \end{bmatrix}.$$

In this model, voltages V_1 and V_2 are expressed in terms of currents I_1 and I_2.

Consider the bridged-T circuit shown in Fig. 7.2. It is a T-circuit with the impedance Z_4 bridged across the top part. Let us find the voltages V_3 and V_2 by nodal analysis, with $V_1 = 1$. The equilibrium equations at the nodes are

$$\frac{V_3 - 1}{Z_1} + \frac{V_3}{Z_3} + \frac{V_3 - V_2}{Z_2} = 0$$

$$\frac{V_2}{Z_5} + \frac{V_2 - V_3}{Z_2} + \frac{V_2 - 1}{Z_4} = 0.$$

Simplifying, we get

$$V_3(Z_1 Z_2 + Z_1 Z_3 + Z_2 Z_3) - V_2 Z_1 Z_3 = Z_2 Z_3$$

$$-V_3 Z_4 Z_5 + V_2(Z_2 Z_4 + Z_4 Z_5 + Z_2 Z_5) = Z_2 Z_5.$$

With

$$Z_1 = 1, \quad Z_2 = 2, \quad Z_3 = 3, \quad Z_4 = 4, \quad Z_5 = 5$$

$$\begin{bmatrix} 11 & -3 \\ -20 & 38 \end{bmatrix} \begin{bmatrix} V_3 \\ V_2 \end{bmatrix} = \begin{bmatrix} 6 \\ 10 \end{bmatrix}.$$

The determinant of the admittance matrix is 358. Solving for V_3 and V_2, we get

$$\begin{bmatrix} V_3 \\ V_2 \end{bmatrix} = \begin{bmatrix} 11 & -3 \\ -20 & 38 \end{bmatrix}^{-1} \begin{bmatrix} 6 \\ 10 \end{bmatrix} = \begin{bmatrix} 0.1061 & 0.0084 \\ 0.0559 & 0.0307 \end{bmatrix} \begin{bmatrix} 6 \\ 10 \end{bmatrix} = \begin{bmatrix} 0.7207 \\ 0.6425 \end{bmatrix}.$$

Now, the terminal currents are

$$I_1 = \frac{V_3}{Z_3} + \frac{V_2}{Z_5} = 0.3687 \quad \text{and} \quad I_2 = -\frac{V_2}{Z_5} = -0.1285.$$

Let us find the impedance and admittance matrices of the two-port circuit and verify that the currents and voltages found by the nodal analysis satisfy the governing relations for the circuit.

Determining the Impedance Matrix

In all the models, the procedure to determine the parameter matrix is the same. An appropriate voltage or current is applied at one port, short-circuit or open-circuit on the other port depending on the parameter of interest, and use regular circuit analysis. Figure 7.3a shows the equivalent circuit in Fig. 7.2 with the Δ part of the bridged-T circuit, indicated by terminals $\{a, b, c\}$, replaced by an equivalent T-circuit and the source and load parts removed. The impedances are obtained, using the conversion formulas, as

$$\frac{Z_1 Z_4}{Z_1 + Z_2 + Z_4} = 0.5714$$

$$\frac{Z_2 Z_4}{Z_1 + Z_2 + Z_4} = 1.1429$$

$$\frac{Z_1 Z_2}{Z_1 + Z_2 + Z_4} = 0.2857.$$

By adding the series connected resistors with values 3 and 0.2857, the value of the bottom branch becomes 3.2857 (Fig. 7.3b). That is,

$$Z_1 = 0.5714 \, \Omega, \quad Z_2 = 1.1429 \, \Omega, \quad Z_3 = 3.2857 \, \Omega.$$

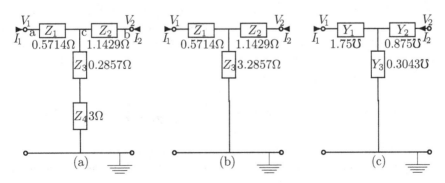

Fig. 7.3 (**a**) The Δ part of the bridged-T circuit, indicated by terminals $\{a, b, c\}$, replaced by an equivalent T-circuit; (**b**) simplified version of (**a**); (**c**) the admittance version

Fig. 7.4 Determination of
the impedance parameters
Z_{11} and Z_{21}

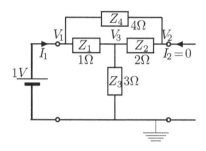

Now,

$$Z_{11} = \frac{V_1}{I_1}|_{I_2=0} = \frac{I_1(0.5714 + 3.2857)}{I_1} = 3.8571$$

$$Z_{12} = \frac{V_1}{I_2}|_{I_1=0} = \frac{I_2(3.2857)}{I_2} = 3.2857$$

$$Z_{21} = \frac{V_2}{I_1}|_{I_2=0} = \frac{I_1(3.2857)}{I_1} = 3.2857$$

$$Z_{22} = \frac{V_2}{I_2}|_{I_1=0} = \frac{I_1(1.1429 + 3.2857)}{I_1} = 4.4286.$$

The first subscript indicates the origin of the voltage and the second indicates the origin of the current. Z_{11} is the impedance, which multiplied by the current at the terminal pair 1, I_1, produces the open-circuit voltage at the terminal pair 11, V_1 with open-circuit condition at terminal pair 2. Z_{12} is the impedance, which multiplied by the current at the terminal pair 2, I_2, produces the open-circuit voltage at the terminal pair 1, V_1 with open-circuit condition at terminal pair 1. Z_{22} is the impedance, which multiplied by the current at the terminal pair 2, I_2, produces the open-circuit voltage at the terminal pair 2, V_2 with open-circuit condition at terminal pair 1. Z_{21} is the impedance, which multiplied by the current at the terminal pair 1, I_1, produces the open-circuit voltage at the terminal pair 2, V_2 with open-circuit condition at terminal pair 2. The impedance matrix is

$$Z = \begin{bmatrix} Z_{11} & Z_{12} \\ Z_{21} & Z_{22} \end{bmatrix} = \begin{bmatrix} Z_1 + Z_3 & Z_3 \\ Z_3 & Z_2 + Z_3 \end{bmatrix} = \begin{bmatrix} 3.8571 & 3.2857 \\ 3.2857 & 4.4286 \end{bmatrix}.$$

The determination of the two-port model parameters looks formidable, in the derivation using symbols. However, it just requires only basic circuit theory. Here, we make a much simplified presentation of deriving the Z parameters using circuits and numerical values. The same approach holds for other models. Consider the circuit, shown in Fig. 7.4, for the determination of the impedance parameter Z_{11}. The output port is open-circuited, $I_2 = 0$. $V_1 = 1$. We have to find I_1. The equivalent impedance of the circuit is

$$Z_{eq} = Z_3 + (Z_1 \parallel (Z_2 + Z_4)) = Z_3 + \frac{Z_1(Z_2 + Z_4)}{(Z_1 + Z_2 + Z_4)} = \frac{V_1}{I_1} = Z_{11}.$$

Substituting the numerical values, we get $Z_{11} = 3.8571$ as obtained earlier.

Consider the circuit, shown in Fig. 7.4, for the determination of the impedance parameter Z_{21}.

$$I_1 = \frac{V_1}{Z_{eq}}$$

Fig. 7.5 Determination of the impedance parameters Z_{22} and Z_{12}

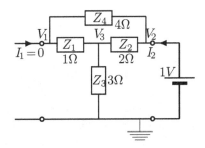

$$I_{z4} = I_1 \frac{Z_1}{(Z_1 + Z_2 + Z_4)}$$

$$V_2 = V_1 - I_{z4} Z_4 \text{ and } Z_{21} = \frac{V_2}{I_1} = Z_{21}.$$

Substituting the numerical values, we get $Z_{21} = 3.2857$ as obtained earlier.

Consider the circuit, shown in Fig. 7.5, for the determination of the impedance parameter Z_{22}. The input port is open-circuited, $I_1 = 0$. $V_2 = 1$. We have to find I_2. The equivalent impedance of the circuit is

$$Z_{eq} = Z_3 + Z_2 \parallel (Z_1 + Z_4) = Z_3 + \frac{Z_2(Z_1 + Z_4)}{(Z_1 + Z_2 + Z_4)} = \frac{V_2}{I_2} = Z_{22}$$

Substituting the numerical values, we get $Z_{22} = 4.4286$ as obtained earlier.

Consider the circuit, shown in Fig. 7.5, for the determination of the impedance parameter Z_{12}.

$$I_2 = \frac{V_2}{Z_{eq}}$$

$$I_{z4} = I_2 \frac{Z_2}{(Z_1 + Z_2 + Z_4)}$$

$$V_1 = V_2 - I_{z4} Z_4 \text{ and } Z_{12} = \frac{V_1}{I_2} = Z_{12}$$

Substituting the numerical values, we get $Z_{12} = 3.2857$ as obtained earlier.

The Impedance Matrix of Circuits Connected in Series

A large circuit is an interconnection of a set of smaller circuits. The interconnection may be series, parallel, or cascade. When they are in series, it is advantageous to characterize them by Z matrices, since the Z matrix of two circuits connected in series is the sum of the individual Z matrices. When they are in parallel, it is advantageous to characterize them by Y matrices, since the Y matrix of two circuits connected in parallel is the sum of the individual Y matrices.

The circuit in Fig. 7.3a can be considered as a series of two circuits with the top three resistors comprising the first part and the resistor with value $3\,\Omega$ being the second part. The input currents are the same and their voltages add up. The common reference point between the two parts is connected together, when connected in series. The Z matrices of the top and bottom parts, respectively, are

$$Z = \begin{bmatrix} Z_{11} & Z_{12} \\ Z_{21} & Z_{22} \end{bmatrix} = \begin{bmatrix} 0.8571 & 0.2857 \\ 0.2857 & 1.4286 \end{bmatrix}$$

and

$$Z = \begin{bmatrix} Z_{11} & Z_{12} \\ Z_{21} & Z_{22} \end{bmatrix} = \begin{bmatrix} 3 & 3 \\ 3 & 3 \end{bmatrix}.$$

The sum of the two matrices is the same as found earlier.

The circuit in Fig. 7.2 has three independent current variables, whereas the corresponding two-port model has only two current variables. That is, the original circuit has been abridged into a two-terminal pair circuit, eliminating the terminal pair that is of no interest. Now, using this model (Eqs. (7.1) and (7.2)), we can determine two of the four of the terminal variables

$$\{V_1, V_2, I_1, I_2\},$$

with the other two known. We have determined the terminal currents $I_1 = 0.3687$ and $I_2 = -0.1285$ by nodal analysis. Using the two-port model, we get

$$\begin{bmatrix} V_1 \\ V_2 \end{bmatrix} = \begin{bmatrix} 3.8571 & 3.2857 \\ 3.2857 & 4.4286 \end{bmatrix} \begin{bmatrix} 0.3687 \\ -0.1285 \end{bmatrix} = \begin{bmatrix} 1 \\ 0.6425 \end{bmatrix}.$$

Along with the defining equations of the two-port network, such as Eqs. (7.1) and (7.2) and the constraints imposed at the input and output, we can determine the parameters such as the input impedance. Form the nodal analysis, we found $V_1 = 1$ and $I_1 = 0.3687$. Now, the input impedance is

$$Z_{in} = \frac{V_1}{I_1} = \frac{1}{0.3687} = 2.7121 \, \Omega.$$

The constraint at the output side is

$$V_2 = -I_2 Z_L.$$

From the characterizing equation of the two-port network, we have

$$Z_{21} I_1 + Z_{22} I_2 = V_2.$$

Eliminating V_2, we get

$$Z_{21} I_1 + Z_{22} I_2 = -I_2 Z_L \quad \text{or} \quad \frac{I_2}{I_1} = -\frac{Z_{21}}{Z_{22} + Z_L}.$$

Substituting for I_2 in the characterizing equation

$$Z_{11} I_1 + Z_{12} I_2 = V_1,$$

we get

$$Z_{in} = \frac{V_1}{I_1} = Z_{11} - \frac{Z_{12} Z_{21}}{Z_{22} + Z_L} = 2.7121$$

as obtained earlier. With $Z_L = \infty$, $I_2 = 0$ and $Z_{in} = Z_{11}$. To determine the output impedance Z_o, which is the Thévenin's Z_{eq} from the terminal pair 2 with the load impedance disconnected and the voltage source short-circuited, $V_1 = 0$. Assuming that the series source impedance is Z_s, from the defining equations of two-port circuit, we have

$$I_1 = -\frac{Z_{12}I_2}{Z_{11} + Z_s} \quad \text{and} \quad I_2 = \frac{V_2 - Z_{12}I_1}{Z_{22}}.$$

Substituting for I_1 in Eq. (7.1), we get

$$Z_o = \frac{V_2}{I_2} = Z_{22} - \frac{Z_{12}Z_{21}}{Z_{11} + Z_s} = 1.6296$$

with $Z_s = 0$. From Thévenin's theorem and Fig. 7.3b,

$$Z_2 + \frac{Z_1 Z_3}{Z_1 + Z_3} = 1.6296.$$

Let us find V_2/V_1.

$$Z_{21}I_1 + Z_{22}(-\frac{V_2}{Z_L}) = V_2 = I_1 \frac{Z_{21}Z_L}{Z_{22} + Z_L}$$

$$V_1 = I_1 \frac{(Z_{11}Z_L + Z_{11}Z_{22} - Z_{12}Z_{21})}{Z_{22} + Z_L}.$$

Therefore,

$$\frac{V_2}{V_1} = \frac{Z_{21}Z_L}{(Z_{11}Z_L + Z_{11}Z_{22} - Z_{12}Z_{21})} = 0.6425.$$

7.2 Admittance Parameters

With nodal analysis, we get

$$Y_{11}V_1 + Y_{12}V_2 = I_1 \tag{7.3}$$

$$Y_{21}V_1 + Y_{22}V_2 = I_2 \tag{7.4}$$

$$Y_{11} = \frac{I_1}{V_1}|_{V_2=0} \quad \text{and} \quad Y_{22} = \frac{I_2}{V_2}|_{V_1=0}$$

are called driving-point admittances. That is, Y_{11} multiplied by the excitation voltage at terminal pair 1 equals the current at the terminal pair 1, when terminal pair 2 is short-circuited, $V_2 = 0$. Similarly, Y_{22} multiplied by the excitation voltage at terminal pair 2 equals the current at the terminal pair 2, when terminal pair 1 is short-circuited, $V_1 = 0$.

$$Y_{12} = \frac{I_1}{V_2}|_{V_1=0} \quad \text{and} \quad Y_{21} = \frac{I_2}{V_1}|_{V_2=0}$$

are called transfer admittances. That is, Y_{12} multiplied by the excitation voltage at terminal pair 2 equals the current at the terminal pair 1, when terminal pair 1 is short-circuited, $V_1 = 0$. Similarly, Y_{21} multiplied by the excitation voltage at terminal pair 1 equals the current at the terminal pair 2, when terminal pair 2 is short-circuited, $V_2 = 0$. In matrix form, we get

$$\begin{bmatrix} Y_{11} & Y_{12} \\ Y_{21} & Y_{22} \end{bmatrix} \begin{bmatrix} V_1 \\ V_2 \end{bmatrix} = \begin{bmatrix} I_1 \\ I_2 \end{bmatrix}.$$

In this model, currents I_1 and I_2 are expressed in terms of voltages V_1 and V_2.

The impedance and admittance matrices are the inverses of each other.

$$\begin{bmatrix} V_1 \\ V_2 \end{bmatrix} = \begin{bmatrix} Y_{11} & Y_{12} \\ Y_{21} & Y_{22} \end{bmatrix}^{-1} \begin{bmatrix} I_1 \\ I_2 \end{bmatrix} \text{ and } \begin{bmatrix} I_1 \\ I_2 \end{bmatrix} = \begin{bmatrix} Z_{11} & Z_{12} \\ Z_{21} & Z_{22} \end{bmatrix}^{-1} \begin{bmatrix} V_1 \\ V_2 \end{bmatrix}.$$

The conversion formulas for the impedance and admittance matrices, from the 2×2 matrix inverse, are

$$Z_{11} = \frac{Y_{22}}{Y_{11}Y_{22} - Y_{12}Y_{21}} \tag{7.5}$$

$$Z_{12} = \frac{-Y_{12}}{Y_{11}Y_{22} - Y_{12}Y_{21}} \tag{7.6}$$

$$Z_{21} = \frac{-Y_{21}}{Y_{11}Y_{22} - Y_{12}Y_{21}} \tag{7.7}$$

$$Z_{22} = \frac{Y_{11}}{Y_{11}Y_{22} - Y_{12}Y_{21}} \tag{7.8}$$

and

$$Y_{11} = \frac{Z_{22}}{Z_{11}Z_{22} - Z_{12}Z_{21}} \tag{7.9}$$

$$Y_{12} = \frac{-Z_{12}}{Z_{11}Z_{22} - Z_{12}Z_{21}} \tag{7.10}$$

$$Y_{21} = \frac{-Z_{21}}{Z_{11}Z_{22} - Z_{12}Z_{21}} \tag{7.11}$$

$$Y_{22} = \frac{Z_{11}}{Z_{11}Z_{22} - Z_{12}Z_{21}}. \tag{7.12}$$

Let us find the admittance matrix of the two-port circuit shown in Fig. 7.3c. The conductances, which are the inverses of the resistances of the elements, are

$$Y_1 = \frac{1}{0.5714} = 1.75\,\text{℧}, \quad Y_2 = \frac{1}{1.1429} = 0.875\,\text{℧}, \quad Y_3 = \frac{1}{3.2857} = 0.3043\,\text{℧}$$

$$Y_{11} = \frac{I_1}{V_1}|_{V_2=0} = \frac{I_1((1.75) \parallel (0.875 + 0.3043))}{I_1} = 0.7045\,\text{℧}$$

$$Y_{12} = \frac{I_1}{V_2}|_{V_1=0} = -\frac{(1.75)(0.875)}{1.75 + 0.875 + 0.3043} = -0.5227$$

$$Y_{21} = \frac{I_2}{V_1}|_{V_2=0} = -\frac{(1.75)(0.875)}{1.75 + 0.875 + 0.3043} = -0.5227$$

$$Y_{22} = \frac{I_2}{V_2}|_{V_1=0} = \frac{I_2((0.875) \parallel (1.75 + 0.3043))}{I_2} = 0.6136\,\text{℧}.$$

The first subscript indicates the origin of the current and the second indicates the origin of the voltage.

Y_{11} is the admittance, the inverse of which multiplied by the current at the terminal pair 1, I_1, produces the open-circuit voltage at the terminal pair 11, V_1 with short-circuit condition at terminal pair 2. Then, conductances 0.875 and 0.3043 are parallel and add up to 1.1793. Now, conductances 1.75 and 1.1793 are in series and

$$Y_{11} = \frac{(1.75)(1.1793)}{1.75 + 1.1793} = 0.7045$$

$$V_1 = I_1(Y_1 \parallel (Y_2 + Y_3)) \text{ and } Y_{11} = \frac{V_1}{I_1} = \frac{(Y_1)(Y_2 + Y_3)}{Y_1 + Y_2 + Y_3}.$$

In terms of impedances,

$$I_1 = \frac{V_1}{Z_1 + (Z_2 \parallel Z_3)}.$$

Simplifying, we get

$$Y_{11} = \frac{I_1}{V_1} = \frac{Z_2 + Z_3}{Z_1 Z_2 + Z_1 Z_3 + Z_3 Z_2}.$$

Y_{12} is the admittance, the inverse of which multiplied by the current at the terminal pair 1, I_1, produces the open-circuit voltage at the terminal pair 22, V_2 with short-circuit condition at terminal pair 1.

$$I_2 = V_2(Y_2 \parallel (Y_1 + Y_3)) \quad \text{and} \quad I_1 = -I_2 \frac{Y_1}{Y_1 + Y_3}.$$

Eliminating I_2, we get

$$Y_{12} = \frac{I_1}{V_2} = \frac{-Y_1 Y_2}{Y_1 + Y_2 + Y_3}$$

$$Y_{12} = -\frac{(1.75)(0.875)}{1.75 + 0.875 + 0.3043} = -0.5227 = Y_{21}.$$

Y_{22} is the admittance, the inverse of which multiplied by the current at the terminal pair 2, I_2, produces the open-circuit voltage at the terminal pair 22, V_2 with short-circuit condition at terminal pair 1. Then, conductances 1.75 and 0.3043 are parallel and add up to 2.0543. Now, conductances 0.875 and 2.0543 are in series and

$$Y_{22} = \frac{(0.875)(2.0543)}{0.875 + 2.0543} = 0.6136$$

$$Y = \begin{bmatrix} Y_{11} & Y_{12} \\ Y_{21} & Y_{22} \end{bmatrix} = \begin{bmatrix} 0.7045 & -0.5227 \\ -0.5227 & 0.6136 \end{bmatrix}.$$

Multiplying the admittance and impedance matrices will result in an identity matrix of size 2×2. The general formula in terms of conductances are

$$Y_{11} = \frac{(Y_1)(Y_2 + Y_3)}{Y_1 + Y_2 + Y_3} \tag{7.13}$$

$$Y_{12} = \frac{-Y_1 Y_2}{Y_1 + Y_2 + Y_3} \tag{7.14}$$

$$Y_{21} = \frac{-Y_1 Y_2}{Y_1 + Y_2 + Y_3} \tag{7.15}$$

$$Y_{22} = \frac{(Y_2)(Y_1 + Y_3)}{Y_1 + Y_2 + Y_3}. \tag{7.16}$$

In general, the formulas for the admittance matrix for a T-circuit, in terms of impedances, are

$$Y_{11} = \frac{Z_2 + Z_3}{Z_1 Z_2 + Z_1 Z_3 + Z_3 Z_2} \tag{7.17}$$

$$Y_{12} = \frac{-Z_3}{Z_1 Z_2 + Z_1 Z_3 + Z_3 Z_2} \tag{7.18}$$

$$Y_{21} = \frac{-Z_3}{Z_1 Z_2 + Z_1 Z_3 + Z_3 Z_2} \tag{7.19}$$

$$Y_{22} = \frac{Z_1 + Z_3}{Z_1 Z_2 + Z_1 Z_3 + Z_3 Z_2}. \tag{7.20}$$

The Admittance Matrix of Circuits Connected in Parallel

A large circuit is an interconnection of a set of smaller circuits. When they are in parallel, it is advantageous to characterize them by Y matrices, since the Y matrix of two circuits connected in parallel is the sum of the individual Y matrices.

The circuit in Fig. 7.2 can be considered as a parallel combination of two circuits with the bottom three resistors comprising the first part and the resistor with value $4\,\Omega$ as the second part. The input voltages are the same and their currents add up. The common reference points between the two parts are tied together, when connected in parallel. The Z matrices of the top and bottom parts, respectively, are

$$Z = \begin{bmatrix} Z_{11} & Z_{12} \\ Z_{21} & Z_{22} \end{bmatrix} = \begin{bmatrix} 4 & 3 \\ 3 & 5 \end{bmatrix}$$

$$Y_1 = Z^{-1} = \begin{bmatrix} Y_{11} & Y_{12} \\ Y_{21} & Y_{22} \end{bmatrix} = \begin{bmatrix} 0.4545 & -0.2727 \\ -0.2727 & 0.3636 \end{bmatrix}$$

$$Y_2 = \begin{bmatrix} Y_{11} & Y_{12} \\ Y_{21} & Y_{22} \end{bmatrix} = \begin{bmatrix} 0.25 & -0.25 \\ -0.25 & 0.25 \end{bmatrix}.$$

The sum of the two matrices is the same as found earlier.

We have determined the input and output terminal voltages as $V_1 = 1$ and $V_2 = 0.6425$ by nodal analysis. Using the two-port admittance model, we get

$$\begin{bmatrix} I_1 \\ I_2 \end{bmatrix} = \begin{bmatrix} 0.7045 & -0.5227 \\ -0.5227 & 0.6136 \end{bmatrix} \begin{bmatrix} 1 \\ 0.6425 \end{bmatrix} = \begin{bmatrix} 0.3687 \\ -0.1285 \end{bmatrix}.$$

7.3   Hybrid Parameters

Some two-port models are convenient or applicable to some physical devices. In this model, V_1 and I_2 are the dependent variables, while I_1 and V_2 are the independent variables. It is especially important in the analysis of amplifying devices and ideal transformers. They are called hybrid parameters, since they are a hybrid combination of ratios.

$$h_{11}I_1 + h_{12}V_2 = V_1 \tag{7.21}$$

$$h_{21}I_1 + h_{22}V_2 = I_2, \tag{7.22}$$

where

$$h_{11} = \frac{V_1}{I_1}|_{V_2=0}\,\Omega, \quad h_{12} = \frac{V_1}{V_2}|_{I_1=0}, \quad h_{21} = \frac{I_2}{I_1}|_{V_2=0}, \quad h_{22} = \frac{I_2}{V_2}|_{I_1=0}\,\mho.$$

They are, respectively, short-circuit input impedance, open-circuit reverse voltage gain, short-circuit forward current gain, and open-circuit output admittance. In matrix form,

$$\begin{bmatrix} h_{11} & h_{12} \\ h_{21} & h_{22} \end{bmatrix} \begin{bmatrix} I_1 \\ V_2 \end{bmatrix} = \begin{bmatrix} V_1 \\ I_2 \end{bmatrix}.$$

The h parameters can be obtained from the corresponding Z parameters by the formula

$$\begin{bmatrix} h_{11} & h_{12} \\ h_{21} & h_{22} \end{bmatrix} = \begin{bmatrix} \frac{\Delta_Z}{Z_{22}} & \frac{Z_{12}}{Z_{22}} \\ -\frac{Z_{21}}{Z_{22}} & \frac{1}{Z_{22}} \end{bmatrix},$$

where $\Delta_Z = Z_{11}Z_{22} - Z_{11}Z_{22}$, the determinant of the Z matrix. With the impedance matrix

$$Z = \begin{bmatrix} Z_{11} & Z_{12} \\ Z_{21} & Z_{22} \end{bmatrix} = \begin{bmatrix} 3.8571 & 3.2857 \\ 3.2857 & 4.4286 \end{bmatrix}$$

and $\Delta_Z = 6.2857$, we get

$$\begin{bmatrix} h_{11} & h_{12} \\ h_{21} & h_{22} \end{bmatrix} = \begin{bmatrix} 1.4194 & 0.7419 \\ -0.7419 & 0.2258 \end{bmatrix}.$$

We have determined the terminal voltages and currents as $V_1 = 1$, $V_2 = 0.6425$, $I_1 = 0.3687$, and $I_2 = -0.1285$ by nodal analysis in an earlier example. Using the two-port hybrid model, we get

$$\begin{bmatrix} V_1 \\ I_2 \end{bmatrix} = \begin{bmatrix} 1.4194 & 0.7419 \\ -0.7419 & 0.2258 \end{bmatrix} \begin{bmatrix} 0.3687 \\ 0.6425 \end{bmatrix} = \begin{bmatrix} 1 \\ -0.1285 \end{bmatrix}.$$

7.4 Transmission Parameters

As there is no restriction in the choice of the independent and dependent variables, we get different models with each choice. This is the fourth of the commonly used models. In this model, V_1 and I_1 are the dependent variables and V_2 and I_2 are the independent variables. This model is particularly useful in the analysis of transmission lines and referred as transmission or $ABCD$ parameters.

$$AV_2 - BI_2 = V_1 \tag{7.23}$$

$$CV_2 - DI_2 = I_1, \tag{7.24}$$

where

$$A = \frac{V_1}{V_2}|_{I_2=0}, \quad B = -\frac{V_1}{I_2}|_{V_2=0}\,\Omega, \quad C = \frac{I_1}{V_2}|_{I_2=0}\,\mho, \quad D = -\frac{I_1}{I_2}|_{V_2=0}.$$

They are, respectively, the open-circuit voltage ratio, the negative short-circuit transfer impedance, the open-circuit transmission admittance and the negative short-circuit current ratio. In matrix form,

$$\begin{bmatrix} A & B \\ C & D \end{bmatrix} \begin{bmatrix} V_2 \\ -I_2 \end{bmatrix} = \begin{bmatrix} V_1 \\ I_1 \end{bmatrix}.$$

The sign of I_2 is changed, as this current is considered to be leaving the circuit. A and D are dimensionless, while B is an impedance and C is an admittance. If $AD - BC = 1$, then the circuit is reciprocal. A circuit is reciprocal, with a single input source, if the points of input and output can be interchanged with the ratio of input and output, remains the same. The parameter matrix is symmetric. This property is known as reciprocity theorem. The ABCD parameter matrix of a circuit, which is a combination of two networks connected in cascade (the output of the first circuit is the input to the second), is the matrix product in that order of their individual parameter matrices.

Let the impedance matrix be

$$Z = \begin{bmatrix} Z_{11} & Z_{12} \\ Z_{21} & Z_{22} \end{bmatrix} = \begin{bmatrix} 3.8571 & 3.2857 \\ 3.2857 & 4.4286 \end{bmatrix}$$

and $\Delta_Z = 6.2857$. The transmission parameters can be obtained from the corresponding Z parameters by the formula

$$\begin{bmatrix} A & B \\ C & D \end{bmatrix} = \begin{bmatrix} \frac{Z_{11}}{Z_{21}} & \frac{\Delta_Z}{Z_{21}} \\ \frac{1}{Z_{21}} & \frac{Z_{22}}{Z_{21}} \end{bmatrix} = \begin{bmatrix} 1.1739 & 1.9130 \\ 0.3043 & 1.3478 \end{bmatrix},$$

where $\Delta_Z = Z_{11}Z_{22} - Z_{11}Z_{22}$, the determinant of the Z matrix.

We have determined the terminal voltages and currents as $V_1 = 1$, $V_2 = 0.6425$, $I_1 = 0.3687$, and $I_2 = -0.1285$ by nodal analysis in an earlier example. Using the two-port ABCD model, we get

$$\begin{bmatrix} V_1 \\ I_2 \end{bmatrix} = \begin{bmatrix} 1.1739 & 1.9130 \\ 0.3043 & 1.3478 \end{bmatrix} \begin{bmatrix} 0.6425 \\ 0.1285 \end{bmatrix} = \begin{bmatrix} 1 \\ 0.3687 \end{bmatrix}.$$

7.5   Examples

7.5.1   Analysis of a π Circuit

Find I_1, I_2, and V_2 for the π circuit shown in Fig. 7.6 by the nodal method. The source voltage is $\cos(t + \pi/3)$ with the positive terminal connected to the ground. Find the admittance parameters and verify the derived model. Let us find the voltage V_2 by nodal analysis. The equilibrium equation at node 2 is

Fig. 7.6 A π circuit

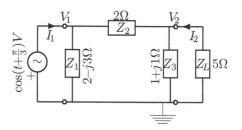

$$\frac{V_2 + (0.5 + j0.866)}{Z_2} + \frac{V_2}{Z_3} + \frac{V_2}{Z_L} = 0.$$

Simplifying, we get

$$V_2(Z_3 Z_L + Z_2 Z_L + Z_2 Z_3) = -Z_L Z_3(0.5 + j0.866).$$

With

$$Z_1 = 2 - j3, \quad Z_2 = 2, \quad Z_3 = 1 + j1, \quad Z_L = 5, \quad V = -(0.5 + j0.866),$$

we get $V_2 = (-0.0494 - j0.3814)$ V. Now, the terminal currents are

$$I_1 = \frac{V_1}{Z_1} + \frac{V_2}{Z_3} + \frac{V_2}{Z_L} = (-0.1024 - j0.4909) \text{ A} \quad \text{and} \quad I_2 = -\frac{V_2}{Z_L} = 0.0099 + j0.0763.$$

Let us find the admittance matrix of the two-port circuit and verify that the currents and voltages found by the nodal analysis satisfy the governing relations for the circuit. The admittance of the branches, in ℧, is the inverse of the impedances shown in the figure. That is,

$$\{Y_1 = 0.1538 + j0.2308, \quad Y(2) = 0.5, \quad Y(3) = 0.5 - j0.5\}.$$

With the output terminals short-circuited,

$$I_1 = V_1(Y_1 + Y_2) \quad \text{and} \quad Y_{11} = \frac{I_1}{V_1} = Y_1 + Y_2.$$

With the input terminals short-circuited,

$$I_1 = -V_2 Y_2 \quad \text{and} \quad Y_{12} = -\frac{I_1}{V_2} = -Y_2 = Y_{21}.$$

With the input terminals short-circuited,

$$I_2 = V_2(Y_3 + Y_2) \quad \text{and} \quad Y_{22} = \frac{I_2}{V_2} = Y_3 + Y_2.$$

Therefore, the admittance matrix is

$$Y = \begin{bmatrix} Y_{11} & Y_{12} \\ Y_{21} & Y_{22} \end{bmatrix} = \begin{bmatrix} Y_1 + Y_2 & -Y_2 \\ -Y_2 & Y_3 + Y_2 \end{bmatrix} = \begin{bmatrix} 0.6538 + j0.2308 & -0.5 \\ -0.5 & 1 - j0.5 \end{bmatrix}.$$

We have determined the input and output terminal voltages $V_1 = -(0.5 + j0.866)$ and $V_2 = (-0.0494 - j0.3814)$ by nodal analysis. Using the two-port model, we get

$$\begin{bmatrix} I_1 \\ I_2 \end{bmatrix} = \begin{bmatrix} 0.6538 + j0.2308 & -0.5 \\ -0.5 & 1 - j0.5 \end{bmatrix} \begin{bmatrix} -(0.5 + j0.866) \\ (-0.0494 - j0.3814) \end{bmatrix} = \begin{bmatrix} -0.1024 - j0.4909 \\ 0.0099 + j0.0763 \end{bmatrix}$$

as found earlier.

7.5.2 Analysis of a Common-Emitter Transistor Amplifier

When we speak in an auditorium, the people sitting in the first few rows can hear the speech. However, with the aid of an amplifier system, all the people can hear. An amplifier amplifies the speech signal. Basic devices used in amplifiers are usually of two types. In one type, called the current-controlled current source, the ratio between the output and input current is high.

The hybrid model of a commonly used transistor amplifier circuit is shown in Fig. 7.7. A source voltage V_s with internal resistance R_s is connected to the input side. A load resistance of R_L is connected to the output side. The parameters h_{11}, h_{12}, h_{21}, and h_{22} are, respectively, short-circuit input impedance, the open-circuit reverse voltage gain, the short-circuit forward current gain, and open-circuit output admittance. Since a combination of different parameters are used, it is called a hybrid model. The principal component is a current-controlled current source. Let us determine the voltage gain, current gain, input and output impedances of the circuit.

In the analysis of practical systems, a mathematical model is formulated. The complexity of the model depends upon the required accuracy of the analysis. Further, it also depends on the operating environment. For example, the complexity of the model may be high for high-frequency operation. For understanding the basic behavior of the system, the simplest model is sufficient. The parameters h_{12} and h_{22} can be, for a first approximation, assumed to be 0. The ratio of the output and input current is high. That is, the device is, basically, a current-controlled current source.

Let us find the current gain I_2/I_1. Since $V_2 = -I_2 Z_L$,

$$I_2 = h_{21} I_1 - h_{22} I_2 Z_L \quad \text{and} \quad A_i = \frac{I_2}{I_1} = \frac{h_{21}}{1 + h_{22} Z_L}.$$

Let us find the input impedance. From the first defining equation of the hybrid model and the current gain A_i, we get

$$V_1 = h_{11} I_1 - h_{12} I_2 Z_L \quad \text{and} \quad Z_{in} = \frac{V_1}{I_1} = h_{11} - \frac{h_{12} h_{21} Z_L}{1 + h_{22} Z_L}.$$

The output impedance is the Thévenin equivalent impedance at the output terminals. The input voltage source is short-circuited, leaving Z_s. Since the circuit contains dependent sources, we apply a 1-V voltage source at the output terminals and disconnect Z_L, $V_2 = 1$. Now, $Z_o = 1/I_2$.

$$I_1 = -\frac{h_{12}}{h_{11} + Z_s}.$$

Fig. 7.7 The hybrid model of a transistor amplifier circuit

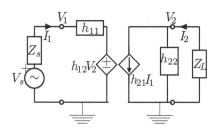

Fig. 7.8 The hybrid
model of a bipolar
transistor amplifier circuit

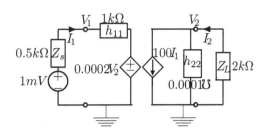

Since $I_2 = h_{22} + h_{21}I_1$ and substituting for I_1, we get

$$Z_o = \frac{1}{I_2} = \frac{h_{11} + Z_s}{(h_{11} + Z_s)h_{22} - h_{12}h_{21}}.$$

Let us find the voltage gain. From the current gain,

$$I_2 = I_1 \frac{h_{21}}{1 + h_{22}Z_L} = I_1 h_{21} + h_{22}V_2 \quad \text{and} \quad I_1 = \frac{h_{22}V_2}{\frac{h_{21}}{1+h_{22}Z_L} - h_{21}}.$$

Since $V_1 = h_{11}I_1 + h_{21}V_2$ and substituting for I_1, we get

$$A_v = \frac{V_2}{V_1} = -\frac{h_{21}Z_L}{h_{11} + Z_L(h_{11}h_{22} - h_{12}h_{21})}.$$

Example

Let us determine the voltage gain, current gain, input and output impedances of the circuit shown in Fig. 7.8. All the required parameter values are shown in the figure. The voltage gain, current gain, input and output impedances, using the formulas derived, are, respectively, -172.4138, 83.3333, $999.9833\,\Omega$ and $11{,}538\,\Omega$. Now, we have to find V_2, I_1 and I_2. Let us find

$$I_1 = \frac{h_{22}V_2}{\frac{h_{21}}{1+h_{22}Z_L} - h_{21}}$$

$$V_1 = \frac{V_2}{A_v}.$$

Applying KVL to the left side of the circuit, we get

$$V_s = Z_s I_1 + V_1.$$

Substituting for I_1 and V_1 in this equation, we get

$$V_s = \left(Z_s \frac{h_{22}}{\frac{h_{21}}{1+h_{22}Z_L} - h_{21}} + \frac{1}{A_v} \right) V_2 \quad \text{and} \quad \frac{V_2}{V_s} = -113.6364,$$

which is the circuit voltage gain. Therefore, this gain multiplied by the input source voltage gives V_2

$$V_2 = (0.001)(-113.6364) = -113.6 \times 10^{-3}\,\text{V} = -113.6\,\text{mV}$$

$$I_2 = \frac{V_2}{Z_L} = \frac{0.1136}{2000} = 56.88 \times 10^{-6} \,\text{A} = 56.8 \,\mu\text{A}$$

$$I_1 = \frac{I_2}{A_i} = \frac{0.0000568}{83.3333} = 0.6818 \times 10^{-6} \,\text{A} = 0.6818 \,\mu\text{A}$$

$$V_1 = \frac{V_2}{A_v} = \frac{-113.6 \,\text{mV}}{-172.4138} = 0.6591 \times 10^{-3} \,\text{V} = 0.6591 \,\text{mV}$$

7.5.3 Analysis of a Bridge Circuit

Let us find the impedance parameters for the bridge circuit shown in Fig. 7.9, assuming that R_4 is the load resistor and DC source. Using the model derived, let us find V_1 and V_2. The parameter definitions are

$$Z_{11} = \frac{V_1}{I_1}\Big|_{I_2=0} \quad \text{and} \quad Z_{22} = \frac{V_2}{I_2}\Big|_{I_1=0}.$$

$$Z_{12} = \frac{V_1}{I_2}\Big|_{I_1=0} \quad \text{and} \quad Z_{21} = \frac{V_2}{I_1}\Big|_{I_2=0}$$

Figure 7.10 is appropriate for finding Z_{11} and Z_{21}. The load circuit is open-circuited, so that the constraint $I_2 = 0$ is met. The DC source is 1V. We have to find the equivalent resistance to find I_1 and V_2.

$$Z_{eq} = Z_{11} = 3 + (1) \parallel (3+1) = \frac{19}{5} \quad \text{and} \quad \frac{1}{Z_{eq}} = I_1 = \frac{5}{19} \,\text{A}.$$

Fig. 7.9 A bridge circuit

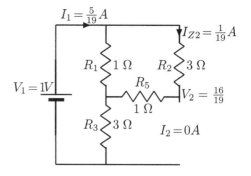

Fig. 7.10 Determination of Z_{11}

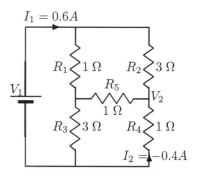

Fig. 7.11 Determination
of Z_{22}

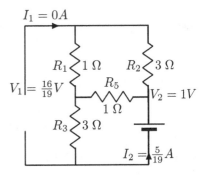

By current division,

$$I_{Z2} = \frac{I_1}{5} = \frac{1}{19} \text{ and } V_2 = 1 - \frac{3}{19} = \frac{16}{19} \text{ V}$$

$$Z_{21} = \frac{V_2}{I_1}\Big|_{I_2=0} = \frac{16}{5}.$$

Figure 7.11 is appropriate for finding Z_{22} and Z_{12}. As the circuit is reciprocal, however, the parameter matrix is symmetric. Therefore,

$$Z = \begin{bmatrix} \frac{19}{5} & \frac{16}{5} \\ \frac{16}{5} & \frac{19}{5} \end{bmatrix}.$$

Using the input–output relations, we get

$$\begin{bmatrix} Z_{11} & Z_{12} \\ Z_{21} & Z_{22} \end{bmatrix} \begin{bmatrix} I_1 \\ I_2 \end{bmatrix} = \begin{bmatrix} V_1 \\ V_2 \end{bmatrix} \text{ and } \begin{bmatrix} \frac{19}{5} & \frac{16}{5} \\ \frac{16}{5} & \frac{19}{5} \end{bmatrix} \begin{bmatrix} 0.6 \\ -0.4 \end{bmatrix} = \begin{bmatrix} 1 \\ 0.4 \end{bmatrix}$$

as found in loop analysis earlier.

From this example, it is clear that the circuit is considered as completely enclosed in a box with access being restricted to the terminals of the input and output ports only. Any two terminal pair of the circuit can be chosen for the two-port model. The model is useful in practice, since we are not usually interested in all the currents and voltages in the circuit. Therefore, in deriving the model, we systematically eliminate all the variables whose accessibility is not called for. Consequently, a N terminal pair circuit is abridged to a two-terminal pair.

7.5.4 Ladder Circuit

Ladder circuits are interconnection of repeating units of impedances with the same structure. Digital-to-analog conversion circuits, transmission lines, and filters are typical applications. As it has a structure, a general loop or nodal analysis is not required. Most of the circuits encountered in practice have some geometrical symmetry. In general, any short cuts applicable to the circuit that is being analyzed should be taken advantage to simplify the analysis.

Figure 7.12 shows a resistive ladder circuit with the resistance values indicated. A 1-V source energizes the circuit. First, let us find the currents and voltages at all parts of the circuit. Then, a two-port model will be derived using the transmission parameters.

The approach is to ignore the source voltage to start with and assume the voltage at the rightmost node to be $v_3 = 1$ V. Then, find the value of the input source voltage that gives v_3, the assumed

Fig. 7.12 A resistive
ladder circuit

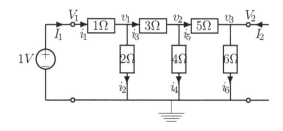

voltage. Now, $i_6 = i_5 = 1/6\,\mathrm{A}$.

$$v_2 = \frac{11}{6}\,\mathrm{V} \quad \text{and} \quad i_4 = \frac{11}{24}\,\mathrm{A}$$

$$v_1 = \frac{11}{6} + \frac{3}{6} + \frac{33}{24} = \frac{89}{24}\,\mathrm{V} \quad \text{and} \quad i_2 = \frac{89}{48}\,\mathrm{A}$$

$$V_1 = \frac{89}{24} + \frac{1}{6} + \frac{11}{24} + \frac{89}{48} = \frac{297}{48}\,\mathrm{V}.$$

To obtain 1 V at the output, an input of 297/48 V is required. For 1 V input, we have to divide all the
currents and voltages by 297/48. Then, we get

$$V_1 = 1,\, v_1 = 0.5993,\, v_2 = 0.2963,\, v_3 = 0.1616$$

and

$$i_1 = 0.4007,\, i_2 = 0.2997,\, i_3 = 0.1010,\, i_4 = 0.0741,\, i_5 = i_6 = 0.0269\,\mathrm{A}.$$

Two-Port Model with Transmission Parameters
The characterizing equations of this model are

$$A = \frac{V_1}{V_2}|_{I_2=0}, \quad B = -\frac{V_1}{I_2}|_{V_2=0}\,\Omega, \quad C = \frac{I_1}{V_2}|_{I_2=0}\,\mho, \quad D = -\frac{I_1}{I_2}|_{V_2=0}.$$

In matrix form,

$$\begin{bmatrix} A & B \\ C & D \end{bmatrix} \begin{bmatrix} V_2 \\ -I_2 \end{bmatrix} = \begin{bmatrix} V_1 \\ I_1 \end{bmatrix}.$$

In deriving this model, we use the property that the transmission parameters of a model of two circuits
connected in cascade is the matrix product of their individual parameters. The order of the matrix
multiplication has to be the same as that of the interconnection.

Consider the part of the circuit with resistors $Z_1 = 1\,\Omega$ and $Y2 = 1/Z_2 = 1/2\,\mho$.

$$V_2 = \frac{V_1 Z_2}{Z_1 + Z_2} \quad \text{and} \quad A = \frac{V_1}{V_2} = 1 + Z_1 Y_2$$

$$B = -\frac{V_1}{-V_1/Z_1} = Z_1$$

$$V_2 = \frac{V_1 Z_2}{Z_1 + Z_2}, \quad I_1 = \frac{V_1}{Z_1 + Z_2} \quad \text{and} \quad C = Y_2$$

$$\begin{bmatrix} A & B \\ C & D \end{bmatrix} = \begin{bmatrix} 1 + Z_1 Y_2 & Z_1 \\ Y_2 & 1 \end{bmatrix}.$$

As three sections are connected, the transmission parameters of the whole network is the product of the three.

$$\begin{bmatrix} A & B \\ C & D \end{bmatrix} = \begin{bmatrix} 1.5000 & 1 \\ 0.5000 & 1 \end{bmatrix} \begin{bmatrix} 1.7500 & 3 \\ 0.2500 & 1 \end{bmatrix} \begin{bmatrix} 1.8333 & 5 \\ 0.1667 & 1 \end{bmatrix} = \begin{bmatrix} 6.1875 & 19.8750 \\ 2.4792 & 8.1250 \end{bmatrix}.$$

Assume that a $Z_L = 1\,\Omega$ load resistor is connected to the output side, making the circuit terminated. Substituting $V_2 = -I_2 Z_L$ in the defining equations of the transmission model, we get the input impedance Z_{in}.

$$AV_2 - BI_2 = V_1 = -I_2 A Z_L - B I_2 = -I_2(AZ_L + B) \tag{7.25}$$

$$CV_2 - DI_2 = I_1 = -I_2 C Z_L - D I_2 = -I_2(CZ_L + D) \tag{7.26}$$

$$Z_{in} = \frac{V_1}{I_1} = \frac{AZ_L + B}{CZ_L + D}.$$

Substituting numerical values, we get $Z_{in} = 2.4578\,\Omega$ and, with 1 V input, $I_{in} = 1/2.4578 = 0.4069$ A. With the load resistor disconnected, the output voltage is 0.1616 V, as found earlier. The Thévenin equivalent resistance, the output resistance of the circuit, with the input source short-circuited and the load resistor disconnected is

$$Z_o = ((((1 \parallel 2) + 3) \parallel 4) + 5) \parallel 6 = 3.2121\,\Omega.$$

The current through the load resistor $1\,\Omega$ is

$$-I_2 = \frac{0.1616}{3.2121 + 1} = 0.0384\,\text{A}.$$

Substituting $V_2 = -I_2 Z_L = 0.0384$ and $-I_2 = 0.0384$ in the defining equation of the model, we get back $V_1 = 1$ V and $I_1 = 0.4069$ A.

$$\begin{bmatrix} 6.1875 & 19.8750 \\ 2.4792 & 8.1250 \end{bmatrix} \begin{bmatrix} 0.0384 \\ 0.0384 \end{bmatrix} = \begin{bmatrix} 1 \\ 0.4069 \end{bmatrix} = \begin{bmatrix} V_1 \\ I_1 \end{bmatrix}.$$

Using the defining equation for V_1 in the model, the voltage gain is

$$\frac{V_2}{V_1} = \frac{Z_L}{AZ_L + B} = 0.0384.$$

7.6 Application

7.6.1 Digital-to-Analog Converter: The $R - 2R$ Ladder Circuit

As digital circuits are advantageous, although most naturally occurring signals are analog, signals are converted to digital form, processed and converted back to analog form. This type of transformation is common in signal processing. We learnt that the mathematically equivalent complex sinusoid is more convenient for analysis, although the real sinusoids occur in practical applications. Similarly, signals naturally occur in the time-domain form, but are transformed to frequency-domain for efficient processing and converted back to time-domain. In using digital systems for processing analog signals, an analog-to-digital converter converts the signal to the digital form. The digital signal is processed and converted back to analog form by a digital-to-analog converter. The $R - 2R$ ladder circuit is a simple converter. The input digital signals are in binary form. The binary digits, called bits, consist of 1 and 0 only.

The Number System

A number is a sequence of digits. In a radix-N number system, there are N distinct digits. In the decimal number system with radix 10, there are 10 distinct digits,

$$\{0, 1, 2, 3, 4, 5, 6, 7, 8, 9\}.$$

The weight of each digit depends on its position. A number is, in general, is written as

$$b_{N-1}, b_{N-2}, \ldots, b_1, b_0.$$

For example,

$$125 = 1 \times 10^2 + 2 \times 10^1 + 5 \times 10^0.$$

While the digital system is suitable for human beings, the binary number system, with only two digits (0,1), is used in digital computers. For example,

$$10 = 1 \times 2^3 + 0 \times 2^2 + 1 \times 2^1 + 0 \times 2^0.$$

Each digit in the binary number system is called a bit, an abbreviation for binary digit. The rightmost digit has the smallest weight and it is called the least significant bit (LSB). The leftmost digit has the highest weight and it is called the most significant bit (MSB). Decimal numbers from 0 to 15 and the corresponding binary numbers are shown, respectively, in the first and second columns of Table 7.1.

Table 7.1 Decimal numbers from 0 to 15 and the corresponding binary numbers

Decimal number	Binary number $b_3 b_2 b_1 b_0$
0	0000
1	0001
2	0010
3	0011
4	0100
5	0101
6	0110
7	0111
8	1000
9	1001
10	1010
11	1011
12	1100
13	1101
14	1110
15	1111

$R - 2R$ Ladder Circuit

The block diagram of a 4-bit digital-to-analog converter with voltage output is shown in Fig. 7.13. The converter converts a 4-bit digital input to an analog output. The 4-bit input can be any of the combination shown in the binary column in Table 7.1. Figure 7.14 shows the $R - 2R$ ladder circuit for 4-bit digital-to-analog conversion. In this circuit, only resistors with two values are used. The binary input is applied in bit-reversed form to the converter. That is, the least significant bit is applied to the leftmost input point. Let us assume that a "1" input implies that 16 V is applied to the circuit. A "0" input implies that 0 V is applied to the circuit (connected to the ground terminal). The converted output will be from 0 to 15 V. The expression for the analog output voltage is given by

$$V_o = \frac{b_0}{16} + \frac{b_1}{8} + \frac{b_2}{4} + \frac{b_3}{2}.$$

The analog output for any of the digital inputs can be determined using Thévenin's and superposition theorems. Let us find the output with the value of the b_0-bit to be "1" and all the other bits input terminal points grounded (zero voltage). Let us replace the b_0-bit stage by its Thévenin equivalent circuit, shown to the left of the circle in Fig. 7.15. The equivalent resistance R is the parallel combination of two resistors, each with value $2R$. The open-circuit voltage is $b_0/2$. At the next stage this voltage further divided by a factor of 2 to become $b_0/4$. In each stage, it gets divided by a factor of 2 and finally the output becomes $b_0/16$. With the input voltage applied at the terminal being 16 V, the output voltage is 1 V. The input voltage can be changed to get the required range of output. In a

Fig. 7.13 Block diagram of a 4-bit digital-to-analog converter with voltage output

Fig. 7.14 $R - 2R$ ladder circuit for 4-bit digital-to-analog conversion

Fig. 7.15 $R - 2R$ ladder circuit with the b_0-bit stage replaced by its Thévenin equivalent circuit

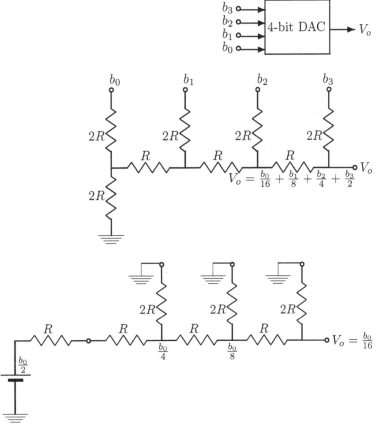

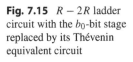

similar way, the output due to b_1 is 2 V. The total output is the some of the voltages contributed by the bits with value "1." Table 7.1 shows the output for all combination of the bits. The output resistance is always equal to R.

7.7   Summary

- A two-port circuit has two separate ports for input and output.
- A two-port circuit may have dependent sources but no independent source. Two-port models are widely used in the study of amplifier, communication and power system circuits.
- In modeling a circuit with impedance parameters, voltages V_1 and V_2 are expressed in terms of currents I_1 and I_2.
- In all the models, the procedure to determine the parameter matrix is the same. An appropriate voltage or current is applied at one port, short-circuit or open-circuit on the other port depending on the parameter of interest, and use regular circuit analysis.
- A large circuit is an interconnection of a set of smaller circuits. The interconnection may be in series, parallel, or cascade. The model for the whole circuit can be expressed as an appropriate combination of its component parts.
- Along with the defining equations of the two-port network, such as Eqs. (7.1) and (7.2) and the constraints imposed at the input and output, we can determine the parameters such as the input impedance.
- In modeling a circuit with admittance parameters, currents I_1 and I_2 are expressed in terms of voltages V_1 and V_2.
- The impedance and admittance matrices are the inverses of each other. Conversion formulas are available to convert one model into another.
- In modeling a circuit with hybrid parameters, V_1 and I_2 are the dependent variables, while I_1 and V_2 are the independent variables. It is especially important in the analysis of amplifying devices and ideal transformers. They are called hybrid parameters, since they are a hybrid combination of ratios.
- In modeling a circuit with transmission parameters, V_1 and I_1 are the dependent variables and V_2 and I_2 are the independent variables. This model is useful in the analysis of transmission lines and referred as transmission or $ABCD$ parameters.

Exercises

* 7.1 Find the impedance parameters for the bridged-T circuit shown in Fig. 7.16 by two methods and verify that both are the same.

Fig. 7.16 A bridged-T circuit

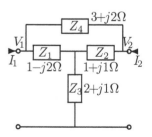

Fig. 7.17 A bridge circuit

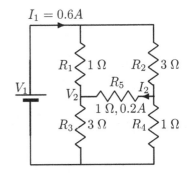

Fig. 7.18 Circuit with a
mutual inductance

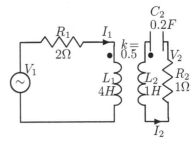

Fig. 7.19 A π circuit

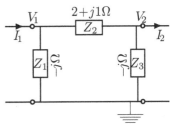

7.2 Find the impedance parameters for the bridge circuit shown in Fig. 7.17, assuming that R_5 is the load resistor and DC source. Using the model derived, find V_1 and V_2.

* 7.3 Find the admittance parameters for the bridged-T circuit shown in Fig. 7.16 by decomposing it into a combination of two parallel circuits with Z_4 constituting one part and the rest another.

7.4 Find the impedance parameters for the mutually coupled inductors shown in Fig. 7.18, assuming that R_2 is the load resistor. The frequency of operation is 2 rad/s. Given that $I_1 = -0.08054 - j0.04362$ and $I_2 = -0.13423 + j0.09396$. Using the model derived, find V_1 and V_2. Use the equivalent T model.

* 7.5 Find the transmission parameters for the circuit shown in Fig. 7.19. Given that $V_2 = (0.1379 - j0.3448)$ V and $I_2 = (0.0690 - j0.1724)$ A. Using the model derived, find V_1 and I_1.

Transform Analysis and Transient Response

8

Transform means change in form. For example, in finding the derivative of the product of two functions, we use the product rule and transform the problem into an easier one. Multiplication operation is more difficult than addition operation. By expressing the numbers in logarithmic form, we reduce the multiplication operation into an addition. For example,

$$8 \times 4 = 2^3 \times 2^2 = 2^5 = 32.$$

Signals occur mostly in the time-domain form, $x(t)$. Time t is the independent variable and $x(t)$, the signal amplitude, is the dependent variable. Practical systems can be mathematically modeled, for analysis and design, by differential equations. For example, the input–output relationship of an inductor is

$$v(t) = L\frac{di(t)}{dt}.$$

Practical systems are composed of large number of elements interconnected. Consequently, the order of the differential equation characterizing the system becomes large and solving it also becomes difficult. The principal transforms, Fourier series, Fourier transform, and Laplace transform used in signal and system analysis reduce the differential equation into an algebraic equation. For example, the input–output relationship of an inductor, in the Laplace transform domain, is

$$V(s) = sLI(s).$$

In the transform domain, frequency becomes the independent variable. Both the time-domain and frequency-domain representation completely specify a signal or a system. As the analysis, in general, is easier in the frequency-domain and the signals occur in the time-domain, we transform the signal to the frequency-domain, find the solution, and transform it back to the time-domain.

Thus far, we mostly studied the response of a circuit to a single frequency sinusoidal voltage and current sources and the DC source with zero frequency. In presenting the linearity property, it was pointed out that the response to an input source, that is a combination of waves of different frequencies, must be individually found and added to find the total response. One reason for the importance of the response to sinusoidal sources with single frequency is that analysis of circuits is easier. Another important reason is that any waveform, encountered in practice, can be decomposed into a set of sinusoidal waveforms by transforms. In applications, the input sources may be periodic or aperiodic

© Springer Nature Switzerland AG 2020
D. Sundararajan, *Introductory Circuit Theory*,
https://doi.org/10.1007/978-3-030-31985-4_8

with arbitrary amplitude profile. The frequency components of the input waveform are first separated using a transform, the response to each frequency component is found separately, and the sum of all the responses is the total response of the circuit. Fourier series is most suitable to decompose periodic waveforms. Therefore, the steps involved in practical circuit analysis are as follows:

1. Decompose the periodic input source into its frequency components by finding the corresponding Fourier series
2. Find the frequency-domain representation of the circuit, as we do in AC circuit analysis
3. Find the response of the circuit to each of the frequency component separately
4. Using the linearity property of linear systems, we add up the individual responses to find the total response

All the tools, such as nodal and mesh analysis and circuit theorems, remain the same with the constraint that each frequency component of the source must be considered individually.

Therefore, the first step in analyzing the response to an arbitrary periodic input is to decompose it in terms of its constitutional sinusoids of various frequencies. While this decomposition, in the mathematical form, looks formidable, it is as simple as finding the amount of a bag of coins with various denominations. The details make it looks difficult. One can easily get used to transform analysis with sufficient paper-and-pencil and programming practice. The amount of coins in a bag can be found by taking one coin at a time, adding its value to a partial sum. After the values of all the coins are added up, we get the total amount. This procedure is preferable if the number of coins is small. For a large number of coins, a better procedure is to sort the coins into various denominations, count the number of coins in each denomination, multiply this number by their corresponding values, and sum all the partial amounts of the various denominations. In the transform analysis of a system, the same procedure is used. The various frequency components are separated by an appropriate transform such as Fourier series, Fourier transform, or Laplace transform. The individual responses of the circuit are found and all the responses are added to find the total response.

8.1 Fourier Series

A periodic signal $x(t)$ satisfies the condition $x(t) = x(t + T)$, for all values of t from $-\infty$ to ∞ and $T > 0$ is a positive constant. The period is the minimum value of T that satisfies the condition. As it repeats its value over a period indefinitely, it remains unchanged by a shift of an integral number of its period. All the signals those do not satisfy $x(t) = x(t + T)$ are called aperiodic signals. That is, the period is infinity.

Each room in a house has a lock and a key. Only a particular key will open the lock of the corresponding room. That is, the key and the lock are a matched pair. Any other key cannot open that lock. Similarly, a signal is composed of a set of frequency components. The frequency components have a property, called the orthogonal property, in that the integral of the product of the signal, over a period, with the conjugate of a certain frequency component will yield the coefficient of that component in the signal alone. This is a similarity test. Repeating this procedure for all the components of interest, we know the frequency content of the signal, called its spectrum.

While practical signal generators, such as an oscillator, generate real sinusoidal waveforms, for analysis purposes, the mathematically equivalent complex exponentials are used due to its compact form and ease of use. In this case, orthogonality of two complex exponentials is that the integral of the product of an exponential and the conjugate of the other exponential over a period is zero or a constant. For the two periodic complex exponentials $e^{j\frac{2\pi}{T}kt}$ and $e^{j\frac{2\pi}{T}lt}$ over a period of T seconds, the orthogonality condition is given by

$$\int_0^T e^{j\frac{2\pi}{T}(k-l)t} dt = \begin{cases} T & \text{for } k = l \\ 0 & \text{for } k \neq l \end{cases}$$

For $k = l$, the integral evaluates to T. For $k \neq l$, the integral evaluates to zero. This is also obvious from the fact that the sine and cosine components of the exponentials are symmetrical about the x-axis. The integral of a cosine or sine waveform, over an integral number of periods, with a nonzero frequency index is always zero. That is, with $k - l = m$,

$$\int_0^T e^{j\frac{2\pi}{T}mt} dt = 0.$$

Fourier analysis problem is to find the coefficients $X(k)$ in the complex exponential polynomial representation of a time-domain function $x(t)$ with period T and the fundamental frequency $\omega_0 = 2\pi/T$

$$x(t) = \sum_{k=-\infty}^{\infty} X(k)e^{jk\omega_0 t}, \tag{8.1}$$

where

$$X(k) = \frac{1}{T} \int_0^T x(t)e^{-jk\omega_0 t} dt, \quad -\infty, \dots, -1, 0, -1, \dots, \infty. \tag{8.2}$$

where k is an integer. As $x(t)$ is periodic with period T, the integral can be evaluated over any continuous interval of duration T. In the second equation, $x(t)$ and the exponentials are known. The coefficients $X(k)$ are found, using the orthogonality property, by integrating the product of $x(t)$ with the conjugate of each of the exponentials, in turn, composing $x(t)$. The Fourier synthesis problem is to find $x(t)$, given $X(k)$ and the exponentials. It is the sum of all the exponentials multiplied by their respective coefficients. The Fourier reconstructed waveform is with respect to the least squares error criterion.

Example 8.1 Find the FS for the signal

$$x(t) = 2 - 2\sin\left(2\frac{2\pi}{5}t + \frac{\pi}{6}\right) + 4\cos\left(4\frac{2\pi}{5}t - \frac{\pi}{3}\right).$$

Solution One period of the signal is shown in Fig. 8.1a. As the signal is given in terms of cosine and sine functions, we just have to express these functions in terms of complex exponentials using Euler's formula. The fundamental frequency of the waveform is $\omega_0 = 2\frac{2\pi}{5}$, as the DC component is periodic with any period. The period is 2.5. All the components of $x(t)$ are rewritten using cosine waveform with positive amplitude for each frequency.

$$x(t) = 2 + 2\cos\left(2\frac{2\pi}{5}t + \frac{2\pi}{3}\right) + 4\cos\left(4\frac{2\pi}{5}t - \frac{\pi}{3}\right)$$

Using Euler's formula, we get

$$x(t) = 2 + 1e^{j\left(2\frac{2\pi}{5}t + \frac{2\pi}{3}\right)} + 1e^{-j\left(2\frac{2\pi}{5}t + \frac{2\pi}{3}\right)} + 2e^{j\left(4\frac{2\pi}{5}t - \frac{\pi}{3}\right)} + 2e^{-j\left(4\frac{2\pi}{5}t - \frac{\pi}{3}\right)}.$$

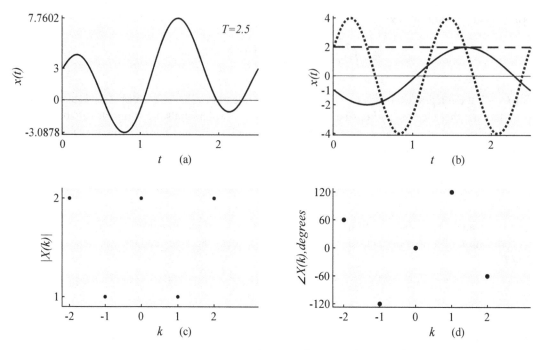

Fig. 8.1 (**a**) A periodic waveform, $x(t) = 2 - 2\sin\left(2\frac{2\pi}{5}t + \frac{\pi}{6}\right) + 4\cos\left(4\frac{2\pi}{5}t - \frac{\pi}{3}\right)$, with period 2.5 s; (**b**) the frequency components of the waveform in (**a**); (**c**) its frequency-domain representation, the magnitude; (**d**) the phase

Comparing this expression with the definition, Eq. (8.1), we get the exponential form of the FS coefficients as

$$\{X(0) = 2, \ X(1) = 1\angle\frac{2\pi}{3}, \ X(-1) = 1\angle-\frac{2\pi}{3}, \ X(2) = 2\angle-\frac{\pi}{3}, \ X(-2) = 2\angle\frac{\pi}{3}.$$

Figure 8.1b shows the three frequency components of the signal. The FS magnitude spectrum and the phase spectrum of the signal are shown, respectively, in Fig. 8.1c, d.   ∎

8.1.1   Fourier Series of a Rectified Sine Wave

Although AC power supply is widely used in the generation, transmission, distribution, and utilization of electrical power, DC power supply is equally important in that it is used in important applications such as traction and most electronic appliances require DC power supply. DC power is mostly derived from the AC supply by rectifiers, followed by filters and regulators. As the AC source is a periodic waveform, Fourier series version of Fourier analysis is appropriate for the analysis and design of DC power supplies.

A rectifier converts the sinusoidal voltage with average value zero into a voltage with a nonzero average value. The negative half-cycle of the sinusoidal waveform is also changed into a positive half-cycle. Then, we have to analyze the rectified voltage using Fourier series representation to find out the magnitude of the various unwanted harmonic components and the extent to which they have to be reduced. Let us find the FS representation of the full-wave rectified waveform given by

$$v(t) = 110|\sin(\omega_0 t)| \ V$$

with the fundamental frequency $f_0 = 60\,\text{Hz}$, $\omega_0 = 2\pi f_0 = 377\,\text{rad/s}$ and period $T = 1/f_0 = 0.0167\,\text{s}$. One period of the waveform is shown in Fig. 8.2a. A waveform is even-symmetric, if the signal is symmetric with respect to the vertical axis at the origin, $v(-t) = v(t)$. As finding the amplitude of each frequency component is the integral of the product and the integral of the product of an even-symmetric signal with an odd-symmetric signal is zero, the rectified waveform $v(t)$ consists of cosine components only. Further, the waveform makes two cycles in its period. That is, it repeats its amplitude profile two times. This type of symmetry is called half-wave symmetry, $v(t) = v(t \pm T/2)$. Consequently, the integral of its product with odd-indexed harmonics is zero. Therefore, the constituent frequency components of $v(t)$ are even-indexed cosine waves only. The Fourier series definition for this type of waveform reduces to finding the integrals of the product of $v(t)$ with even-indexed cosine waves only. The average value, with the harmonic index $k = 0$, is

$$X(0) = 110 \left(\frac{2}{T}\right) \int_0^{\frac{T}{2}} \sin\left(\frac{2\pi}{T}t\right) dt = 110 \left(\frac{2}{\pi}\right) = 70.0282.$$

For other even-indexed k, we get

$$X(2k) = \frac{4}{T} \int_0^{\frac{T}{2}} 110 \sin(\omega_0 t) \cos(2k\omega_0 t) dt$$

$$= 2\frac{110}{T} \int_0^{\frac{T}{2}} (\sin((1-2k)\omega_0 t) + \sin((1+2k)\omega_0 t)) dt$$

$$= -2\frac{110}{T} \left(\frac{\cos((1-2k)\omega_0 t)}{(1-2k)\omega_0} + \frac{\cos((1+2k)\omega_0 t)}{(1+2k)\omega_0}\right) \Big|_0^{\frac{T}{2}}$$

$$= \frac{220}{\pi} \left(\frac{1}{1-4k^2}\right)$$

$$v(t) = \frac{220}{\pi} + \sum_{k=1}^{\infty} \frac{440}{\pi(1-4k^2)} \cos(2k(377)t)$$

$$= \frac{220}{\pi} - \frac{440}{3\pi} \cos(2\omega_0 t) - \frac{4}{15\pi} \cos(4\omega_0 t) - \frac{4}{35\pi} \cos(6\omega_0 t) + \cdots.$$

Figure 8.2b shows its DC component and its reconstruction with the DC and the second harmonic. Figure 8.2c shows its reconstruction with the DC and the second and fourth harmonics. Figure 8.2d shows its reconstruction with the DC and the second, fourth, and sixth harmonics. The decomposition of waveforms in Fourier analysis basically serves two purposes. One is to reconstruct the waveform by truncating the series (for compression). In this type of application, all the harmonics are of interest. In another type, the harmonic content has to be reduced. For example, they are unwanted in making a DC power supply. Sometimes, they may represent noise components of a signal. In Fourier reconstruction of the waveform, the waveform resembles the original waveform as more and more harmonics are used in the reconstruction, as shown in Fig. 8.2b–d. In applications, where the magnitude of the harmonics has to be reduced, the harmonic content is reduced to a desired level. Figure 8.2e, f shows, respectively, the fourth and sixth harmonics of the rectified voltage waveform. For filtering, harmonics have to be reduced to an acceptable level.

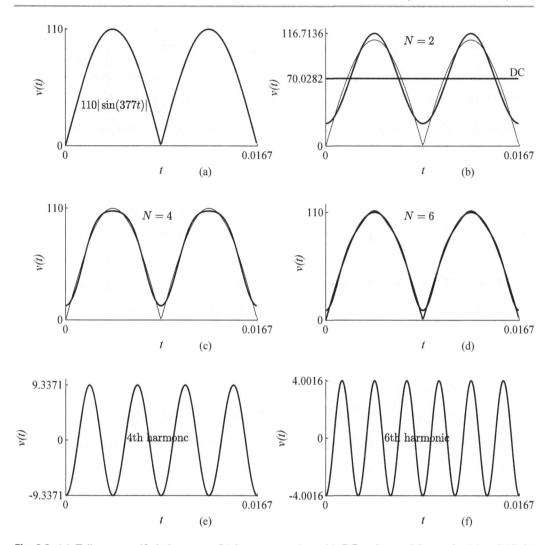

Fig. 8.2 (**a**) Full-wave rectified sine wave; (**b**) its reconstruction with DC and second harmonic; (**c**) and (**d**) its reconstruction with the DC, second, fourth, and sixth harmonics, respectively; (**e**) its fourth harmonic; (**f**) its sixth harmonic

Fourier decomposition of the square wave is usually presented as the first example in teaching Fourier analysis. One reason is that it is important in the operation of digital circuits. Another reason is that it brings out a minor shortcoming of the Fourier analysis. The sinusoidal waveforms are used in Fourier reconstruction of waveforms. It is not possible to reconstruct a square pulse with sharp edges exactly by smooth sinusoids, as presented later.

Example 8.2 Find the FS for a square wave defined over one period as

$$v(t) = \begin{cases} 1 & \text{for } 0 < t < \pi \\ 0 & \text{for } \pi < t < 2\pi \end{cases}$$

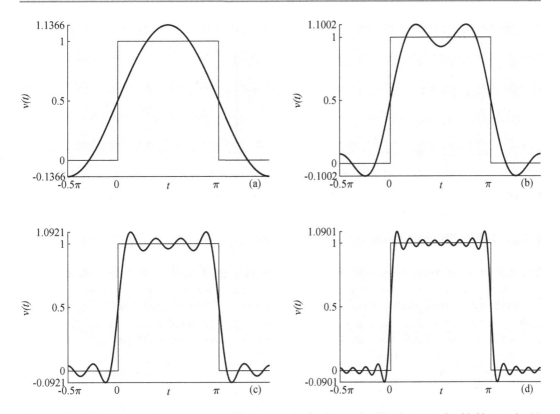

Fig. 8.3 The FS reconstructed square wave. (**a**) Using up to the first harmonic; (**b**) using up to the third harmonic; (**c**) using up to the seventh harmonic; (**d**) using up to the fifteenth harmonic

Solution The waveform is shown in Fig. 8.3. The period of the waveform is $T = 2\pi$ and the fundamental frequency is $\omega_0 = 1$. In contrast to the rectified waveform in the last example, this waveform is odd- and odd half-wave symmetric. Therefore, it is composed of odd-indexed sine components only, in addition to the DC component.

$$X(0) = \frac{1}{2\pi} \int_0^\pi dt = \frac{1}{2}$$

$$X(k) = \frac{2}{2\pi} \int_0^\pi \sin(k\,t)dt = \begin{cases} \frac{2}{k\pi} & \text{for } k \text{ odd} \\ 0 & \text{for } k \text{ even and } k \neq 0 \end{cases}$$

The coefficient $X(0)$ is the average value of $v(t)$.

$$v(t) = \frac{1}{2} + \frac{2}{\pi}(\sin(t) + \frac{1}{3}\sin(3t) + \frac{1}{5}\sin(5t) - \cdots). \tag{8.3}$$

In exponential form, the magnitude of the components, except DC, gets divided by 2. The FS magnitude spectrum and the phase spectrum of the signal in exponential form are shown, respectively, in Fig. 8.4a, b. ∎

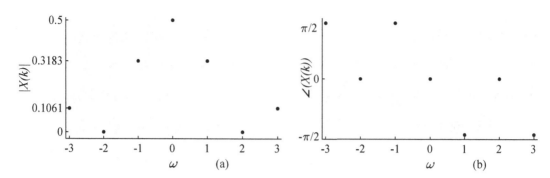

Fig. 8.4 (**a**) The FS magnitude spectrum and (**b**) the phase spectrum of the square wave in exponential form

8.1.2 Gibbs Phenomenon

The reconstructed square waveforms using up to the first, third, seventh, and fifteenth harmonics are shown in Fig. 8.3a–d, respectively. The magnitude of the harmonics decreases with increasing frequency. The more smoother the waveform, the faster is the decay. For waveforms with discontinuities, such as the square wave, the magnitude decrease is inversely proportional to the harmonic number. For waveforms without any discontinuity, the reconstructed waveform converges to the original waveform with more number of harmonics.

At any discontinuity, the reconstructed waveform never converges even in the limit. There is a 8.69% deviation, called the Gibbs phenomenon. However, the area under the deviation tends to zero. Let p and q be the values of the given waveform at either side of a discontinuity. The reconstructed waveform converges to a value r. As the criterion of convergence is in a least squares error sense, at the point of discontinuity,

$$(p-r)^2 + (q-r)^2$$

must be minimum. By differentiating the expression with respect to r and then equating it to zero, we get $p - r + q - r = 0$ and $r = (p+q)/2$. The expression for the FS reconstructed waveform up to the first harmonic is $v(t) = \frac{1}{2} + \frac{2}{\pi}\sin(t)$. Differentiating this expression with respect to t and equating it to zero, we get $\cos(t) = 0$. The point $t = \pi/2$ is a solution to this equation. Substituting $t = \pi/2$ in the expression for $v(t)$, we get the value of the peak as 1.1366, as shown in Fig. 8.3a. The maximum overshoots in other cases can be found similarly.

8.2 Fourier Transform

The FT is the limiting case of the FS as the period of the waveform tends to infinity. The frequency increment tends to zero. Then, we get a transform pair in which both the time-domain signal and its spectrum are aperiodic and continuous. The process of finding the spectrum remains the same in that we find the integral of products of the signal with complex exponentials. The limit of the integral is from $-\infty$ to ∞ and there are an infinite number of frequency components in the continuous spectrum. The FT $X(j\omega)$ of $v(t)$ is defined as

$$X(j\omega) = \int_{-\infty}^{\infty} v(t)e^{-j\omega t}dt. \tag{8.4}$$

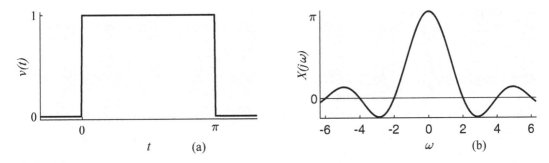

Fig. 8.5 (a) The pulse $v(t) = u(t) - u(t - \pi)$ and (b) its FT spectrum

The inverse FT $v(t)$ of $X(j\omega)$ is defined as

$$v(t) = \frac{1}{2\pi} \int_{-\infty}^{\infty} X(j\omega) e^{j\omega t} d\omega. \tag{8.5}$$

Example 8.3 Find the FT of the rectangular pulse $v(t) = u(t) - u(t - a)$, where $u(t)$ is the unit-step signal.

Solution

$$X(j\omega) = \int_{0}^{a} e^{-j\omega t} dt = \frac{2 \sin(0.5\omega a)}{\omega} e^{-j0.5\omega a}$$

$$u(t) - u(t - a) \leftrightarrow \frac{2 \sin(0.5\omega a)}{\omega} e^{-j0.5\omega a}.$$

The pulse and its FT are shown, respectively, in Fig. 8.5a, b with $a = \pi$. ∎

Fourier transform is the most generalized version of the Fourier analysis. The FS of signal, which is a repetition of an aperiodic signal, can be deduced from the FT. The relationship is given by

$$X(k) = \frac{1}{T} X(jk\omega_0).$$

The discrete samples of the FT spectrum at intervals of the fundamental frequency, ω_0, divided by the period are the FS spectrum. The reason for the FT spectrum to be continuous is that it has to reconstruct the pulse, in addition to the infinite extent zeros on either side of the pulse.

Let us determine the FS for the example from the corresponding FT. Since $X(k) = \frac{1}{T} X(jk\omega_0)$, with $T = 2\pi$ and $\omega = k\omega_0 = k\frac{2\pi}{2\pi} = k$, we get

$$X(k) = \frac{2}{2\pi} \frac{\sin(0.5\pi k)}{k} e^{-j0.5\pi k} = \frac{1}{\pi} \frac{\sin(\frac{\pi}{2} k)}{k} e^{-j\frac{\pi}{2} k}, \ k \neq 0 \quad \text{and} \quad X(0) = \frac{1}{2}.$$

The FS is

$$x(t) = \frac{1}{2} + \sum_{k=-\infty}^{\infty} \frac{1}{\pi} \frac{\sin(\frac{\pi}{2} k)}{k} e^{-j\frac{\pi}{2} k} e^{jkt}, \ k \neq 0,$$

as found earlier. The even-indexed coefficients are zero and the coefficients of the sine component of the complex exponentials are $2/(k\pi)$. For example, with $k = 1$ and $k = -1$ and leaving the factor π

in the denominator, we get

$$(1)(-j1)e^{jt} = -je^{jt} = \frac{e^{jt}}{j}, \quad (1)(j1)e^{-jt} = je^{-jt} = -\frac{e^{-jt}}{j},$$

$$\frac{e^{jt} - e^{-jt}}{j} = 2\sin(t).$$

Since t or ω is zero, the values $X(j0)$ and $v(0)$ can be easily determined and can be used to verify the closed-form expressions for $X(j\omega)$ or $v(t)$.

$$X(j0) = \int_{-\infty}^{\infty} v(t)dt \quad \text{and} \quad v(0) = \frac{1}{2\pi} \int_{-\infty}^{\infty} X(j\omega)d\omega.$$

For the example waveform, the area of $v(t)$ is π, which is $X(j0)$. The area of the spectrum is the area inscribed by the main lobe of the oscillating spectrum. The area is 2π, which divided by 2π is $v(0) = 1$.

8.2.1   The Transfer Function and the Frequency Response

The FT extends Fourier analysis for the analysis of circuits with aperiodic source signals, which is important in that most signals in applications are aperiodic. The transfer function is a frequency-domain model of systems that relates the input and output. As multiplication with it transfers input to output, it is called as a transfer function, $H(j\omega)$. It is defined as the ratio of the output response, $Y(j\omega)$, to the input excitation, $V(j\omega)$, where $V(j\omega)$ and $Y(j\omega)$ are, respectively, the FT of input $v(t)$ and output $y(t)$. That is,

$$H(j\omega) = \frac{Y(j\omega)}{V(j\omega)}.$$

The transfer function is also called the frequency response, as it is the response to the input $e^{j\omega t}$ with the frequency continuously varying from $-\infty$ to ∞. The frequency response clearly depicts the response of the circuit for various frequency components.

The sinusoidal and the impulse signals are the most important waveforms in the analysis of signals and systems. The sinusoid is the basis signal in the frequency-domain and the impulse is the basis signal in the time-domain. The unit-impulse, as presented in detail later, is characterized by its unit area enclosed at $t = 0$.

Example 8.4 Determine the FT of the unit-impulse signal $v(t) = \delta(t)$.

Solution As it encloses unit area at $t = 0$, we get

$$X(j\omega) = \int_{-\infty}^{\infty} \delta(t)e^{-j\omega t}dt = e^{-j\omega 0} \int_{-\infty}^{\infty} \delta(t)dt = 1 \qquad \text{and} \qquad \delta(t) \leftrightarrow 1.$$

Since the spectrum is 1, the impulse is composed of components of all frequencies from $\omega = -\infty$ to $\omega = \infty$ in equal proportion. That is,

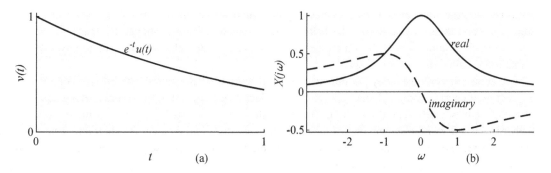

Fig. 8.6 (a) $v(t) = e^{-t}u(t)$ and (b) its FT spectrum. The real part of the FT is shown by the continuous line and the imaginary part is shown by the dashed line

$$\delta(t) = \frac{1}{2\pi} \int_{-\infty}^{\infty} e^{j\omega t} d\omega = \frac{1}{\pi} \int_{0}^{\infty} \cos(\omega t) d\omega.$$

∎

The real exponential signal is also important in that the natural response of systems is of that form and it occurs in problems involving exponential growth and decay of signals, such as capacitor discharge and computation of compound interest.

Example 8.5 Find the FT $X(j\omega)$ of the exponential signal $v(t) = e^{-t}u(t)$, where $u(t)$ is the unit-step signal.

Solution

$$X(j\omega) = \int_{0}^{\infty} e^{-t} e^{-j\omega t} dt = \int_{0}^{\infty} e^{-(1+j\omega)t} dt = -\frac{e^{-(1+j\omega)t}}{1+j\omega}\bigg|_{0}^{\infty} = \frac{1}{1+j\omega}$$

$$e^{-t}u(t) \leftrightarrow \frac{1}{1+j\omega}.$$

Figure 8.6a, b show, respectively, the signal $e^{-t}u(t)$ and its FT spectrum. The real part of the spectrum (continuous line) is an even function with a peak value of 1 at $\omega = 0$ and the imaginary part (dashed line) is an odd function with peaks of value ± 0.5 at $\omega = \mp 1$. ∎

8.3 Transient Response

So far, we considered the steady-state response of circuits. In applications, such as control systems, both the steady-state and transient responses are important. The response of systems can be considered as composed of two parts. One is the steady-state response, which is the state of the system after equilibrium conditions prevail upon the application of a source, whose form is similar to the input source. The other one is the transient response, which dies down in relatively short time for stable systems. If we start a water pump, we do not get the full flow of water instantly. The connecting pipes in the system have to be filled up first. Therefore, at the instant we start, there is no water flow. After a short while, the water flow is as expected. Again, when we put the pump off, the water flows for some time and then only it stops. The behavior of a system, before it attains its steady state after the excitation is applied or to become dead from its steady-state behavior after the removal of the excitation is called its transient response.

The transient response may be due to the input source or the initial conditions. For circuits consisting of energy storage elements, such as an inductor or a capacitor, the characterizing equation of the system is a differential equation. The more the number of such elements in the circuit, the higher is the order of the system and of the characterizing differential equation. Therefore, transform analysis is required for solving higher-order differential equations. Consequently, we use transform methods to solve the differential equations.

8.3.1   The Unit-Impulse and Unit-Step Signals

In addition to the sinusoid, the unit-impulse and the unit-step signals, shown in Fig. 8.7a, b respectively, are often required in the analysis of signals and systems, particularly for transient analysis. Switching on a power supply to a system is a common operation. While practical switches require finite time in transition from one state to the other state, for mathematical convenience, we assume that the switching occurs instantaneously. Although it has a discontinuity and no derivative in the normal function theory, the impulse is derived by a limiting process to be its derivative.

The unit-impulse is characterized by its unit area enclosed at $t = 0$. It is defined by an integral of the product of a continuous signal and itself as

$$\int_{-\infty}^{\infty} v(t)\delta(t)\, dt = v(0).$$

The impulse with zero width and infinite area is difficult to visualize. In practice, an approximation with a suitable finite width and unit area is adequate for understanding and measurement purposes. The integral of the product of a function with unit-area pulse evaluates to the average value of the function during the width of the pulse. As the width is reduced, the pulse resembles more like an impulse and the accuracy improves. In the limit as the width tends to zero, the exact value of the function is returned.

The pulse, $\delta_q(t)$ which approximates the impulse, and the signal $v(t) = e^t$ are shown in Fig. 8.8a over a width 2. Their product $v(t)\delta_q(t)$ is shown by the dotted line. The integral of the product is 1.1752 with four-digit precision. $v(t)$ is approximated by, in terms of the pulse, $(1.1752)(2)\delta_q(t)$ during the width.

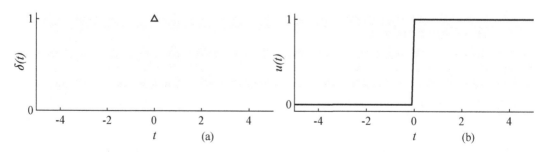

Fig. 8.7 (**a**) The unit-impulse signal; (**b**) the unit-step signal

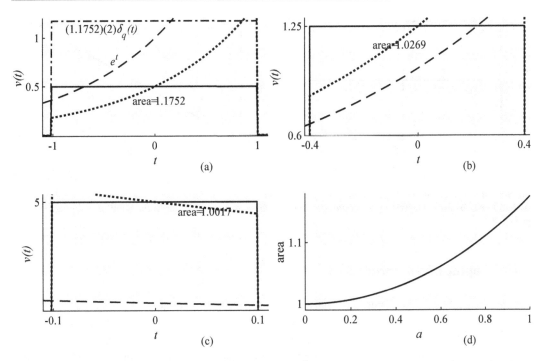

Fig. 8.8 (**a**) The pulse $\delta_q(t)$ (solid line) with width 2 and height $\frac{1}{2} = 0.5$. The function $v(t) = e^{-t}$ (dashed line) and the product $\delta_q(t)v(t)$ (dotted line). (**b**) Same as (**a**) with width 0.8. (**c**) Same as (**a**) with width 0.2. (**d**) The area enclosed by the product $\delta_q(t)v(t)$ for various values of width a

Figure 8.8b, c shows the functions with widths 0.8 and 0.2, respectively. As the pulse width is reduced, the variation in the amplitude of the function $v(t) = e^{-t}$ is also reduced and the integral of the product $\delta_q(t)v(t)$, which is the local average of $v(t)$, approaches the value $v(0) = 1$, as shown in Fig. 8.8d.

The Unit-Impulse as the Derivative of the Unit-Step

The unit-impulse has an area of 1 at $t = 0$. If we integrate it from $t = -\infty$ to ∞, the integral will evaluate to zero for $t < 0$ and to 1 for $t \geq 0$, which is the unit-step function. Consider the unit-area pulse of width 1 and height 1, shown in Fig. 8.9a by a solid line, which is an approximation to the unit-impulse. The integral of it is a ramp function with slope 1 from $t = 0$ to 1 and remains 1 after that, which is an approximation to the unit-step function. As we reduce the width of the unit-area pulse, as shown by the dashed line, the pulse becomes a better approximation to the impulse and, consequently, its integral approximates the unit-step function better, as shown in Fig. 8.9b. As the width tends to zero, the pulse becomes the unit-impulse and its integral becomes the unit-step. The result is obtained by limiting process. Therefore,

$$\frac{du(t)}{dt} = \delta(t) \qquad \text{and} \qquad \int_{-\infty}^{t} \delta(\tau)d\tau = u(t).$$

For example, the current through an inductor is proportional to the integral of the voltage across it. Therefore, a unit-impulse voltage applied across an inductor of one henry will produce a unit-step current through it.

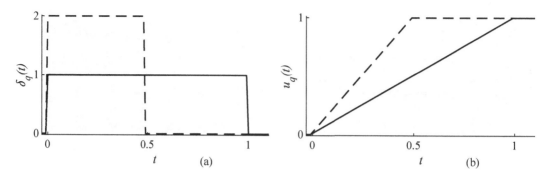

Fig. 8.9 (a) The quasi-impulse $\delta_q(t)$ with width 1 and height $\frac{1}{1} = 1$ (solid line), and with width 0.5 and height $\frac{1}{0.5} = 2$ (dashed line); (b) their integrals $u_q(t)$, approximating the unit-step function better as the width of the quasi-impulse tends to zero

8.4   Laplace Transform

For system analysis, where both the transient and steady-state responses are required, the generalization of Fourier analysis, called the Laplace transform, is more convenient. There are three
differences from Fourier analysis. One is that it is specifically designed for causal signals, for signals
$v(t) = 0, t < 0$, which enables the easier analysis of systems with initial conditions, such as a charged
capacitor. Another difference is that it can be used to study unstable systems also, as the basis signals
include exponentially growing and decaying sinusoids in contrast to constant amplitude sinusoids
used in Fourier analysis. Another difference is that the frequency variable is s rather than $j\omega$ making
it easier to manipulate expressions. These differences make it highly suitable for system analysis. The
basic principle of transforms, correlation of the signal with the basis signals to find the frequency
content remains the same.

The one-sided or unilateral Laplace transform $X(s)$ of the time-domain signal $x(t)$ is defined as

$$X(s) = \int_{0^-}^{\infty} x(t)e^{-st}dt,$$

where $s = (\sigma + j\omega)$. If $\sigma = 0$, the definition resembles that of the FT. The lower limit is 0^-, which
implies that the condition of the signal immediately before $t = 0$ is taken into account. That is, any
jump discontinuities or impulses at $t = 0$ are included in the analysis. Further, with this definition,
handling of the initial conditions at $t = 0^-$ becomes easier.

Example 8.6 Determine the Laplace transform of the unit-impulse signal, $\delta(t)$ from the definition.

Solution

$$X(s) = \int_{0^-}^{\infty} \delta(t)e^{-st}dt = 1, \text{ for all } s \qquad \text{and} \qquad \delta(t) \leftrightarrow 1, \text{ for all } s.$$

Remember that the impulse is characterized by its unit area at $t = 0$.   ∎

Example 8.7 Determine the Laplace transform of the real exponential signal, $e^{-at}u(t)$ from the definition. Substitute $a = 0$ in the transform obtained and get the Laplace transform of the unit-step signal, $u(t)$.

Solution

$$X(s) = \int_{0^-}^{\infty} e^{-at}u(t)e^{-st}dt = \int_{0^-}^{\infty} e^{-at}e^{-st}dt$$

$$= \int_{0^-}^{\infty} e^{-(s+a)t}dt = -\frac{e^{-(s+a)t}}{s+a}\Big|_{0^-}^{\infty} = \frac{1}{s+a} - \frac{e^{-(s+a)t}}{s+a}\Big|_{t=\infty}.$$

While the Laplace transform is applicable to a larger class of signals, for some signals the transform does not exist. Therefore, the condition, called the region of convergence, has to be mentioned for each transform. The Laplace transform pair for the exponential signal becomes

$$e^{-at}u(t) \leftrightarrow \frac{1}{s+a}, \quad \text{Re}(s) > -a.$$

For complex-valued a, the convergence condition is $\text{Re}(s) > \text{Re}(-a)$.

Substituting $a = 0$, we get the transform pair for the unit-step signal $u(t)$ as

$$u(t) \leftrightarrow \frac{1}{s}, \quad \text{Re}(s) > 0.$$

∎

With $a = \mp j\omega_0$,

$$e^{j\omega_0 t}u(t) \leftrightarrow \frac{1}{s - j\omega_0}, \quad \text{and} \quad e^{-j\omega_0 t}u(t) \leftrightarrow \frac{1}{s + j\omega_0}, \quad \text{Re}(s) > 0.$$

Then,

$$\sin(\omega_0 t)u(t) = -0.5j(e^{j\omega_0 t}u(t) - e^{-j\omega_0 t}u(t)) \leftrightarrow \frac{\omega_0}{s^2 + \omega_0^2}, \quad \text{Re}(s) > 0.$$

8.4.1 Properties of the Laplace Transform

The use of properties makes the analysis easier. Signals and their transforms can be decomposed in terms of simpler components. This makes it easy to find the transform of complicated signals. Further, the inverse transform is easily obtained.

Linearity
The transform of a linear combination of signals is the same linear combination of their individual transforms. This property is the basis for transform analysis. Complex signals and their transforms can be easily obtained by decomposing them into a linear combination of simpler signals.

8.4.2 Time-Differentiation

This property expresses the transform of the time derivative of a signal in terms of its transform. If $v(t) \leftrightarrow X(s)$, then

$$\frac{d\,v(t)}{dt} \leftrightarrow sX(s) - v(0^-),$$

where $v(0^-)$ the initial value of the signal at $t = 0^-$. As the signal is expressed in terms of exponentials of the form $X(s)e^{st}$, the transform of the derivative is $sX(s)$, as given by the first term. The second term is the derivative at $t = 0$. This property makes the analysis of systems with initial conditions easier.

This property can be extended, by repeated application, to find the transform of higher-order derivatives. For example,

$$\frac{d}{dt}\left(\frac{d\,v(t)}{dt}\right) = \frac{d^2 v(t)}{dt^2} \leftrightarrow$$

$$s(sX(s) - v(0^-)) - \frac{d\,v(t)}{dt}\Big|_{t=0^-} = s^2 X(s) - sv(0^-) - \frac{d\,v(t)}{dt}\Big|_{t=0^-}$$

The entity

$$\frac{d\,v(t)}{dt}\Big|_{t=0^-}$$

is the value of the derivative of $v(t)$ at $t = 0^-$.

In circuit analysis, this property is used to model inductors with initial conditions. The input–output relationship of an inductor is

$$v(t) = L\frac{di(t)}{dt},$$

where $v(t)$ is the voltage across the inductor and $i(t)$ is the current flowing through it. The value of the inductor is L henries. Let the initial value of current in the inductor be $i(0^-)$. From the time-differentiating property, we get the value of the voltage across the inductor as, in the frequency-domain,

$$V(s) = L(sI(s) - i(0^-)).$$

In common with other transforms, the linearity and time-differentiation properties reduce the differential equation characterizing a system into algebraic equations, making the analysis simpler. In addition, transform methods bring out the salient properties of signals and systems.

8.4.3 Integration

If $v(t) \leftrightarrow X(s)$, then

$$\int_{0^-}^{t} v(\tau)\,d\tau \leftrightarrow \frac{1}{s}X(s).$$

As the integration can be considered as the inverse of the derivative property for a signal with zero DC component, the variable s appears in the denominator in the transform of the integral of the signal.

Of particular interest to circuit analysis is that this property enables easier analysis of components with initial conditions, such as a capacitor with some initial charge. The input–output relationship of a capacitor is

$$v(t) = \frac{1}{C}\int_{0^-}^{t} i(\tau)\,d\tau + v(0^-),$$

where $v(0^-)$ is the initial value of voltage across the capacitor, $v(t)$ is the voltage across the capacitor, and $i(t)$ is the current flowing through it. The value of the capacitor is C farads. From the time-integration property, we get the value of the voltage across capacitor as, in the frequency-domain,

$$V(s) = \frac{I(s)}{sC} + \frac{v(0^-)}{s}.$$

8.4.4 Initial Value

The initial and final values of a function can be easily found using the initial and final value properties of the transform. These values can be used to check the transforms of the signals. If $v(t) \leftrightarrow X(s)$ and the degree of the numerator polynomial of $X(s)$ is less than that of the denominator polynomial, then

$$v(0^+) = \lim_{s \to \infty} sX(s).$$

8.4.5 Final Value

If $v(t) \leftrightarrow X(s)$ and the ROC of sX(s) includes the $j\omega$ axis, then

$$\lim_{t \to \infty} v(t) = \lim_{s \to 0} sX(s).$$

The use of these properties is demonstrated in the following examples.

8.4.6 Circuit Analysis in the Frequency-Domain

Circuits are composed of sources, resistors, capacitors, and inductors. In the time-domain, circuits are characterized by differential equations. We can write down the differential equation and use the Laplace transform to solve it. However, it is easier to represent the circuits in the frequency-domain directly and solve it, as though it is a resistive circuit. We used this procedure in Chap. 3 to find the steady-state response of circuits. Now, we study the Laplace transform method of analyzing circuits with or without initial conditions to find the complete response. The complete response consists of both the transient and steady-state responses. Using the Laplace transform to represent the circuit elements, the analysis of circuits becomes algebraic, as is the case with other transforms also. It is no longer necessary to solve the differential equations to find the response. The Laplace transform is a generalization of the Fourier analysis and is particularly useful to analyze circuits with initial conditions. Further, the response can be found to unbounded signals also. In the transform domain representation also, all the analysis methods, such as loop and nodal, and all the circuit theorems are equally applicable.

The input–output relationship of an inductor with no initial current is

$$v(t) = L\frac{di(t)}{dt},$$

where $v(t)$ and $i(t)$ are, respectively, the voltage across and the current through the inductor of value L henries. In the transform domain, the independent variable is the frequency and the basis functions are of the form e^{st}. The derivative of this function with respect to t, se^{st}, is of the same form and multiplied by the frequency variable s. In taking the Laplace transform, the input waveform is decomposed in terms of frequency components of all frequency. Therefore, the representation of an inductor is similar to that of a resistor. That is,

$$V(s) = LsI(s).$$

If we replace s by $j\omega$, then it becomes a Fourier-domain representation. The difference is that the Laplace transform is a generalized version of the Fourier analysis. This generalization brings the advantages of analyzing circuits with initial conditions and also with unbounded signals. Each transform is more suitable for certain types of circuit analysis.

For a capacitor with no initial voltage, the input–output relationship is

$$v(t) = \frac{1}{C} \int_0^t i(t)\,dt,$$

where $v(t)$ and $i(t)$ are, respectively, the voltage across and the current through the capacitor of value C farads. The representation of a capacitor, in the transform domain, is similar to that of a resistor. That is,

$$V(s) = \frac{1}{Cs} I(s).$$

For a resistor of value $R\,\Omega$,

$$V(s) = RI(s).$$

The resistance offered by a resistor for the flow of current through it is independent of the frequency of excitation. Summarizing, in the frequency-domain, the voltage–current relationship of all the elements is algebraic, with their impedance values R, sL, and $1/Cs$. The impedance of an element is the ratio $V(s)/I(s)$, assuming zero initial conditions.

Resistor is not a storage device. The current through and the voltage across it are instantaneous. Capacitors and inductors have the capability to store energy. These devices can be represented as a device with zero initial condition and an appropriate source in addition. For a capacitor with an initial voltage $v(0^-)$,

$$i(t) = C\frac{dv(t)}{dt}.$$

Taking the Laplace transform of this expression, we get

$$I(s) = C(sV(s) - v(0^-)) \quad \text{or} \quad V(s) = \frac{I(s)}{sC} + \frac{v(0^-)}{s}.$$

The charged capacitor is represented as an impedance $\frac{1}{sC}$ in series with an ideal voltage source $\frac{v(0^-)}{s}$. The additional voltage source in series becomes a short-circuit with no initial charge.

By rearranging the expression for $V(s)$, we get an alternative representation as

$$V(s) = \frac{1}{sC}\left(I(s) + Cv(0^-)\right).$$

The voltage across the capacitor is due to the current $(I(s) + Cv(0^-))$ flowing through it, impedance multiplied by the current. This form of representation implies an uncharged capacitor in parallel with an impulsive current source $Cv(0^-)$. There is an additional current source in parallel, which becomes an open-circuit with no initial charge.

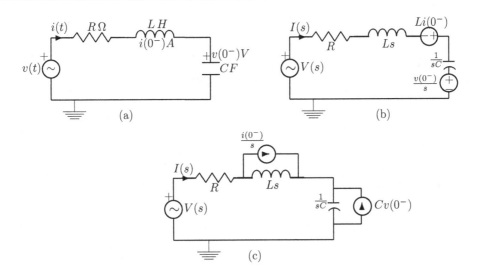

Fig. 8.10 Modeling of inductors and capacitors with initial conditions in the frequency-domain

In the time-domain, an inductor is characterized by

$$v(t) = L\frac{di(t)}{dt}.$$

In the frequency-domain, we get

$$V(s) = L(sI(s) - i(0^-)) = LsI(s) - Li(0^-) = sL\left(I(s) - \frac{i(0^-)}{s}\right).$$

The inductor is modeled as an impedance sL in series with an ideal impulsive voltage source $-Li(0^-)$, which becomes a short-circuit with no initial current.

Alternatively, the voltage across the inductor is due to the current $(I(s) - \frac{i(0^-)}{s})$ flowing through it. An inductor with initial current $i(0^-)$ is modeled as an inductor, with no initial current, in parallel with a current source $-\frac{i(0^-)}{s}$, which becomes an open-circuit with no initial current. Modeling of inductors and capacitors with initial conditions in the frequency-domain is shown in Fig. 8.10. Figure 8.10a shows a RLC series circuit in the time-domain with initial current $i(0^-)$ in the inductor and initial voltage $v(0^-)$ in the capacitor. Figure 8.10b, c shows, respectively, the circuit in the frequency-domain with series and parallel initial condition generators with impedances Ls and $1/sC$.

Example 8.8 The input voltage is the unit-step function, $u(t)$. Assume that the series circuit consists of only the resistor with value $R = 2\,\Omega$ and the inductor with value $L = 3\,$H, shown in Fig. 8.11a. Let the initial current through the inductor be zero. Find the current through the circuit and the voltage across the inductor after the excitation is applied. Deduce the impulse response of the circuit.

Solution The excitation and the impedance, in the Laplace transform domain, are

$$V(s) = \frac{1}{s} \quad \text{and} \quad Z(s) = R + Ls$$

Therefore,

$$I_L(s) = \frac{V(s)}{Z(s)} = \frac{1}{s(R + Ls)} = \frac{1}{Ls(s + R/L)} = \frac{1}{R}\left(\frac{1}{s} - \frac{1}{(s + R/L)}\right)$$

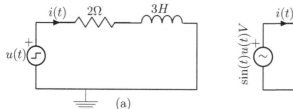
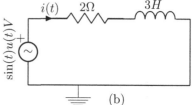

Fig. 8.11 A RL circuit in time-domain

Applying initial and final value theorems, we get

$$i(0^+) = \lim_{s \to \infty} s \frac{1}{s(R + Ls)} = 0$$

$$\lim_{t \to \infty} i(t) = \lim_{s \to 0} s \frac{1}{s(R + Ls)} = \frac{1}{R}$$

For the example, $R = 2$ and $\lim_{t \to \infty} i(t) = 0.5$. These values can be used to check the complete response derived. Taking the inverse Laplace transform, we get

$$i_L(t) = \frac{1}{R}(1 - e^{-\frac{R}{L}t})u(t).$$

With $Z_L(s) = Ls$,

$$V_L(s) = I_L(s)Z_L(s) = \frac{1}{(s + R/L)} \quad \text{and} \quad v_L(t) = e^{-\frac{R}{L}t}u(t).$$

The unit-step responses are

$$v_L(t) = e^{-\frac{R}{L}t}u(t) \quad \text{and} \quad i_L(t) = \frac{1}{R}(1 - e^{-\frac{R}{L}t})u(t).$$

Since the impulse is the derivative of the unit-step function, as the circuit is linear, the impulse response is the derivative of the response to the unit-step excitation. The derivative of these expressions, the unit-impulse responses are

$$v_L(t) = -\frac{R}{L}e^{-\frac{R}{L}t}u(t) \quad \text{and} \quad i_L(t) = \frac{1}{L}e^{-\frac{R}{L}t}u(t).$$

∎

Figure 8.12a, b shows, respectively, the voltage across the inductor and the current through for the unit-step source voltage for the example. The voltage is 1 at $t = 0$ and asymptotically approaches zero as $t \to \infty$. For larger values of inductance, the rate of fall is low. The current is 0 at $t = 0$ and reaches the limit $1/2 = 0.5$. The inductor influences the nature of the response only during the initial period, called the transient interval. In steady state, the impedance of the inductor is zero, as frequency of excitation is zero. That is, the excitation is constant. Figure 8.13a, b shows, respectively, the voltage across the inductor and the current through for the unit-impulse source voltage. These responses are

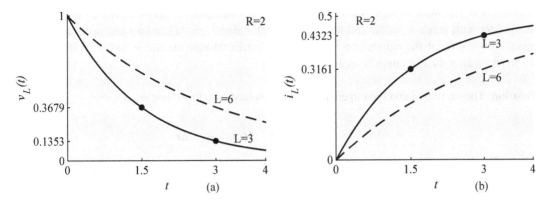

Fig. 8.12 The unit-step response of the series RL circuit. (**a**) the voltage across the inductor; (**b**) the current

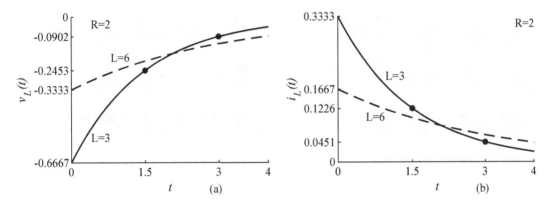

Fig. 8.13 The unit-impulse response of the series RL circuit. (**a**) the voltage across the inductor; (**b**) the current

the derivative of the responses for the unit-step excitation. The voltage across the resistor and that induced in the inductor are of the same magnitude with opposite polarities, since the excitation is zero for $t > 0$ and KVL has to be satisfied.

With the source connected, the inductor stores energy $0.5Li^2(t)$ and the resistor dissipates energy at the rate of $Ri^2(t)$. When the source disconnected, the stored energy is eventually dissipated by the resistor. The rate of energy dissipation, which is controlled by the ratio L/R, decreases with the time and finally both the energy and the current become zero. The larger the value of L/R, the smaller will be the rate of decay.

The value of the exponential signal $e^{-kt}u(t)$ is 1 at $t = 0$. At $t = 1/k$, the value gets reduced to $e^{-1} = 1/e = 0.3679$. At $t = 2/k$, the value gets reduced to $e^{-2} = e^{-1}e^{-1} = 0.1353$. The value $1/k$ is called time constant of the corresponding circuit. The time constant is useful for comparison of different responses. The time constant for the RL circuit is L/R. The transient response is zero, only when $t \to \infty$. For practical purposes, the duration of four time constants is considered as the duration of the transient response, as the response increases to 98% of the final value for a growing exponential and the response reduces to below 2% for a decaying exponential.

AC or DC and transient or steady state, the constraint is that KVL and KCL are to be satisfied at any part of a linear circuit at any instant. The usual difference of impedances replacing resistors has to be taken into account. The DC excitation is replaced by an AC excitation.

Example 8.9 The input voltage is $v(t) = \sin(t)u(t)$. Assume that the series circuit consists of only the resistor with value $R = 2\,\Omega$ and the inductor with value $L = 3\,\text{H}$, shown in Fig. 8.11b. Let the initial current through the inductor be zero. Find the current through the circuit and the voltage across the inductor after the excitation is applied.

Solution The excitation and the impedance, in the Laplace transform domain, are

$$V(s) = \frac{1}{s^2 + 1} \quad \text{and} \quad Z(s) = R + Ls.$$

Therefore,

$$I_L(s) = \frac{V(s)}{Z(s)} = \frac{1}{(s^2 + 1)(R + Ls)} = \frac{1}{L(s^2 + 1)(s + R/L)}$$

$$= \frac{1}{3}\left(\frac{-9(0.5 + j/3)/13}{s - j} + \frac{-9(0.5 - j/3)/13}{s + j} + \frac{(0.6923)}{(s + 2/3)}\right)$$

$$= \frac{1}{3}\left(\frac{-0.3462 - j0.2308}{s - j} + \frac{-0.3462 + j0.2308}{s + j} + \frac{(0.6923)}{(s + 2/3)}\right).$$

For this example, applying the initial value theorem, we get $i(0^+) = 0$. The final value theorem is not applicable, as the response is oscillatory. Taking the inverse Laplace transform, we get

$$i_L(t) = (0.8320\cos(t - 2.5536) + 0.6923e^{-\frac{2}{3}t})/3u(t) = (0.2773\cos(t - 2.5536) + 0.2308e^{-\frac{2}{3}t})u(t)$$

Since $v_L(t) = L(di_L(t)/dt)$, by differentiating, we get

$$v_L(t) = (0.8320\sin(t - 2.5536) - 0.4615e^{-\frac{2}{3}t})u(t) = (0.8320\cos(t - 0.9828) - 0.4615e^{-\frac{2}{3}t})u(t)$$

∎

Figure 8.14a, b shows, respectively, the voltage across the inductor and the current through for the $v(t) = \sin(t)u(t)$ source voltage. The steady-state response is given by the sinusoidal

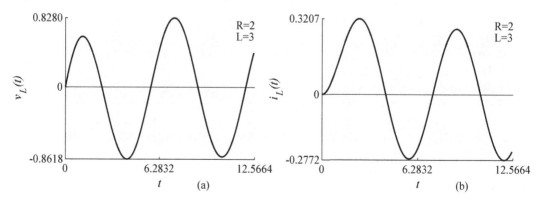

Fig. 8.14 The response of the series RL circuit to the input $v(t) = \sin(t)u(t)$. (**a**) the voltage across the inductor; (**b**) the current

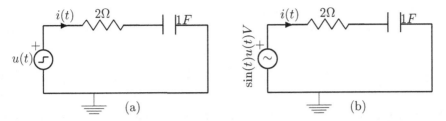

Fig. 8.15 A *RC* circuit in time-domain

component while the exponential component is the transient. The magnitude of the steady-state current component is

$$\left|\frac{-j}{2+j3}\right| = 0.2774.$$

The transient response eventually dies down. For $\cos(t)u(t)$ input, since the cosine function is the derivative of the sine function, we have to differentiate the expressions.

Example 8.10 The input voltage is the unit-step function, $u(t)$. Assume that the series circuit consists of only the resistor with value $R = 2\Omega$ and the capacitor with value $C = 1\,\text{F}$, shown in Fig. 8.15a. Let the initial voltage across the capacitor be zero. Find the current through the circuit and the voltage across the capacitor after the excitation is applied. Deduce the impulse response of the circuit.

Solution The excitation and the impedance, in the Laplace transform domain, are

$$V(s) = \frac{1}{s} \quad \text{and} \quad Z(s) = R + (1/Cs).$$

Therefore,

$$I_C(s) = \frac{V(s)}{Z(s)} = \frac{1}{s(R+(1/Cs))} = \frac{1}{R}\left(\frac{1}{s+(1/RC)}\right).$$

Applying initial and final value theorems, we get

$$i(0^+) = \frac{1}{R} \quad \text{and} \quad \lim_{t\to\infty} i(t) = 0.$$

With $R = 2$, $1/R = 1/2 = 0.5$. Taking the inverse Laplace transform, we get

$$i_C(t) = \frac{1}{R}e^{-\frac{t}{RC}}u(t).$$

With $Z_C = 1/Cs$,

$$V_C(s) = I(s)Z_C(s) = \frac{1}{RCs(s+(1/RC))} = \left(\frac{1}{s} - \frac{1}{(s+1/RC)}\right) \quad \text{and} \quad v_C(t) = (1 - e^{-\frac{t}{RC}})u(t)$$

The unit-step response is

$$v_C(t) = (1 - e^{-\frac{t}{RC}})u(t) \quad \text{and} \quad i_C(t) = \frac{1}{R}e^{-\frac{t}{RC}}u(t).$$

The derivative of these expressions, the unit-impulse response is

$$v_C(t) = \frac{1}{RC}e^{-\frac{t}{RC}}u(t) \quad \text{and} \quad i_C(t) = -\frac{1}{R^2C}e^{-\frac{t}{RC}}u(t).$$

■

Initially the impedance of the capacitor is zero and the current is limited by the resistor. In steady state, the excitation becomes DC and the current in the circuit becomes 0 as the impedance of the capacitor is infinite. The excitation voltage appears across the capacitor. With the source connected, the capacitor stores energy $0.5Cv^2(t)$ and the resistor dissipates energy at the rate of $Ri^2(t)$. When the source disconnected, the stored energy is eventually dissipated by the resistor. The rate of energy dissipation, which is controlled by the product RC, decreases with the time and finally both the energy and the current become zero. The larger the value of RC, the smaller will be the rate of decay. The time constant for the circuit is RC. Figure 8.16a, b shows, respectively, the voltage across the capacitor and the current through for the unit-step source voltage. Figure 8.17a, b shows, respectively, the voltage across the capacitor and the current through for the unit-impulse source voltage.

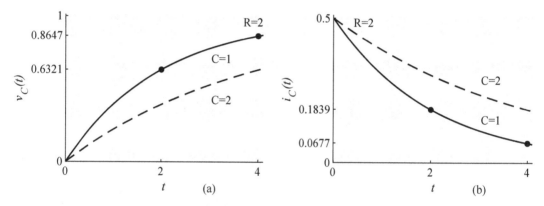

Fig. 8.16 The unit-step response of the series RC circuit. (**a**) the voltage across the capacitor; (**b**) the current

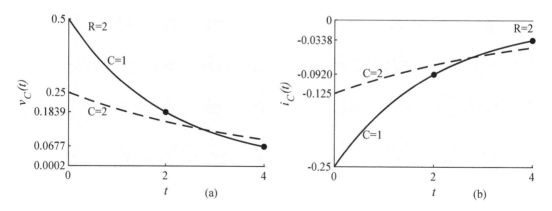

Fig. 8.17 The unit-impulse response of the series RC circuit. (**a**) the voltage across the capacitor; (**b**) the current

Example 8.11 The input voltage is $v(t) = \sin(t)u(t)$. Assume that the series circuit consists of only the resistor with value $R = 2\,\Omega$ and the capacitor with value $C = 1\,\text{F}$, shown in Fig. 8.15b. Let the initial current through the capacitor be zero. Find the current through the circuit and the voltage across the capacitor after the excitation is applied.

Solution The excitation and the impedance, in the Laplace transform domain, are

$$V(s) = \frac{1}{s^2 + 1} \quad \text{and} \quad Z(s) = R + (1/Cs).$$

Therefore,

$$I(s) = \frac{V(s)}{Z(s)} = \frac{s}{R(s^2 + 1)(s + (1/RC))} = \frac{0.5s}{(s^2 + 1)(s + 0.5)}$$

$$= \left(\frac{0.1 - j0.2}{s - j} + \frac{0.1 + j0.2}{s + j} + \frac{(-0.2)}{(s + 0.5)} \right).$$

Taking the inverse Laplace transform, we get

$$i_C(t) = (0.4472 \cos(t - 1.1071) - 0.2e^{-0.5t})u(t).$$

Since $v_C(t) = \frac{1}{C} \int i_C(t)\, dt$, by integrating, we get

$$v_C(t) = (0.4472 \sin(t - 1.1071) + 0.4e^{-0.5t})u(t).$$

∎

Figure 8.18a, b shows, respectively, the voltage across the capacitor and the current through for $v(t) = \sin(t)u(t)$ source voltage. The steady-state response is given by the sinusoidal component while the exponential component is the transient. The magnitude of the steady-state current component is

$$\left| \frac{-j}{2 - j1} \right| = 0.4472.$$

For $\cos(t)u(t)$ input, since the cosine function is the derivative of the sine function, we have to differentiate the expressions.

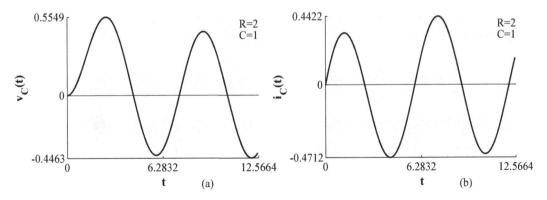

Fig. 8.18 The response of the series RC circuit to the input $v(t) = \sin(t)u(t)$. (**a**) the voltage across the capacitor (**b**) the current

Analysis of Circuits with Arbitrary Initial Conditions

The initial condition of a capacitor or an inductor can be represented by an appropriate equivalent voltage or current source. The sum of the responses due to these sources and the response due to some specific excitation, all computed with zero initial conditions, yields the total response of the circuit. Therefore, the procedure for analyzing a circuit with zero initial conditions is adequate to analyze circuits with arbitrary initial conditions.

Example 8.12 Assume that the series circuit consists of only the resistor with value $R = 2\,\Omega$ and the capacitor with value $C = 0.1\,\text{F}$. Let the initial voltage across the capacitor be $v_0^- = 1\,\text{V}$. Find the current through the circuit and the voltage across the capacitor for $t \geq 0$.

Solution The circuit is shown in Fig. 8.19a in the time-domain. The equivalent frequency-domain circuit is shown in Fig. 8.19b, with a voltage source representing the initial condition. The excitation and the impedance, in the Laplace transform domain, are

$$\frac{v_0^-}{s} \quad \text{and} \quad R + (1/Cs).$$

Therefore,

$$I(s) = \frac{v_0^-}{s(R + (1/Cs))} = \frac{v_0^-}{R}\left(\frac{1}{s + (1/RC)}\right).$$

Applying the initial and final value theorems, we get initial and final values of current, respectively, as

$$\frac{v_0^-}{R} \quad \text{and} \quad 0.$$

Taking the inverse Laplace transform, we get

$$i(t) = \frac{v_0^-}{R}e^{-\frac{t}{RC}}u(t).$$

The initial and final currents are $i_0^+ = v_0^-/R$ and $i_\infty = 0$. Multiplying current $i(t)$ by the resistance R, we get

$$v(t) = v_0^-(e^{-\frac{t}{RC}})u(t).$$

∎

The response of the series RC circuit, for example, with $R = 2\,\Omega$ and $C = 0.1\,\text{F}$, to the initial capacitor voltage $1\,\text{V}$ is shown in Fig. 8.20. The voltage across the capacitor is shown in (a) and the current in (b).

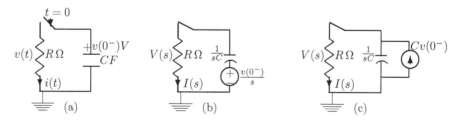

Fig. 8.19 Modeling of capacitors with initial conditions in the frequency-domain

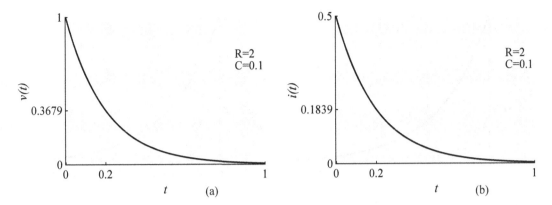

Fig. 8.20 The response of the series RC circuit to the initial capacitor voltage 1 V. (**a**) the voltage across the capacitor; (**b**) the current

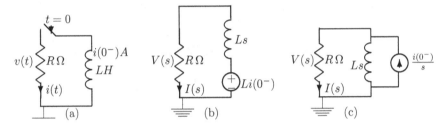

Fig. 8.21 Modeling of inductors with initial conditions in the frequency-domain

Using the alternative model shown in Fig. 8.19c, we get

$$Cv_0^- = \frac{V(s)}{R} + V(s)Cs \quad \text{and} \quad V(s) = \frac{v_0^-}{s + (1/RC)}.$$

This model is the same as obtained by source transformation. The equivalent source is a current source model of the voltage source.

Example 8.13 Assume that the series circuit consists of only the resistor with value $R = 2\,\Omega$ and the inductor with value $L = 1\,\text{H}$. Let the initial current in the inductor be $i_0^- = 1\,\text{A}$. Find the current through the circuit and the voltage across the inductor.

Solution The circuit is shown in Fig. 8.21a in the time-domain. The equivalent frequency-domain circuit is shown in Fig. 8.21b, with a voltage source representing the initial condition. The excitation and the impedance, in the Laplace transform domain, are

$$Li_0^- \quad \text{and} \quad R + Ls.$$

Therefore,

$$I(s) = \frac{Li_0^-}{R + Ls} = \frac{i_0^-}{(R/L) + s}.$$

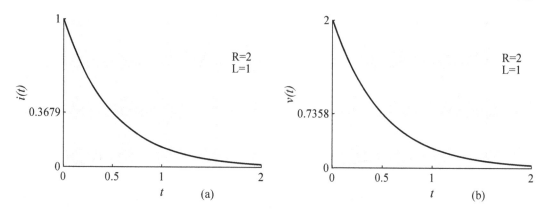

Fig. 8.22 The response of the series RL circuit to the initial inductor current 1 A. (**a**) the current through the inductor; (**b**) the voltage across

Fig. 8.23 A RLC circuit in time-domain

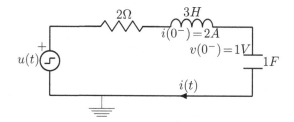

Taking the inverse Laplace transform, we get

$$i(t) = i_0^- e^{-\frac{R}{L}t} u(t).$$

The initial and final currents are $i_0^+ = Li_0^-$ and $i_\infty = 0$. Multiplying current $i(t)$ by the resistance R, we get

$$v(t) = i_0^- R e^{-\frac{R}{L}t} u(t).$$

The response of the series RL circuit, for example, to the initial current in the inductor $i_0^- = 1$ A is shown in Fig. 8.22. The voltage across the inductor is shown in (b) and the current in (a).

The source in the model in Fig. 8.21c is a current source model of the voltage source in Fig. 8.21b.   ∎

Example 8.14 Determine the current in the RLC circuit, shown in Fig. 8.23. The initial current through the inductor $i(0^-) = 2A$ and the initial voltage across capacitor $v(0^-) = 1$ V and, the input $x(t) = u(t)V$, the unit-step signal.

Solution The RLC circuit is shown in Fig. 8.24 in the frequency-domain. The net voltage in the circuit is

$$\frac{1}{s} + 6 - \frac{1}{s} = 6.$$

The impedance of the circuit is

$$2 + 3s + \frac{1}{s} = \frac{3s^2 + 2s + 1}{s}.$$

Fig. 8.24 The RLC
circuit in
frequency-domain

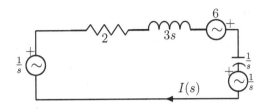

We get the current in the circuit by dividing the voltage by the impedance

$$I(s) = 6\frac{s}{3s^2 + 2s + 1} = \frac{2s}{s^2 + \frac{2}{3}s + \frac{1}{3}}.$$

Applying the initial and final value theorems, we get the currents as 2 and zero. The excitation voltage source gets cancelled with the equivalent voltage source due to the initial capacitor charge. Therefore, the only source that is effective is the equivalent impulsive voltage source due to the initial current in the inductor. As the rate of current raise is infinite, the impedances of the resistance and capacitor become negligible. Therefore, the current at $t = 0$ is controlled by the inductor, $6/3 = 2$.

Expanding into partial fractions, we get

$$I(s) = \frac{1 + j0.7071}{s + (0.3333 - j0.4714)} + \frac{1 - j0.7071}{s + (0.3333 + j0.4714)}.$$

Expressing the numerators in polar form, we get

$$I(s) = \frac{1.2247\angle 0.6155}{s + (0.3333 - j0.4714)} + \frac{1.2247\angle - 0.6155}{s + (0.3333 + j0.4714)}.$$

If two terms are conjugate in a partial fraction expansion, then the inverse is the twice the real part of any one of the inverses. For example,

$$I(s) = \frac{r\angle \theta}{s + (a - jb)} + \frac{r\angle - \theta}{s + (a + jb)}.$$

Taking the inverse Laplace transform, we get the current in the circuit as

$$i(t) = 2re^{-at}\cos(bt + \theta)u(t).$$

For this example, we get

$$i(t) = 2.4495e^{-0.3333t}\cos(0.4714t + 0.6155)u(t).$$

The current through the series RLC circuit with initial conditions is shown in Fig. 8.25.   ∎

Let us find the response with zero initial conditions. The net voltage in the circuit is

$$\frac{1}{s}.$$

The impedance of the circuit is

$$2 + 3s + \frac{1}{s} = \frac{3s^2 + 2s + 1}{s}.$$

Fig. 8.25 The current
through the series RLC
circuit with initial
conditions

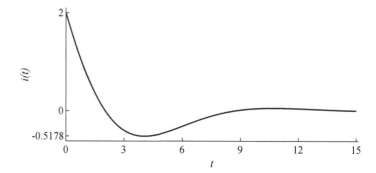

Fig. 8.26 The current
through the series RLC
circuit with zero initial
conditions

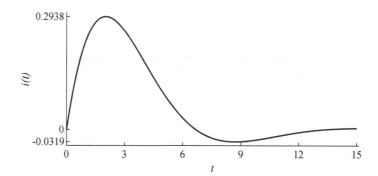

We get the current in the circuit by dividing the voltage by the impedance

$$I(s) = \frac{1}{3s^2 + 2s + 1} = \frac{(1/3)}{s^2 + \frac{2}{3}s + \frac{1}{3}}.$$

Applying the initial and final value theorems, we get the currents as 0 and 0. Expanding into partial
fractions, we get

$$I(s) = \frac{-j0.3536}{s + (0.3333 - j0.4714)} + \frac{j0.3536}{s + (0.3333 + j0.4714)}.$$

Expressing the numerators in polar form, we get

$$I(s) = \frac{0.3536\angle - \pi/2}{s + (0.3333 - j0.4714)} + \frac{0.3536\angle\pi/2}{s + (0.3333 + j0.4714)}.$$

Taking the inverse Laplace transform, we get the current in the circuit as

$$i(t) = 0.7071e^{-0.3333t} \cos\left(0.4714t - \frac{\pi}{2}\right) u(t).$$

The current through the series RLC circuit with zero initial conditions is shown in Fig. 8.26.

8.5 Application

To design a suitable filter, Fourier analysis is required to analyze the waveforms before and after filtering. Let the full-wave rectified waveform, presented earlier, be applied to the lowpass filter circuit, shown in Fig. 8.27. The filter is a resistor-capacitor series circuit, with the resistor $10\,\Omega$ and capacitor $0.01\,\mathrm{F}$. The output voltage is across the capacitor, $v_o(t)$. The FS for the full-wave rectified sine wave, with $\omega_0 = 377\,\mathrm{rad/s}$, is

$$v(t) = \frac{220}{\pi} - \frac{440}{3\pi}\cos(2\omega_0 t) - \frac{4}{15\pi}\cos(4\omega_0 t) - \frac{4}{35\pi}\cos(6\omega_0 t) + \cdots.$$

An ideal lowpass filter will pass the DC component only, suppressing rest of the components. However, practical filters can attenuate the unwanted components to any desired level. We learn the filter concept through the RC filter. More complex, both active and passive, filters are used in practice. The RC filter is a frequency-dependent voltage divider, since the reactance offered by the capacitor to the flow of current depends on the frequency of the constituent components of the rectified input waveform. The resistor is invariant with respect to frequency and, therefore, remains constant. The reactance of the capacitor, $-j/\omega C$, is very high at low frequencies and low at high frequencies. In particular, at $\omega = 0$, its reactance is ∞ and at $\omega = \infty$, its reactance is zero. Therefore, most of the voltage drop across the capacitor occurs at low frequencies. The voltage across the capacitor in the frequency-domain, by voltage division, is

$$V_o(j\omega) = V(j\omega)\frac{1/(j\omega C)}{R + 1/(j\omega C)} = V(j\omega)\frac{1}{1 + j\omega RC}.$$

Note that, with $\omega RC = 1$, the magnitude of the attenuation becomes 0.7071. For DC, with the fundamental frequency $\omega_0 = 377$, $k = 0$ and $\omega = 0$, we get

$$v_o(t) = \frac{220}{\pi}.$$

The capacitor is an open-circuit to DC, so that the output voltage is equal to the input voltage in steady state.

The cutoff frequency of a filter is defined as the frequency at which its magnitude response is 0.7071 of its nominal passband value. Beyond this point the attenuation increases rapidly. For RC filter, the cutoff frequency is defined, in radians, as

$$\omega_c = \frac{1}{RC}.$$

Let us fix $\omega_c = 10$. The cutoff frequency has to be fixed to just satisfy the attenuation requirements. Otherwise, the filter cost and complexity will increase unnecessarily. Then, $RC = 0.1$. Let $R = 10\,\Omega$.

Fig. 8.27 A resistor-capacitor lowpass filter circuit

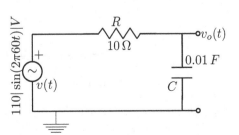

Then, $C = 0.01$ F. For the second harmonic, with the fundamental frequency $\omega_0 = 377$, $k = 2$, and $\omega = 2(377)$, we get

$$H(j\omega) = \frac{1}{1 + j2(377)0.1} = |H(j\omega)|\angle(H(j\omega)$$

$$v_o(t) = -\frac{440}{3\pi}|H(j\omega)|\cos(2(377)t + \angle(H(j\omega)).$$

With $R = 10\,\Omega$ and capacitor $C = 0.01$ F, the magnitude of the output is

$$\left|\frac{440}{3\pi}\frac{1}{1 + j2(377)(0.1)}\right| = \frac{440}{3\pi}0.0133 = 0.6191 \text{ V}.$$

The magnitude of the second harmonic has been attenuated from 46.6854 to 0.6191. The reduction of the harmonic components makes the output waveform closer to the desired, which is DC. In a similar way, the other frequency components are attenuated. The higher the frequency, the higher is the attenuation. The total output is the sum of the response to all the infinite harmonics in theory. In practice, after a relatively small number of harmonics, the amplitude of the harmonics becomes negligible, as it reduces in proportion to the square of the frequency. Note the k^2 term in the denominator of the amplitude of the frequency components. As the circuit attenuates high frequency components more, it is called a lowpass filter circuit. The magnitude response, $|H(j\omega)|$ of the lowpass filter and its phase response, versus the frequency ω are shown in Fig. 8.28.

A resistor-inductor series circuit, which is a highpass filter, is shown in Fig. 8.29. It is also a frequency-dependent voltage divider circuit. The input–output relationship is, in the frequency-domain, given by voltage division

$$H(j\omega) = \frac{j\omega L}{1 + j\omega RL} = |H(j\omega)|\angle(H(j\omega)$$

$$v_o(t) = v(t + \angle(H(j\omega)))\left|\frac{j\omega L}{1 + j\omega RL}\right|.$$

The voltage across the inductor is the output of the circuit. The reactance of the inductor increases with frequency. Therefore, most of the voltage drop occurs across the inductor at high frequencies. For DC, the reactance is zero and, hence, the output is zero. As it passes high frequency components readily, it is a highpass filter. For DC input, from the input–output relationship with $\omega = 0$,

$$v_o(t) = 0.$$

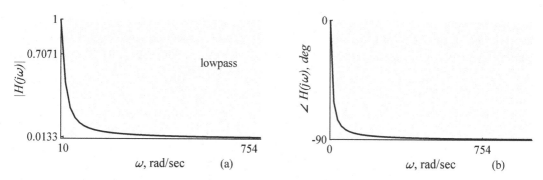

Fig. 8.28 (a) The magnitude response, $|H(j\omega)|$ of the lowpass filter; (b) the phase response

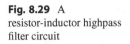

Fig. 8.29 A resistor-inductor highpass filter circuit

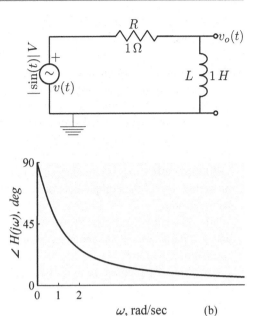

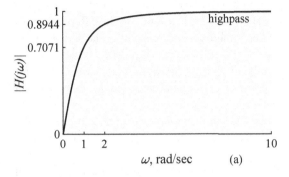

Fig. 8.30 (a) The magnitude response, $|H(j\omega)|$ of the highpass filter; (b) the phase response

For the second harmonic, with $\omega = 2$,

$$v_o(t) = |-\frac{j2L}{R + j2L}|\frac{4}{3\pi}\cos(2t + \angle(H(j\omega))$$

$$\omega_c = \frac{R}{L} = 1.$$

With $R = 1\,\Omega$ and $L = 1\,$H, the magnitude of the output is

$$\left|\frac{j2 \times 1}{1 + j2 \times 1}\right| = 0.8944$$

times that of the input, assuming the input is 1. As the frequency increases, the attenuation reduces and, hence, it is called a highpass filter. The total response of the circuit is the sum of the responses to all the harmonic components of the input. The magnitude response, $|H(j\omega)|$ of a highpass filter and its phase response versus the frequency ω are shown in Fig. 8.30.

8.6   Summary

- Transform means change in form.
- The principal transforms, Fourier series, Fourier transform, and Laplace transform used in signal and system analysis reduce the differential equation, which is a model for linear systems, into an algebraic equation.
- In the transform domain, frequency becomes the independent variable. We transform the signal to the frequency-domain, find the solution, and transform it back to the time-domain.

- Any waveform, encountered in practice, can be decomposed into a set of sinusoidal waveforms by transforms.
- All the tools, such as nodal and mesh analysis and circuit theorems, remain the same with the constraint that each frequency component of the source must be considered individually.
- The various frequency components are separated by an appropriate transform such as Fourier series, Fourier transform, and Laplace transform. The individual responses of the circuit are found and all the responses are added to find the total response.
- The frequency components have a property, called the orthogonal property, in that the integral of the product of the signal and the conjugate of a certain frequency component, over a period, will yield the coefficient of that component in the signal alone. This is a similarity test. Repeating this procedure for all the components of interest, we know the frequency content of the signal, called its spectrum.
- While practical signal generators, such as an oscillator, generate real sinusoidal waveforms, for analysis purposes, the mathematically equivalent complex exponentials are used due to its compact form and ease of use.
- Fourier analysis problem is to find the coefficients $X(k)$ in the complex exponential polynomial representation of a time-domain function $x(t)$ with period T and the fundamental frequency $\omega_0 = 2\pi/T$
- The Fourier synthesis problem is to find $x(t)$, given $X(k)$ and the exponentials. It is the sum of all the exponentials multiplied by their respective coefficients. The Fourier reconstructed waveform is with respect to the least squares error criterion.
- At any discontinuity, the reconstructed waveform never converges even in the limit. There is a 8.69% deviation, called the Gibbs phenomenon. However, the area under the deviation tends to zero.
- In the Fourier transform, which is the limiting case of the FS as the period of the waveform tends to infinity, both the time-domain signal and its spectrum are aperiodic and continuous. The process of finding the spectrum remains the same in that we find the integral of products of the signal with complex exponentials.
- The transfer function is a frequency-domain model of systems that relates the input and output, in the frequency-domain.
- The behavior of a system, before it attains its steady-state after the excitation is applied or to become dead from its steady-state behavior after the removal of the excitation is called its transient response.
- The unit-impulse signal is characterized by its unit area enclosed at $t = 0$. It is the derivative of the unit-step signal.
- For system analysis with or without initial conditions, where both the transient and steady-state responses are required, the generalization of Fourier analysis, called the Laplace transform, is more convenient.
- We can write down the differential equation circuit model and use the Laplace transform to solve it. However, it is easier to represent the circuits in the frequency-domain and solve it, as though it is a resistive circuit.
- In most of the signal and system analysis, it is found that transform methods are found to be more advantageous.

Exercises

* **8.1.1** Find the FS for the signal

$$x(t) = -1 + \cos\left(\frac{2\pi}{8}t - \frac{\pi}{3}\right) + 3\sin\left(3\frac{2\pi}{8}t + \frac{\pi}{3}\right).$$

8.1.2 Find the FS for the signal

$$x(t) = 3 + 3\cos\left(2\frac{2\pi}{8}t + \frac{\pi}{6}\right) + \sin\left(5\frac{2\pi}{8}t - \frac{\pi}{6}\right).$$

8.1.3 Find the FS for the signal

$$x(t) = 1 + 2\cos\left(3\frac{2\pi}{8}t + \frac{\pi}{4}\right) + 2\sin\left(\frac{2\pi}{8}t + \frac{\pi}{3}\right).$$

* **8.2** Determine the FS expansion for the periodic sawtooth waveform, one period of which is shown in Fig. 8.31. Use the time-differentiation property

$$\frac{d^n x(t)}{dt^n} \leftrightarrow (jk\omega_0)^n X(k).$$

8.3 Determine the FT of the aperiodic sawtooth pulse waveform, shown in Fig. 8.31. From the FT obtained, deduce the FS expansion of the periodically extended periodic signal. Verify that it is same as that obtained in Exercise 8.2. Use the time-differentiation property

$$\frac{d^n x(t)}{dt^n} \leftrightarrow (j\omega)^n X(j\omega).$$

8.4 Let the impulse be approximated by a rectangular pulse, centered at $t = 0$, of width $2w$ and height $\frac{1}{2w}$. The signal $x(t)$ is sampled by this quasi-impulse. Compare the sample values of $x(t)$ at t with $w = 1$, $w = 0.1$, $w = 0.01$ with the exact value of $x(t)$ at t.

8.4.1 $x(t) = e^{-t}$, $t = 1$.

8.4.2 $x(t) = \sin(t)$, $t = -\pi/6$.

* 8.4.3 $x(t) = \cos(t)$, $t = \pi/6$.

8.4.4 $x(t) = \frac{\sin(t)}{t}$, $t = 0$. Use numerical integration.

8.4.5 $x(t) = e^{jt}$, $t = \pi/3$.

8.5 The input current is the unit-step function, $u(t)$ A. Assume that the parallel circuit consists of only the resistor with value $R = 2\,\Omega$ and the inductor with value $L = 1$ H, shown in Fig. 8.32.

Fig. 8.31 One period of the sawtooth waveform

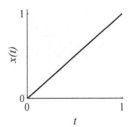

Fig. 8.32 A parallel RL
circuit

Fig. 8.33 A parallel RC
circuit

Fig. 8.34 A parallel RLC
circuit

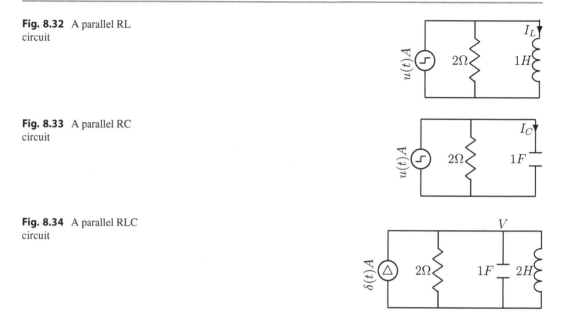

Let the initial current through the inductor be zero. Find the current through and the voltage across the inductor after the excitation is applied. Deduce the corresponding impulse response of the circuit.

* **8.6** The input current is the unit-step function, $u(t)\ A$. Assume that the parallel circuit consists of only the resistor with value $R = 2\,\Omega$ and the capacitor with value $C = 1\,F$, shown in Fig. 8.33. Let the initial voltage across the capacitor be zero. Find the current through and the voltage across the capacitor after the excitation is applied. Deduce the corresponding the impulse response.

* **8.7** The input current is the unit-impulse function, $u(t)\ A$. Assume that the parallel circuit consists of only the resistor with value $R = 2\,\Omega$, inductor with value $L = 2$ and the capacitor with value $C = 1\,F$, shown in Fig. 8.34. Let the initial conditions be zero. Find the voltage across the capacitor after the excitation is applied. Deduce the corresponding unit-step response.

Matrices

A

Linear algebra is an important branch of mathematics that includes the theory of systems of linear equations, matrices, determinants and linear transformations. For solving circuit analysis problems, matrix multiplication, inversion and determinants are often required.

In both mesh and nodal analysis, the problem is formulated in terms of a set of linear equations. These equations are solved to find the independent currents or voltages. The solution can be found using the equations themselves by the substitution method, as we learnt from our high school mathematics. For example, consider the two equations with variables V_2 and V_3, we came across in Chap. 2.

$$-7V_2 + 3V_3 = -3 \tag{A.1}$$

$$3V_2 - 7V_3 = -1. \tag{A.2}$$

Solving Eq. (A.1) for V_3, we get

$$V_3 = \frac{-3 + 7V_2}{3}$$

Substituting for V_3 in Eq. (A.2), we get

$$3V_2 - 7\left(\frac{-3 + 7V_2}{3}\right) = -1 \quad \text{or} \quad 9V_2 - 7(-3 + 7V_2) = -3$$

and $V_2 = 0.6$.

$$V_3 = \frac{-3 + 7V_2}{3} = \frac{-3 + 7(0.6)}{3} = 0.4$$

These values satisfy the given equations, verifying that the solution is correct.

Using matrices, these equations are written as

$$\begin{bmatrix} -7 & 3 \\ 3 & -7 \end{bmatrix} \begin{bmatrix} V_2 \\ V_3 \end{bmatrix} = \begin{bmatrix} -3 \\ -1 \end{bmatrix}. \tag{A.3}$$

This form is very compact and leads to efficient algorithms for solving problems, such as solving systems of simultaneous equations. Matrix is any rectangular array of numbers. Any matrix consists of horizontal rows and vertical columns, numbered, respectively, from top to bottom and left to right. A general 2×2 matrix (containing two rows and two columns) is written as

$$\begin{bmatrix} a_{11} & a_{12} \\ a_{21} & a_{22} \end{bmatrix}$$

© Springer Nature Switzerland AG 2020
D. Sundararajan, *Introductory Circuit Theory*,
https://doi.org/10.1007/978-3-030-31985-4

This matrix with equal number of rows and columns is called a square matrix. A general 2×1 matrix (containing two rows and one column, called column vector) is written as

$$\begin{bmatrix} a_1 \\ a_2 \end{bmatrix}$$

Equation (A.3) can represent the two equations only if the product of the left two matrices get back the equations. That requires the rows of the 2×2 matrix is multiplied, elementwise, by the column of the 2×1 matrix and the partial products added. That is matrix multiplication. The product of matrices is defined if and only if the number of columns in the first matrix is equal to number of rows in the second matrix. That is, the product of $m \times n$ and $n \times p$ matrices, A and B, is of size $m \times p$, the number of rows in the first matrix times the number of columns in the second matrix. The product of A and B, AB, is the $m \times p$ matrix, whose ijth entry is given by

$$a_{i1}b_{1j} + a_{i2}b_{2j} + \cdots + a_{in}b_{nj}$$

and AB, in general, is not equal to BA. The main diagonal of a $n \times n$ matrix consists of the elements

$$\{a_{11}, a_{22}, \ldots, a_{nn}\}$$

The (i, j)-th minor of a 3×3 matrix, denoted M_{ij} is the determinant of the 2×2 matrix that is left after the ith row and jth column are deleted. Let us find M_{ij} for the 2×2 matrix

$$\begin{bmatrix} -7 & 3 \\ 3 & -7 \end{bmatrix}$$

$$\begin{bmatrix} |-7| \end{bmatrix} \begin{bmatrix} |3| \end{bmatrix} \begin{bmatrix} |3| \end{bmatrix} \begin{bmatrix} |-7| \end{bmatrix}$$

As the determinant of a 1×1 matrix is itself,

$$M_{ij} = \begin{bmatrix} -7 & 3 \\ 3 & -7 \end{bmatrix}$$

The given matrix happens to be symmetric. For a nonsymmetric matrix, the elements get swapped. The (i, j)th cofactor of the matrix, denoted by C_{ij} is $(-1)^{i+j} M_{ij}$. Therefore,

$$C_{ij} = \begin{bmatrix} -7 & -3 \\ -3 & -7 \end{bmatrix}$$

The adjoint of A, denoted by adj A, is defined as

$$\text{adj } A(i, j) = C(j, i)$$

The determinant of $|A|$ is

$$|A| = a_{11}a_{22} - a_{12}a_{21}$$

The inverse of the square matrix A is defined as

$$A^{-1} = \frac{\text{adj } A}{|A|}$$

$$\begin{bmatrix} V_2 \\ V_3 \end{bmatrix} = \begin{bmatrix} -7 & 3 \\ 3 & -7 \end{bmatrix}^{-1} \begin{bmatrix} -3 \\ -1 \end{bmatrix} = \frac{1}{40} \begin{bmatrix} -7 & -3 \\ -3 & -7 \end{bmatrix} \begin{bmatrix} -3 \\ -1 \end{bmatrix} = \begin{bmatrix} 0.6 \\ 0.4 \end{bmatrix}$$

The product of a matrix and its inverse must be an identity matrix. An identity matrix has 1s on the main diagonal with the rest 0s. In the circuit analysis problems, the inverse always exists. In that case, the determinant of the matrix will be nonzero. If the determinant is zero, most probably the equilibrium equations are not independent and must be checked for errors. For the example matrix

$$\begin{bmatrix} -7 & 3 \\ 3 & -7 \end{bmatrix} \frac{1}{40} \begin{bmatrix} -7 & -3 \\ -3 & -7 \end{bmatrix} = \begin{bmatrix} 1 & 0 \\ 0 & 1 \end{bmatrix}$$

A.1   Determinants

A matrix is an orderly arrangement of elements, whereas a determinant has a numerical value.

$$\det \begin{bmatrix} a_{11} & a_{12} & a_{13} \\ a_{21} & a_{22} & a_{23} \\ a_{31} & a_{32} & a_{33} \end{bmatrix} = \begin{vmatrix} a_{11} & a_{12} & a_{13} \\ a_{21} & a_{22} & a_{23} \\ a_{31} & a_{32} & a_{33} \end{vmatrix}$$

In terms of minors

$$\begin{vmatrix} a_{11} & a_{12} & a_{13} \\ a_{21} & a_{22} & a_{23} \\ a_{31} & a_{32} & a_{33} \end{vmatrix} = a_{11}M_{11} - a_{12}M_{12} + a_{13}M_{13}$$

In terms of cofactors

$$\begin{vmatrix} a_{11} & a_{12} & a_{13} \\ a_{21} & a_{22} & a_{23} \\ a_{31} & a_{32} & a_{33} \end{vmatrix} = a_{11}C_{11} + a_{12}C_{12} + a_{13}C_{13}$$

In general, the determinant of a matrix can be obtained by multiplying each entry in any column or row by the corresponding cofactor and adding the products. For a 2×2 matrix,

$$\begin{vmatrix} a_{11} & a_{12} \\ a_{21} & a_{22} \end{vmatrix} = a_{11}a_{22} - a_{12}a_{21}$$

For a 3×3 matrix,

$$\begin{vmatrix} a_{11} & a_{12} & a_{13} \\ a_{21} & a_{22} & a_{23} \\ a_{31} & a_{32} & a_{33} \end{vmatrix} = a_{11} \begin{vmatrix} a_{22} & a_{23} \\ a_{32} & a_{33} \end{vmatrix} - a_{12} \begin{vmatrix} a_{21} & a_{23} \\ a_{31} & a_{33} \end{vmatrix} + a_{13} \begin{vmatrix} a_{21} & a_{22} \\ a_{31} & a_{32} \end{vmatrix}$$

Let A be a square matrix. If a matrix B of the same size exists such that $AB = I$, then B is said to be the inverse of A and denoted as $B = A^{-1}$. I is the identity matrix of the same size as A with the main diagonal elements equal to 1 and the rest of the elements equal to zero.

The inverse of an arbitrary 3×3 matrix A is

$$A = \begin{bmatrix} a_{11} & a_{12} & a_{13} \\ a_{21} & a_{22} & a_{23} \\ a_{31} & a_{32} & a_{33} \end{bmatrix} \quad \text{and} \quad A^{-1} = \frac{1}{|A|} \begin{bmatrix} \begin{vmatrix} a_{22} & a_{23} \\ a_{32} & a_{33} \end{vmatrix} & -\begin{vmatrix} a_{21} & a_{23} \\ a_{31} & a_{33} \end{vmatrix} & \begin{vmatrix} a_{21} & a_{22} \\ a_{31} & a_{32} \end{vmatrix} \\ -\begin{vmatrix} a_{12} & a_{13} \\ a_{32} & a_{33} \end{vmatrix} & \begin{vmatrix} a_{11} & a_{13} \\ a_{31} & a_{33} \end{vmatrix} & -\begin{vmatrix} a_{11} & a_{12} \\ a_{31} & a_{32} \end{vmatrix} \\ \begin{vmatrix} a_{12} & a_{13} \\ a_{22} & a_{23} \end{vmatrix} & -\begin{vmatrix} a_{11} & a_{13} \\ a_{21} & a_{23} \end{vmatrix} & \begin{vmatrix} a_{11} & a_{12} \\ a_{21} & a_{22} \end{vmatrix} \end{bmatrix}^T$$

The transpose A^T of an $m \times n$ matrix A is the $n \times m$ matrix, obtained by interchanging the rows and columns in A. That is, the ith row of A becomes the ith column of A^T. A square matrix A is symmetric, if $a_{ij} = a_{ji}$. If the matrix is symmetric, its inverse is also symmetric. Therefore, the determination of about one half of the elements is sufficient.

Let us find the inverse of the matrix

$$A = \begin{bmatrix} 4 & -1 & -3 \\ -1 & 5 & -1 \\ -3 & -1 & 5 \end{bmatrix} \quad \text{and} \quad A^{-1} = \begin{bmatrix} 0.6000 & 0.2000 & 0.4000 \\ 0.2000 & 0.2750 & 0.1750 \\ 0.4000 & 0.1750 & 0.4750 \end{bmatrix}$$

which appears in an example in Chap. 2. The determinant of A is

$$4(24) - 8 - 3(16) = 40$$

$$\begin{bmatrix} 4 & -1 & -3 \\ -1 & 5 & -1 \\ -3 & -1 & 5 \end{bmatrix} \begin{bmatrix} 0.6000 & 0.2000 & 0.4000 \\ 0.2000 & 0.2750 & 0.1750 \\ 0.4000 & 0.1750 & 0.4750 \end{bmatrix} = \begin{bmatrix} 1 & 0 & 0 \\ 0 & 1 & 0 \\ 0 & 0 & 1 \end{bmatrix}$$

Let us find the inverse of the matrix

$$A = \begin{bmatrix} 4 & 0 & 3 \\ 0 & -2 & 2 \\ 2 & 0 & 1 \end{bmatrix} \quad \text{and} \quad A^{-1} = \begin{bmatrix} -0.50 & 0 & 1.50 \\ 1.00 & -0.50 & -2.00 \\ 1.00 & 0 & -2.00 \end{bmatrix}$$

The determinant of A is

$$4(-2) - 0 + 3(4) = 4$$

$$\begin{bmatrix} 4 & 0 & 3 \\ 0 & -2 & 2 \\ 2 & 0 & 1 \end{bmatrix} \begin{bmatrix} -0.50 & 0 & 1.50 \\ 1.00 & -0.50 & -2.00 \\ 1.00 & 0 & -2.00 \end{bmatrix} = \begin{bmatrix} 1 & 0 & 0 \\ 0 & 1 & 0 \\ 0 & 0 & 1 \end{bmatrix}$$

Consider the equation $5x = 15$. Solving for x, we get

$$x = \frac{1}{5}(15) = 3$$

$\frac{1}{5}$ is the multiplicative inverse of 5. The multiplicative inverse of a number x is a number which when multiplied by x yields 1. Similarly,

$$AX = V \quad \text{and} \quad X = A^{-1}V$$

Matrix algebra is an extension of finding the solution of a single equation to a set of equations. It provides compactness and efficiency.

The inverse of an arbitrary 2×2 matrix A exists if

$$|A| = (a_{11}a_{22} - a_{12}a_{21}) \neq 0$$

Then, A^{-1} is given by

$$A = \begin{bmatrix} a_{11} & a_{12} \\ a_{21} & a_{22} \end{bmatrix} \quad \text{and} \quad A^{-1} = \frac{1}{(a_{11}a_{22} - a_{12}a_{21})} \begin{bmatrix} a_{22} & -a_{12} \\ -a_{21} & a_{11} \end{bmatrix}$$

Let us find the inverse of the matrix

$$A = \begin{bmatrix} -10.0000 + j9.0000 & 10.0000 + j0.0000 \\ 0.3861 + j0.8614 & -1.6762 - j1.6624 \end{bmatrix}$$

$$A^{-1} = \begin{bmatrix} -0.0423 - j0.0704 & -0.3372 - j0.0856 \\ -0.0056 - j0.0323 & -0.4142 + j0.2178 \end{bmatrix}$$

which appears in an example in Chap. 3. The determinant is

$$(-10 + j9)(-1.6762 - j1.6624) - (10)(0.3861 + j0.8614) = 27.8626 - j7.0758$$

and

$$\begin{bmatrix} -10.0000 + j9.0000 & 10.0000 + j0.0000 \\ 0.3861 + j0.8614 & -1.6762 - j1.6624 \end{bmatrix} \begin{bmatrix} -0.0423 - j0.0704 & -0.3372 - j0.0856 \\ -0.0056 - j0.0323 & -0.4142 + j0.2178 \end{bmatrix}$$

$$= \begin{bmatrix} 1 & 0 \\ 0 & 1 \end{bmatrix}$$

Note that, only with higher precision, the product will be exactly equal to the identity matrix.

Complex Numbers

B

The resistance of a resistor is completely specified by a single number. However, the impedance is characterized by an ordered pair of real numbers. A 2-element vector is an ordered pair of elements, (a, b). Several entities in science and engineering are characterized by vectors, such as velocity, sinusoid, point in a plane and color. For such entities, analysis using vector form of representation is advantageous. The complex number system is an extension of the real number system. A complex number is an ordered pair of real numbers, a 2-element vector. The complex number $z = 2 + j1$, called its rectangular form, is shown in Fig. B.1. The two real numbers a and b are called, respectively, the real and imaginary parts of the complex number $z = a + jb$ and $j = \sqrt{-1}$ is the imaginary unit. The necessity for complex numbers is that it is more efficient to represent related entities in the vector form. For example, at a given frequency, a sinusoid is characterized by its amplitude and phase. In signal analysis, the complex form of representing the amplitude and phase of a sinusoid is more convenient than by two scalars. In a Cartesian coordinate system, a point is represented by its distance from a set of perpendicular lines that intersect at the origin of the system. A Cartesian coordinate system in which the horizontal and vertical axes represent, respectively, the real and imaginary parts of a complex number is called a complex plane. Complex number $z = a + jb$ and $p = c + jd$ are equal, if and only if $a = c$ and $b = d$.

A complex number $z = a + jb$ can also be written in its polar form $A\angle\theta$ or exponential form $Ae^{j\theta}$. The exponential form of representing the complex number $z = 2 + j1$ is $\sqrt{5}e^{j26.5651}$ using degree measure for the angle, as shown in Fig. B.1. The magnitude A and angle θ are, respectively,

$$A = +\sqrt{a^2 + b^2} \qquad \text{and} \qquad \theta = \tan^{-1}\frac{b}{a}$$

Fig. B.1 The complex plane with some complex numbers

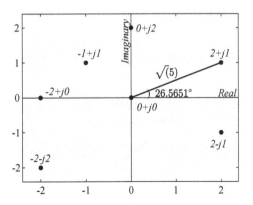

© Springer Nature Switzerland AG 2020
D. Sundararajan, *Introductory Circuit Theory*,
https://doi.org/10.1007/978-3-030-31985-4

The correct angle has to be determined, depending on the polarities of a and b. Assuming a and b are positive, the angle of

- $z = a + jb$ lies in the first quadrant of the complex plane.
- $z = -a + jb$ lies in the second quadrant
- $z = -a - jb$ lies in the third quadrant
- $z = a - jb$ lies in the fourth quadrant

As $A\angle\theta = A\angle(\theta + 2N\pi)$, where N is an integer, the value of $\theta = -\pi < \theta \le \pi$ is called its principal value.

The inverse relations are

$$a = A\cos(\theta) \quad \text{and} \quad b = A\sin(\theta)$$

The real number system is a subset of the complex number system. Therefore, all the operations, if the imaginary parts are zero, reduce to real arithmetic operations.

Addition and Subtraction

Let the two numbers be $z = a + jb$ and $p = c + jd$. Then,

$$q = z \pm p = (a \pm c) + j(b \pm d)$$

With $z = 2 + j3$ and $p = 1 - j4$, $q = z + p = 3 - j1$ and $q = z - p = 1 + j7$. The sum of two complex numbers is another complex number, in which the real part is the sum of their real parts and the imaginary part is the sum of their imaginary parts. The addition and subtraction operations are shown in the complex plane in Fig. B.2a, b, respectively.

Multiplication

Let the two numbers be $z = a + jb$ and $p = c + jd$. Then,

$$q = (z)(p) = (a + jb)(c + jd) = (ac - bd) + j(ad + bc),$$

where $j^2 = -1$. With $z = 2 + j3$ and $p = 1 - j4$, $q = (z)(p) = 14 - j5$.

In exponential form,

$$q = (z)(p) = (a + jb)(c + jd) = Ae^{j\theta}Ce^{j\phi} = ACe^{j(\theta + \phi)}$$

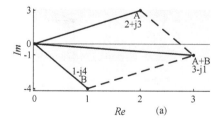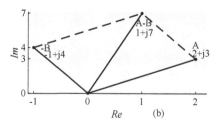

Fig. B.2 (**a**) Addition; (**b**) subtraction

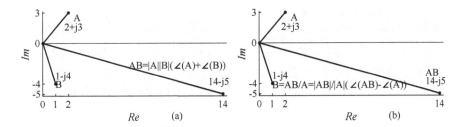

Fig. B.3 (a) Multiplication; (b) division

$$(2 + j3)(1 - j4) = 3.6056e^{j0.9828}4.1231e^{-j1.3258} = 14.8661e^{-j0.3430} = 14 - j5$$

using radian measure. The product of two complex numbers is another complex number, whose magnitude is the product of their magnitudes and the angle is the sum of their angles. The multiplication and division operations are shown in the complex plane in Fig. B.3a, b. respectively.

Complex Conjugate

The conjugate of a complex number $z = a + jb$ is $z^* = a - jb$, obtained by replacing j by $-j$. z^* is the mirror image of z about the real axis in the complex plane. In polar form, the conjugate of $Ae^{j\theta}$ is $Ae^{-j\theta}$. Obviously, the product of a complex number with its conjugate is its magnitude squared, A^2. That is,

$$(z)z^* = (a + jb)(a - jb) = a^2 + b^2$$

$$z + z^* = 2a \quad \text{and} \quad z - z^* = j2b$$

Division

With $z = a + jb$ and $p = c + jd$.

$$q = \frac{z}{p} = \frac{zp^*}{pp^*} = \frac{zp^*}{|p|^2} = \frac{ac + bd}{c^2 + d^2} + j\frac{bc - ad}{c^2 + d^2}$$

In exponential form,

$$q = \frac{z}{p} = \frac{Ae^{j\theta}}{Ce^{j\phi}} = \frac{A}{C}e^{j(\theta - \phi)}$$

For example,

$$\frac{(14 - j5)}{(2 + j3)} = \frac{14.8661e^{-j0.3430}}{3.6056e^{j0.9828}} = 4.1231e^{-j1.3258} = (1 - j4)$$

Powers and Roots of Complex Numbers

Since $x^2 \geq 0$ for all real numbers, the quadratic equation $x^2 = -1$ has no solution in the real number system. In the complex number system, the two roots are j and $-j$ and, in fact, every polynomial equation does have a solution.

$$z^N = (Ae^{j\theta})^N = A^n e^{jN\theta}$$

Replacing N by $1/N$ and adding $2k\pi$ to θ, we get

$$z^{\frac{1}{N}} = +\sqrt[N]{A}e^{\frac{j(\theta+2k\pi)}{N}} = +\sqrt[N]{A}\left(\cos\left(\frac{(\theta+2k\pi)}{N}\right) + j\sin\left(\frac{(\theta+2k\pi)}{N}\right)\right), k = 0, 1, 2, \ldots, N-1$$

With $A = 1$ and $\theta = 0$, we get the Nth roots of unity, which form the DFT basis functions.

$$1^{\frac{1}{N}} = \cos\left(\frac{2k\pi}{N}\right) + j\sin\left(\frac{2k\pi}{N}\right), \quad k = 0, 1, 2, \ldots, N-1$$

For example, with $N = 4$, we get the roots as $\{1, j, -1, -j\}$. Each root raised to the power of 4 will yield 1. Since the magnitude of the roots is 1, their angles add to $\{0, 2\pi, 4\pi, -2\pi\}$. The complex number with these arguments is 1.

Answers to Selected Exercises

Chapter 1

1.1.2

$$R_{eq} = 15, I = -3/15 = -0.2$$

$$V_{R1} = (-0.2)1 = -0.2, V_{R2} = (-0.2)2 = -0.4, V_{R3} = (-0.2)3 = -0.6,$$

$$V_{R4} = (-0.2)4 = -0.8, V_{R5} = (-0.2)5 = -1$$

1.2.3

$$V_{R1} = (-4)2 = -8, V_{R2} = (-4)1 = -4, V_{R3} = (-4)4 = -16,$$

$$V_{R4} = (-4)3 = -12, V_{R5} = (-4)6 = -24$$

1.3.3

$$Y_{eq} = 2.25, I_{R1} = -4/(2Y_{eq}) = -0.8889, I_{R2} = -4/(1Y_{eq}) = -1.7778,$$

$$I_{R3} = -4/(4Y_{eq}) = -0.4444, I_{R4} = -4/(3Y_{eq}) = -0.5926, I_{R5} = -4/(6Y_{eq}) = -0.2963 \text{ A}$$

The voltage across the resistors is -1.7778 V.

1.4.2

$$Y_{eq} = 2.2833, I = V Y_{eq} = -6.85, I_{R1} = I/(1Y_{eq}) = -3, I_{R2} = I/(2Y_{eq}) = -1.5,$$

$$I_{R3} = I/(3Y_{eq}) = -1, I_{R4} = I/(4Y_{eq}) = -0.75, I_{R5} = I/(5Y_{eq}) = -0.6 \text{ A}$$

1.5.2 The equivalent admittance of the four parallelly connected resistors is $Y_{eq} = 2.0833$. The equivalent impedance of the circuit is $Z_{eq} = 5 + (1/2.0833) = 5.48 \, \Omega$.

$$I = V/Z_{eq} = -0.5474$$

$$I_{R1} = I/(1Y_{eq}) = -0.2628, I_{R2} = I/(2Y_{eq}) = -0.1314,$$

$$I_{R3} = I/(3Y_{eq}) = -0.0876, I_{R4} = I/(4Y_{eq}) = -0.0657$$

$$I_{R5} = I$$

The voltage across the parallel resistors is -0.2628. The voltage across R_5 is -2.7372.

© Springer Nature Switzerland AG 2020
D. Sundararajan, *Introductory Circuit Theory*,
https://doi.org/10.1007/978-3-030-31985-4

1.6.2 The equivalent admittance of the four parallelly connected resistors is $Y_{eq} = 2.0833$.

$$I_{R1} = I/(1Y_{eq}) = -1.44, I_{R2} = I/(2Y_{eq}) = -0.72,$$

$$I_{R3} = I/(3Y_{eq}) = -0.48, I_{R4} = I/(4Y_{eq}) = -0.36$$

$$I_{R5} = I$$

The voltage across the parallel resistors is -1.44. The voltage across R_5 is -15.

Chapter 2

2.1.2

$$I_1 = 0.4730, \quad I_2 = 0.3514, \quad I_3 = -0.1351$$

$$\{I_{R1} = I_1 - I_2 = 0.1216, \quad I_{R4} = I_2 + I_3 = 0.2162, \quad I_{R3} = I_1 - I_2 - I_3 = 0.2568\}$$

The power consumed is $0.4730\,\text{W}$.

2.2.2

$$I_1 = -0.2750, \quad I_2 = 0.2500, \quad I_3 = -0.7250$$

$$\{I_{R1} = -0.5250, \quad I_{R3} = 0.2, \quad I_{R4} = -0.4750\}$$

The power consumed is $1.45\,\text{W}$.

2.3.2

$$I_1 = 0.1186, \quad I_2 = 0.2203, \quad I_3 = 0.2712$$

Power consumed is $0.4915\,\text{W}$.

2.4.2

$$I_1 = -1/2. \quad I_2 = -1, \quad I_3 = -3/7$$

Power consumed is

$$1(1/2)^2 + 3(1/2)^2 + 2(1)^2 + 3(3/7)^2 + 4(4/7)^2 = 4.8571\,\text{W}$$

2.5.2

$$I_1 = 1/2, I_2 = -3/2$$

The power consumed is $17.5\,\text{W}$.

2.6.2

$$I_1 = 1.8333, \quad I_2 = -1.1667, \quad I_3 = 0.3333$$

Power consumed is $18.0553\,\text{W}$.

2.7.2

$$I_1 = -3, \quad I_2 = -2, \quad I_3 = -4$$

Power consumed is $54\,\text{W}$.

2.9.2 The maximum power transferred is $0.1246\,\text{W}$.

2.10.1 The current through R_3 is $1/3$.

Chapter 3

3.1.3 $x(t) = 5\cos(2\pi t + \frac{5\pi}{6})$, $t = -\frac{5}{12}$.

3.2.4 $x(t) = 4\sin(2\pi t - \frac{\pi}{4}) = -2\sqrt{2}\cos(2\pi t) + 2\sqrt{2}\sin(2\pi t)$, $\quad t = \frac{3}{8}, \frac{11}{8}, \frac{19}{8}$.

3.3.4

$$x(t) = 3\sin\left(\frac{2\pi}{6}t + \frac{\pi}{6}\right) = 1.5(e^{j\left(\frac{2\pi}{6}t - \frac{\pi}{3}\right)} + e^{-j\left(\frac{2\pi}{6}t - \frac{\pi}{3}\right)})$$

3.4.2

$$i(t) = 0.9969\cos\left(\frac{2\pi}{4}t - 2.6964\right)$$

3.5.2

$$I = \frac{V}{Z_{eq}} = 0.1246 - j0.3418$$

$$V_{Z1} = -1.1393 - j0.4153, \, V_{Z2} = 0.3738 - j1.0254, \, V_{Z3} = 0.2087 + j0.0121,$$

$$V_{Z4} = 0.2492 - j0.6836, \, V_{Z5} = 0.3076 + j0.1121$$

3.6.2

$$V_{Z1} = -5, \, V_{Z2} = -j2, \, V_{Z3} = -2.6471 - j4.4118,$$

$$V_{Z4} = -j8, \, V_{Z5} = 0.4$$

3.7.3

$$I_{Z1} = -0.0333 + j0.0946, \, I_{Z2} = 0.9460 + j0.3332, \, I_{Z3} = -0.0167 + j0.0473,$$

$$I_{Z4} = -0.0189 - j0.0067, \, I_{Z5} = -0.0111 + 0.0315$$

3.8.2

$$I_{Z1} = 0.4714 - j0.4714, \, I_{Z2} = -2.3570 - j2.3570, \, I_{Z3} = 0.7071 - j0.7071,$$

$$I_{Z4} = 0.5657 + j0.5657, \, I_{Z5} = 0.4714 - j0.4714$$

3.9.2

$$I_{Z1} = 0.0026 + j0.0095, \, I_{Z2} = 0.0475 - j0.0129, \, I_{Z3} = -0.1293 - j0.4750, \, I_{Z4} = 0.0317 - j0.0086$$

$$V_{Z5} = -0.0950 - j0.9741$$

3.10.2

$$I_{Z1} = -0.0477 + j0.0408, \, I_{Z2} = 0.0680 + j0.0795, \, I_{Z3} = 0.7947 - j0.6799, \, I_{Z4} = 0.0510 + j0.0596$$

$$V_{Z5} = 0.8660 - j0.5$$

3.11.2

$$I_1 = -0.4272 - j0.5434, \quad I_2 = -0.5130 - j0.5303, \quad I_3 = 0.2276 + j0.1187$$

3.12.2

$$I_1 = -0.0344 + j0.0598, \quad I_2 = -0.0255 - j0.2326, \quad I_3 = -0.0105 + j0.2538$$

3.13.2

$$I_1 = -0.0260 - j0.1236, \quad I_2 = 0.1753 - j0.2164, \quad I_3 = 0.1928 - j0.1237$$

3.14.2 See Tables B.1 and B.2

Table B.1 Voltages due to each source in circuit in Exercise 3.14.2

Source	V_1	V_2	V_3
V source, $\omega = 1$	1	$1.0038 + j0.0421$	$0.3357 + j0.0586$
DC, $\omega = 0$	0	0	$-\frac{1}{3}$
I source, $\omega = 2$	0	$0.0694 + j0.0116$	$0.1976 - j0.6799$

Table B.2 Currents due to each source in circuit in Exercise 3.14.2

Source	I_1	I_2	I_3
V source	$0.4206 - j0.0382$	$0.3341 - j0.0083$	$0.3341 - j0.0083$
DC	$-\frac{1}{3}$	$-\frac{1}{3}$	$-\frac{1}{3}$
I source	$-0.0578 + j0.3470$	$-0.0641 + j0.3457$	$-0.0641 - j0.6543$

3.15.2

$$I_1 = 2.0200 + j0.3990 \quad \text{and} \quad I_3 = I_1 - 2 = 0.0200 + j0.3990$$

$$V_2 = 0.0399 - j0.2020$$

3.16.2

$$I_1 = -1.7024 + j1.3750, \quad I_2 = 3.4048 - j1.7499, \quad I_3 = 1.3208 - j0.6567$$

3.17.2

$$I_1 = 0.0236 + j0.1376, \quad I_2 = -0.0253 - j0.4250, \quad I_3 = 0.0483 + j0.4668$$

3.19.2

$$P_m = 0.0646 \, \text{W}$$

3.20.1

$$I = (-0.5000 + j0.5000) \, \text{A}$$

Chapter 4

4.2

$$L = 1.25 \, \text{H}, \, pf = 0.8$$

$$L = 2 \, \text{H}, \, pf = 1$$

4.4

$$C = 0.0186 \, \text{F}$$

Chapter 5

5.1.2

$$I_1 = 0.16 - j0.12 \quad I_2 = 0.14 + j0.02$$

$$S = P_{av} + jQ_{av} = 0.08 + j0.06 \text{ va.}$$

The average value of the energy stored is 0.0150.
The energy stored at $t = 1$ is 0.08 J.

5.2.2

$$I_1 = -0.1891 - j0.1317 \quad I_2 = -0.0289 - j0.0875$$

$$S = P_{av} + jQ_{av} = 0.0659 + j0.0946 \text{ va.}$$

The average value of the energy stored is 0.0236.
The energy stored at $t = 1$ is 0.0316 J.

5.3.2

$$I_1 = 0.4268 + j0.4854, \quad I_2 = 0.4576 + j0.1912, \quad I_3 = 0.2722 + j0.2673$$

$$S = P_{av} + jQ_{av} = 0.3062 - j0.1035 \text{ va.}$$

The average value of the magnetic fields is 0.0577.
The average value of the electric field is 0.1094.
At $t = 1$, the values of the magnetic fields and the electric field are 0.0153 and 0.0443, respectively.

Chapter 6

6.1.1 Positive sequence
The voltages are

$$v_a(t) = 2\cos\left(\frac{2\pi}{15}t\right), \quad v_b(t) = 2\cos\left(\frac{2\pi}{15}t - \frac{2\pi}{3}\right), \quad v_c(t) = 2\cos\left(\frac{2\pi}{15}t - \frac{4\pi}{3}\right)$$

Negative sequence

$$v_a(t) = 2\cos\left(\frac{2\pi}{15}t\right), \quad v_b(t) = 2\cos\left(\frac{2\pi}{15}t + \frac{2\pi}{3}\right), \quad v_c(t) = 2\cos\left(\frac{2\pi}{15}t + \frac{4\pi}{3}\right)$$

6.2.1

$$I_a = 2/(1 + j3) = 0.2 - j0.6, \quad I_{bn} = -0.6196 + j0.1268, \quad I_{cn} = 0.4196 + j0.4732$$

The power stored and consumed by the circuit in phase a is $0.2 + j0.6$ va.

$$pf = 0.3162, \quad C = 0.1125 \text{ F}$$

6.3.2

$$I_a = 3/(1 + j4) = 0.1765 - j0.7059, \quad I_{bn} = -0.6995 + j0.2001, \quad I_{cn} = 0.5231 + j0.5058$$

The power stored and consumed by the circuit in phase a is $0.2647 + j1.0588$ va.

$$pf = 0.2425, \quad C = 0.0876\,\text{F}$$

6.4.1 Same as for Exercise 6.2.1
6.5.1 Same as for Exercise 6.2.1

Chapter 7

7.1

$$Z = \begin{bmatrix} Z_{11} & Z_{12} \\ Z_{21} & Z_{22} \end{bmatrix} = \begin{bmatrix} 3.7308 - j0.3462 & 2.5385 + j0.6923 \\ 2.5385 + j0.6923 & 2.9231 + j1.6154 \end{bmatrix}$$

7.3

$$Y = \begin{bmatrix} Y_{11} & Y_{12} \\ Y_{21} & Y_{22} \end{bmatrix} = \begin{bmatrix} 0.5692 + j0.1385 & -0.4615 + j0.0000 \\ -0.4615 + j0.0000 & 0.6154 - j0.2308 \end{bmatrix}$$

7.5

$$\begin{bmatrix} A & B \\ C & D \end{bmatrix} = \begin{bmatrix} 1 & 0 \\ Y_1 & 1 \end{bmatrix} \begin{bmatrix} 1 & Z_1 \\ 0 & 1 \end{bmatrix} \begin{bmatrix} 1 & 0 \\ Y_3 & 1 \end{bmatrix} = \begin{bmatrix} j2 & 2 + j1 \\ -2 + j1 & j2 \end{bmatrix}$$

$$\begin{bmatrix} V_1 \\ I_1 \end{bmatrix} = \begin{bmatrix} j2 & 2 + j1 \\ -2 + j1 & j2 \end{bmatrix} \begin{bmatrix} 0.1379 - j0.3448 \\ 0.0690 - j0.1724 \end{bmatrix} = \begin{bmatrix} 1 \\ 0.4138 + j0.9655 \end{bmatrix}$$

Chapter 8

8.1.1

$$X(0) = 1\angle\pi, \ X(1) = 0.5\angle-\frac{\pi}{3}, \ X(-1) = 0.5\angle\frac{\pi}{3}, \ X(3) = 1.5\angle-\frac{\pi}{6}, \ X(-3) = 1.5\angle\frac{\pi}{6}$$

8.2

$$x(t) = \frac{1}{2} - \frac{1}{\pi}\left(\sin(2\pi t) + \frac{\sin(2(2\pi)t)}{2} + \frac{\sin(3(2\pi)t)}{3} + \cdots\right)$$

8.4.3 The sample values are

$$0.7287, \ 0.8646, \ 0.8660$$

The exact value is 0.8660.

8.6 The unit-step response is

$$v_C(t) = R(1 - e^{-\frac{t}{RC}})u(t) \quad \text{and} \quad i_C(t) = e^{-\frac{t}{RC}}u(t)$$

The unit-impulse response is

$$v_C(t) = \frac{1}{C}e^{-\frac{t}{RC}}u(t) \quad \text{and} \quad i_C(t) = -\frac{1}{RC}e^{-\frac{t}{RC}}u(t)$$

8.7 The unit-impulse response is

$$v(t) = 1.069e^{-0.25t}\cos(0.6614t + 0.3614)u(t)$$

Since the unit-step response is the integral of that of the unit-impulse response, we integrate $v(t)$ to get

$$y(t) = \frac{1.069e^{-0.25t}}{0.25^2 + 0.6614^2}(\, 0.9354((-0.25)\cos(0.6614t) + (0.6614)\sin(0.6614t)) +$$

$$(-0.3536)((-0.25)\sin(0.6614t) - (0.6614)\cos(0.6614t))\,)$$

Bibliography

Alexander, C. K., & Sadiku, M. N. O. (2013). *Fundamentals of electric circuits*. New York: McGraw Hill.

Boylestad, R. L. (2012). *Introductory circuit analysis*. New York: Pearson.

Carlson, A. B. (2002). *Circuits*. New York: Thomson.

Guillemin, E. A. (1963). *Introductory circuit theory*. New York: Wiley.

Irwin, J. D., & Nelms, R. M. (2006). *Basic engineering circuit analysis*. New York: Wiley.

Sundararajan, D. (2008). *Signals and systems – a practical approach*. Singapore: Wiley.

The MathWorks. (2019). *Matlab/Simulink user's guide*. Natick: The MathWorks, Inc.

The MathWorks. (2019). *Matlab signal processing tool box user's guide*. Natick: The MathWorks, Inc.

Van Valkenburg, M. E. (1974). *Network analysis*. Englewood Cliffs: Prentice-Hall.

Index

A

Admittance, 28, 81
    matrix, 28
    parameters, 221
    connected in parallel, 81
Alternating current (AC), 19, 77
    advantages of, 77
    DC circuit analysis
        differences, 85
    circuit analysis, 80
        advantages of sinusoidal input, 80
        circuit elements, 80
        frequency dependence, 85
        importance of, 86
        the importance of using sinusoids, 81
        Ohm's law, 81
        superposition theorem, 134
Ampere, 3
Amplitude, 78
Analysis of circuits with one or two variables, 113
Aperiodic signal, 240
Approximation of, 250
Average apparent power, 182
Average energy stored, 150

B

Balanced three-phase circuits, 205
    $\Delta - \Delta$ connection, 211
    $Y - Y$ connection, 205
    $Y - \Delta$ connection, 207
    $\Delta - Y$ connection, 210
    advantages, 203
    instantaneous power, 205
    line voltage, 206
    negative sequence, 204
    phase voltage, 205
    positive sequence, 203
    power-factor improvement, 207, 208
    sum of voltages or currents, 204
    typical applications, 212
    voltages, 203
Branch, 19, 20

C

Bridge circuit, 58, 59, 94
Bridged-T circuit, 216

C

Capacitor, 80
    with initial voltage, 256
    input-output relationship in frequency domain, 81
    input-output relationship in time domain, 80
Circuit, 1, 4, 19
    transform analysis, 84
Circuit analysis, 4, 19
    complete response, 255
    equilibrium conditions, 20
    in the frequency-domain, 255
    practical, 240
Circuit analysis (AC)
    superposition method, 114, 116
Circuit theorems, 59, 122
Column vector, 276
Common-emitter transistor amplifier, 228
Complex amplitude, 79
Complex and real sinusoids
    equivalency, 80
Complex numbers, 281
    addition and subtraction, 282
    angle, 281
    complex plane, 281
    conjugate, 283
    division, 283
    equality, 281
    exponential form, 281
    imaginary unit, 281
    magnitude, 281
    multiplication, 282
    polar form, 281
    real and imaginary parts, 281
    rectangular form, 281
Complex plane representation of voltages and currents, 89
Complex sinusoids, 79
    advantages, 79
Conductance, 3

© Springer Nature Switzerland AG 2020
D. Sundararajan, *Introductory Circuit Theory*,
https://doi.org/10.1007/978-3-030-31985-4

Conductively coupled, 177
Controlled sources, 40
Coulomb, 1
Coupling coefficient, 179
Current, 1, 3
    directions, 25
    source, 6, 35
        arrow sign, 35
        in parallel, 36
        in series, 36
    through a capacitor, 89
    through an inductor, 89
Current-controlled current source, 40
Current-controlled voltage source, 40
Cutoff frequency, 269

D
Δ-circuit
    defined, 55
$\Delta - \Delta$ circuit, 211
$\Delta - Y$ circuit, 210
$\Delta - Y$ transformation, 55, 57, 207
$\Delta - Y$ transformation (AC), 132
Determinant, 96, 98–102, 108, 110, 112, 113, 134, 191,
            195, 197, 277, 295
Difference between AC and DC circuit analysis,
            80
Digital-to-analog converter, 233
Direct current (DC), 19
Driving-point admittances, 221
Driving-point impedances, 215
Duality, 56
Dual nature, 12

E
Electricity, 1
Energy in reactive elements with sinusoidal sources,
            150
Energy stored in the magnetic fields, 179
Equilibrium conditions, 19
Equilibrium equations, 22
Equivalent circuit
    magnetically coupled inductors, 180, 183
Euler's identity, 79
Even-symmetric, 243

F
Filters, 135
    RC circuit, 136
    RL circuit, 135
Fourier and Laplace transforms
    differences, 256
Fourier series
    Gibbs phenomenon, 246
    of a rectified sine wave, 242
    signal analysis, 241

signal synthesis, 241
    of a square wave, 244
Fourier transform, 246
    definition, 246
    of exponential, 249
    of impulse, 248
    inverse, 247
    of pulse, 247
    relation to FS, 247
Frequency domain, 84
Frequency response, 248

G
Gibbs phenomenon, 246
Ground node, 27

H
Highpass filter, 271
Hybrid parameters, 224

I
Ideal current source, 89
Ideal voltage source, 87
Identity matrix, 24
Impedance, 23, 81
    matrix, 23, 24, 217
    parameters, 215
    connected in series, 81
    in parallel, 91
        combined admittance, 91
        defined, 91
        voltage across, 91
    in series, 85
        combined impedance, 86
        current through, 86
        defined, 86
        voltage across, 87
Impedances in series and parallel, 93
    combined impedance, 93
Impulse representation of signals, 250
Impulse signal
    as the derivative of step signal, 251
Induction motor, 198, 212
Inductor, 80, 177
    with initial voltage, 257
    input-output relationship in frequency domain, 81
    input-output relationship in time domain, 80
Instantaneous power, 205

K
Kilowatthour (kWh), 173
Kirchhoff's current law (KCL)
    defined, 12
Kirchhoff's voltage law (KVL), 11, 89
    defined, 11

L
Ladder circuit, 231
Laplace transform
  definition, 252
  of an exponential, 253
  properties
    final value, 255
    initial value, 255
    integration, 254
    time-differentiation, 254
  of sine signal, 253
  of unit-impulse, 252
  of unit-step, 253
Linearity, 37
Link, 21
Link currents, 21
Loop, 20
Loop analysis, 95, 103, 111, 112
Loop analysis (DC), 20

M
Magnetically coupled, 177
Magnetically coupled circuits, 177
Magnitude, 78
Matrices, 275
  adjoint, 276
  cofactor, 276
  complex, 279
  definition, 275
  determinant of, 277
  identity, 277
  inverse, 276, 278
  minor, 276
  multiplication, 276
  square, 276
  symmetric, 278
  transpose, 278
Maximum power transfer theorem, 65
Maximum power transfer theorem (AC), 131
Mesh analysis, 21, 95, 103, 111, 112
  controlled voltage source, 107
Mesh analysis (AC), 104, 114, 116, 119, 120
  bridge circuit, 94, 98, 102
  current-controlled current source, 112
  current source, 103
  determinant of the impedance matrix, 96
  differences from DC analysis, 94
  equilibrium equations, 95
  impedances, 95
  solving the equilibrium equations, 96
  source, 94
  superposition theorem, 105
  verifying the solutions, 96
  voltage and current sources of different frequencies,
    105
Mesh analysis (DC), 20, 21, 34, 39, 40, 45, 47, 49–55,
    118, 121
  bridge circuit, 21, 29
  current directions, 25

determinant of the impedance matrix, 24
equilibrium equations, 21–23
independent and dependent variables, 22
loop independence, 20
negative value for the current, 26
power consumed, 26, 33, 35, 38, 39, 44, 46
solving equilibrium equations, 24
supermesh, 36
supernode, 33
superposition method, 37, 48, 49
symmetry of the impedance matrix, 24
verifying the solutions, 24
Mesh and nodal analysis
  comparison, 111
Mutual inductance, 177
Mutual induction, 177
  coupling coefficient, 179
Mutually induced voltage
  polarity, 180

N
Negative sequence, 204
Nodal analysis, 97, 99, 104, 109, 111, 127
  controlled voltage source, 109
Nodal analysis (AC), 104, 113, 115, 118, 119, 121
  bridge circuit, 97, 99
  current-controlled current source, 111
  current source, 102
  determinant of the admittance matrix, 100
  supernode, 100
Nodal analysis (DC), 21, 27, 36, 39, 42, 44, 46, 49–54,
    118, 122
  bridge circuit, 27, 31, 33
  determinant of the admittance matrix, 29
  equilibrium equations, 28
    matrix form, 28
  independent and dependent variables, 27
  steps of, 27
Node, 20
Nortan's theorem (AC), 123, 130
Norton's theorem, 59, 61
  steps of, 61

O
Odd-symmetric, 245
Ohm's law, 3
  in AC circuit analysis, 81
Orthogonality, 241

P
Parallel circuit, 7
Parallel DC circuits, 7
  combined resitance, 7
  current, 7
  current division, 8
  definition, 7
  power, 8

Parallel DC circuits (*cont.*)
   symbol, 8
   voltage, 7
Periodic signal, 240
Pf, 152
Phasor, 85, 88
Phasor and real sinusoids, 88
π circuit, 226
Polarity of the mutually induced voltage, 180
Port, 215
Positive sequence, 203
Power, 4, 149
   active, 152
   apparent, 152
   factor, 152
   factor correction, 159
   reactive, 152
   relations in a circuit, 151
   sign convention, 47
   steady state, 149
Power-factor correction, 152

R

R-2R ladder circuit, 233
RC circuit, 82
RC circuit with initial voltage, 264
RC lowpass filter, 269
Reciprocal, 226
Relation between stored energy and reactive power, 156
Representation of sources in time and frequency, 89
Resistance, 3
Resistor, 21
RLC circuit analysis, 266, 267
RL circuit analysis, 81
RL circuit with initial current, 266
RMS value, 155

S

Self-inductance, 177
Series and parallel DC circuits, 9
Series circuit, 4
Series DC circuits
   combined resistance, 5
   current, 4
   defined, 4
   element interchange, 5
   power, 6
   resistance, 5
   voltage, 5
   voltage division, 5
Series RC Circuit
   frequency-domain analysis, 84
   time-domain analysis, 82
Signal
   aperiodic, 240
   periodic, 240

Sinusoidal analysis
   importance, 239
Sinusoids, 77
   advantages of, 77
   amplitude, 78
   angular frequency, 78
   complex, 79
   cosine and sine waveforms, 78
   cosine waveform in terms of sine, 79
   cyclic frequency, 78
   odd and even components, 79
   period, 78
   the periodic nature, 79
   phase, 78
   polar form, 78
   polar form to rectangular, 79
   rectangular form, 79
   rectangular form to polar, 79
   sine waveform in terms of cosine, 79
Sources
   controlled, 40
Source transformation, 65
Source transformation (AC), 133
Spectrum
   magnitude, 242
   phase, 242
Steady-state analysis, 85
Steady-state response, 249
Storage elements, 149
Stored energy, 178, 182, 185, 187, 189, 193
   capacitor, 150
   discharge, 259
   electric field, 157
   magnetic field, 155
   sinusoidal source, 179, 180
Stored energy, inductor, 149
Strain gauge measurement, 67
Substitution method, 275
Supermesh, 36, 104
Supernode, 33, 34, 100, 101
Superposition, 37
Symmetrical components, 205
Symmetry of the impedance matrix, 24

T

Thévenin's theorem, 59
   bucking voltage method, 60
   equivalent resistance, 59
   equivalent voltage, 59
   steps of, 59
Thévenin's theorem (AC), 122
   bucking voltage method, 124
   current source, 125
   sources with different frequencies, 126
   voltage-controlled voltage source, 127
Three-phase circuits, 203
   line voltages, 209
   unbalanced, 205
Three-phase voltages, 203

Time constant, 259
Time domain, 84
Transfer admittances, 221
Transfer function, 248
Transfer impedances, 216
Transform, 239
Transform analysis of circuits
   steps of, 84
Transformers, 198
   ideal, 198
Transient response, 249
Transmission parameters, 225
Tree, 20
Two-port circuit, 215
   applications, 215
   input impedance, 220
   necessity, 215
   output impedance, 220
   parallel circuits, 224
   parameter matrix, 217
   series circuits, 219
Two-port model
   bridge circuit, 230
Two-port networks, 215

U
Unit-impulse signal, 250
   the integral of, 251

Unit-step responses
   RC circuit, 261, 290
   RL circuit, 258
Unit-step signal, 250
   the derivative of, 251

V
Volt, 3
Voltage, 3
Voltage-controlled current source, 40
Voltage-controlled voltage source, 40
Voltage source, 5, 20
   in parallel, 20
   + sign, 20
   in series, 20

W
Watt, 4

Y
Y-circuit
   defined, 55
$Y - \Delta$ and $\Delta - Y$ transformations, 55
$Y - \Delta$ circuit, 207
$Y - \Delta$ transformation, 56
$Y - Y$ circuit, 206

Printed in the United States
By Bookmasters